T0185902

FLUID DYNAMICS WITH A COMPUTATIONAL PERSPECTIVE

Modern fluid dynamics is a combination of traditional methods of theory and analysis and newer methods of computation and numerical simulation. Underlying both are the principles of fluid flow. *Fluid Dynamics with a Computational Perspective* synthesizes traditional theory and modern computation. It is neither a book on methods of computation, nor a book on analysis; it is about fluid dynamics – consistent with the state of the art in that field. The book is ideal for a course on fluid dynamics. The early chapters review the laws of fluid mechanics and survey computational methodology and the subsequent chapters study flows where the Reynolds number increases from creeping flow to turbulence, followed by a thorough discussion of compressible flow and interfaces. Although all significant equations and their solutions are presented, their derivations are informal. References for detailed derivations are provided. A chapter on intermediate Reynolds number flows provides illustrative case studies by pure computation. Elsewhere, computations and theory are interwoven.

Paul A. Durbin is the Martin C. Jischke professor of aerospace engineering at Iowa State University. He was previously a professor in mechanical engineering at Stanford University. His research interests are in turbulence and transition, including computation, theory, and analytical modeling. He is a member of AIAA and ASME and a Fellow of APS. He is an associate editor of the ASME *Journal of Fluids Engineering*. He has extensive experience in teaching fluid dynamics and has written *Statistical Theory and Modeling for Turbulent Flow* and numerous articles.

Gorazd Medic is a research associate at the Center for Integrated Turbulence Simulations of the Mechanical Engineering Department at Stanford University. His research interests are in turbulence, numerical methods, and high-performance computing. He is a member of AIAA, ASME, APS, and SIAM. He has extensive experience in computational fluid dynamics for a variety of applications ranging from aircraft engines to biomechanical systems.

Fluid Dynamics with a Computational Perspective

PAUL A. DURBIN

Iowa State University

GORAZD MEDIC

Stanford University

CAMBRIDGE
UNIVERSITY PRESS

CAMBRIDGE
UNIVERSITY PRESS

32 Avenue of the Americas, New York NY 10013-2473, USA

Cambridge University Press is part of the University of Cambridge.

It furthers the University's mission by disseminating knowledge in the pursuit of education, learning and research at the highest international levels of excellence.

www.cambridge.org
Information on this title: www.cambridge.org/9781107699311

First published 2007
First paperback edition 2013

A catalogue record for this publication is available from the British Library

Library of Congress Cataloguing in Publication data

Durbin, Paul A.
Fluid dynamics with a computational perspective / Paul A. Durbin, Gorazd Medic.
   p. cm.
ISBN 9780521850179
Includes bibliographical references and index.
1. Fluid dynamics – Mathematics. I. Medic, Gorazd, 1970–   II. Title.
TA357.D855 2007
620.1´0640151 – dc22   2007012221

ISBN 978-0-521-85017-9 Hardback
ISBN 978-1-107-69931-1 Paperback

To Cinian, Seth, and Sanja, who were patient and supportive while we wrote, and to Professor Enoch Durbin, for a lifetime of encouragement

Contents

Preface

This is a book on fluid dynamics. It is not a book on computation. Many excellent books on fluid dynamics are available: why is another needed?

In recent decades, numerical algorithms and computer power have advanced to the point that computer simulations of the Navier–Stokes equations have become routine. This vastly expands our ability to solve these equations, further extending our understanding of fluid flow and providing a tool for engineering analysis. Computer simulations are solutions of a different nature from classical exact and approximate solutions. They are numerical data rather than formulas. One of our objectives in this text is to relate computer solutions to theoretical fluid dynamics. Indeed, it is this goal, rather than computation as a tool for complex engineering analysis, that provides the guideline for this text. Computer solutions can reproduce closed-form and approximate solutions; they can illuminate the merits and limits of simple analyses; and they can provide entirely new solutions of varying degrees of complexity. The time is ripe to integrate computer solutions into fluid dynamics education.

From a pedagogical perspective, readily available, commercial computational fluid dynamics (CFD) software provides a new resource for teaching fluid dynamics. This software converts CFD from a technique used by researchers and engineers in industry into a readily accessible facility. It is a challenge to integrate such software packages into the educational structure. Most of the examples in this book have been computed with commercial software, and exercises to be solved with such software have been suggested. How far to go in this direction was a true quandary. We started out ambitiously, intending an intimate use of commercial tools, but backed off, deciding on a text that provides the reader with illustrative computations. Grids and specifications needed to effect computer simulation are described, but far short of the level of detail found in tutorials. The endeavor was to make the book useable either on its own or as an adjunct to an integrated course on fluid dynamics and computation. A lecturer might supplement this text with a computer laboratory, instructing students via tutorials.

This certainly is not a book on methods of computational fluid dynamics. We assume that a student who uses this text in conjunction with computational analysis has access to CFD software, including both flow solver and mesh generator. Methods and algorithms are mentioned only to the extent that they bear on the fidelity of computations. In other words, the perspective is that of a code user, not of a developer. One impetus for this book was the observation by many of our colleagues that students are increasingly proficient at using software, but often without the understanding of fluid dynamics needed to use it effectively. We hope this text will be a resource for educating prospective software users.

Despite the power of computer simulation, it has not supplanted theory and analysis in fluid dynamics teaching, even at advanced levels; nor have theoretical efforts diminished within the research community. No wholesale displacement of theory by computation is remotely on the horizon. The complexity and intricacy of flow phenomena are too great, and the range of applications are too vast, for any single approach to be sufficient. Hence, in this book, we endeavor both to present some basics of the theory of fluid flow and to explore them by computer simulation. The intent is a marriage between classical theory and modern computer-aided analysis.

Deciding how much mathematics to include was another quandary. A great many texts detailing analyses and solutions to governing equations are available. It is not our intent to reproduce such material. At the same time, there is a need to establish a framework for computer-aided analysis. This means that equations and solutions must be cited. To a large extent, we have quoted results with informal derivations, providing references where detailed developments can be found. In some cases, formulas might appear out of the blue, but in almost all cases, some basis has been provided, without rigor.

The prerequisite to this book is a course on basic fluid dynamics, including elementary viscous flow. The book is self-contained, but the pace might seem fast without the prerequisite. It is directed to students at advanced undergraduate level or to graduate students at master's level. It is also meant for scientists and engineers who want background in viscous flow phenomena.

After two chapters that provide background on the laws of fluid mechanics and survey computational methodology, the next four chapters increase the Reynolds number from creeping flow to turbulence. They are followed by a chapter on compressible flow and a final chapter on interfaces. We have been guided by the content of standard curricula in fluid mechanics.

Acknowledgments

We are extremely grateful to Frank Albina, George Constantinescu, Gianluca Iaccarino, Tomoaki Ikeda, Georgi Kalitzin, Yang Liu, Milovan Peric, Jorg Schluter, Meng Wang, Edwin van der Weide, and Xiaohua Wu for providing figures, as well as to the Institute for Mathematical Sciences, Singapore National University, for its hospitality during part of the writing. Peter Gordon of Cambridge University Press has been wonderfully encouraging.

1 Introduction to Viscous Flow

1.1 Why Study Fluid Dynamics?

Fluid dynamics is a branch of classical physics. It is an instance of continuum mechanics. A fluid is a continuous, deformable material. It is a material that flows in response to imposed forces. This is embodied in the everyday experience of draining water from a sink. The water flows under the action of gravity. It does not have a fixed shape; it fills the sink, conforming to its shape. The water flows with variable velocity, depending on its distance from the drain. All these distinguish fluid motion from solid dynamics. As another example, a pump propels water through a pipe or through the cooling system of a car. How does the reciprocating movement of the pump produce directed flow, extending to distant parts of the cooling circuit? One way or the other, the pump must be exerting forces on the fluid; one way or the other, these forces are communicated to distant portions of the fluid and sets them in motion. It is far from obvious what the nature of that flow will be, especially in a complex geometry. It may be laminar, it may be turbulent; it may be unidirectional, it may be recirculating.

Recirculation is the occurrence of backflow, opposite to the direction of the primary stream. This can be seen behind the pedestals supporting a bridge in a swift river. Despite the strong current, the flow direction reverses, and a circulating eddy forms in a region behind the pedestal. How is such behavior understood and predicted? An understanding requires knowledge of viscous action, of vorticity, of turbulence, and of the governing equations.

As a fluid flows around an obstruction, different points in the fluid flow with different velocities. The motion is continuous; hence, the velocity varies smoothly with position. The fluid might be imagined to be divided into infinitesimal volumes – which can be referred to as fluid particles. Then the fluid flow involves relative movement of the particles. Because the flow varies continuously with position, these particles must influence each other to maintain the smoothness of the flow field. That influence is via the forces of pressure and of viscous friction. It is a scientific triumph that when equations are devised to describe these forces acting on the fluid particles, the rich variety of fluid mechanics emerges. That success is most evident in computer simulation of fluid flow.

The trajectories of fluid particles are the streamlines of motion. These streamlines can be curves stretching from inflow to outflow, or they can be closed curves, corresponding to eddying flow. The flow over a rock in a stream or past a spoon stirred through a cup of coffee are examples containing eddying flow. It is not difficult to observe phenomenology of fluid motion: hold a lighted candle, or tissue paper, at arm's length and puff very quickly. A vortex ring travels from your mouth to the candle. There is a time delay before it flickers; that is because the vortex travels with a finite speed. Now suck air into your mouth; the candle shows no evidence of flow. The vortex flows into your mouth. The flow outside is then quite different from when you blew the air out.

Some interesting and challenging scientific and engineering questions already suggest themselves. What causes the eddies and vortices? Why is the outflow so different from the inflow? What drives the flow through a conduit or over an obstacle? How is motion communicated to distant portions of the fluid? How can the flow rate be predicted? How are the streamlines of a fluid flow determined? In any but the simplest cases, these are challenging questions. There is a body of knowledge that can be called on; but one is rapidly struck with how difficult it is to answer even fairly simple questions about fluid flow.

Computer simulation has changed this, rather substantially. Flow in complex geometry can be solved numerically. The phenomena seen in laboratory experiments, and in more casual experience, can be reproduced with quantitative accuracy. To an extent, older theories and analyses have been enlarged by computation. The laws governing fluid flow have been found to be extremely accurate; one wants only sufficient computer power and efficient algorithms to produce solutions. Computer simulation becomes a method of solution, complementary to paper-and-pencil analysis. It provides further understanding of fluid flow and a tool for engineering analysis. However, computer simulations are solutions of a different nature from classical, exact, or approximate solutions. They are numerical data rather than formulas. Traditional theory is not displaced; its role evolves and it provides the understanding needed to formulate and make sense of computer-aided analysis. The additional understanding of fluid dynamics that stems from simulation should be developed in concert with theory. The motive to study fluid dynamics is to understand its phenomena. The approach is to devise and solve the laws of fluid motion.

1.2 Viscosity

At the root is the governing laws. These are the equations of conservation; conservation of mass and momentum, at the present stage. Friction is an important element of fluid flow. In the absence of friction, a flat plate, dragged tangentially across the surface of a tank of water, would slide freely and induce no movement in the water. But, in the presence of friction, motion is communicated to the water adjacent to the plate and thence a circulation is established in the tank.

Friction internal to a fluid flow is characterized by viscosity. The viscosity coefficient, μ, is an empirical property of the fluid. For instance, a liquid is more viscous than a gas. As temperature increases, liquids flow more easily (think of tar, for example) so viscosity decreases with temperature. Gases have the opposite property,

Table 1.1. *Viscosities at room temperature* $(20°C)$

Fluid	μ g/cm · s	ρ g/cm^3	ν cm^2/s
Air	1.8×10^{-4}	1.2×10^{-3}	0.15
Water	0.01	1.0	0.01
CO_2	1.37×10^{-4}	1.79×10^{-3}	0.077
Engine oil	10	0.89	11.2
Glycerin	7.99	1.26	6.34
Kerosene	0.024	0.78	0.031
Methyl alcohol	0.0055	0.785	0.007

that viscosity increases with temperature. That behavior is less intuitive. It originates in the increased molecular agitation as temperature increases. Clearly, μ must be measured as a function of the fluid and of the temperature. In gases, the coefficient increases approximately as the square root of temperature; in liquids, it falls as the exponential of one over temperature $[\exp(E/kT)]$. Detailed formulas need not be discussed here. Values of μ for many fluids are available in computational fluid dynamics (CFD) codes and in handbooks. The magnitude of the viscosity is essential to determining flow regimes. Table 1.1 contains a few representative values at room temperature.

The coefficient, μ, is the *dynamic* viscosity. For future reference, the *kinematic viscosity* is defined as dynamic viscosity divided by density as follows:

$$\nu = \mu/\rho. \tag{1.1}$$

Kinematic viscosity has dimensions of *length^2/time*; dynamic viscosity has dimensions of density times this or *mass/length · time*. Kinematic viscosity is most relevant to constant density, incompressible flow. Oddly enough, the kinematic viscosity of liquids is often lower than that of gasses. For instance, air at $20°C$ and 1 atmosphere, has a kinematic viscosity of 0.15 cm^2/s; for water it is 0.01 cm^2/s. This is a consequence of the higher density of water.

Viscosity produces forces as a consequence of the relative motion of fluid particles. That might be thought of as friction associated with the particles rubbing across one another. A more correct statement is that viscous *stress* is a consequence of the fluid *rate of strain*.

The need to distinguish rate of strain from simply relative motion is because a fluid in solid body rotation experiences no viscous stress. At a macroscopic level that is clear: if the entire fluid is in solid body rotation, then in a frame rotating with the fluid there is no motion and hence no viscous stress. In a fixed frame, solid body motion means that the velocity is Ωr in the angular direction. There is relative motion in the sense that fluid at $r = 0$ is at rest, whereas at $r > 0$ it is in motion, but there is no viscous stress.

The same concept applies, less obviously, at any point in a nonrotating fluid. Relative motion can be separated into rotation and rate of strain; only the latter produces viscous stress. The velocity is a field: at any point $\boldsymbol{x} = (x, y, z)$, three components of velocity (u, v, w) can be measured. Rate of strain is a measure of how this velocity varies from point to point within the vicinity of $\boldsymbol{x}$. Mathematically, if two points are separated by a distance $d\boldsymbol{x}$, their relative velocity is

$$u_i(\boldsymbol{x} + d\boldsymbol{x}) - u_i(\boldsymbol{x}) \approx d\boldsymbol{x} \cdot \nabla u_i,$$

where $i = 1, 2,$ or 3 corresponding to $u, v,$ or w. The last term expands to

$$dx \frac{\partial u_i}{\partial x} + dy \frac{\partial u_i}{\partial y} + dz \frac{\partial u_i}{\partial z}. \tag{1.2}$$

The convention of summation on repeated subscripts permits this to be written equivalently as

$$dx_j \frac{\partial u_i}{\partial x_j}.$$

Because the same index, j, appears twice in this product, the convention is that j is summed from 1 through 3, so that this is exactly the same expression as Eq. (1.2). This is a rather terse introduction to index notation. The uninitiated reader might want to write out corresponding formulas in index notation and in Cartesian components. That exercise is illustrated in the next paragraph.

We now introduce the separation of the velocity gradient into rate of strain and rotation; it is equivalent to a separation into symmetric and antisymmetric components, respectively. Specifically,

$$\frac{\partial u_i}{\partial x_j} = \frac{1}{2} \left(\frac{\partial u_i}{\partial x_j} + \frac{\partial u_j}{\partial x_i} \right) + \frac{1}{2} \left(\frac{\partial u_i}{\partial x_j} - \frac{\partial u_j}{\partial x_i} \right). \tag{1.3}$$

The first term on the right is the rate of strain, which can be denoted S_{ij}; the second is minus the rate of rotation, which can be denoted $-\Omega_{ij}$:

$$S_{ij} = \frac{1}{2} \left(\frac{\partial u_i}{\partial x_j} + \frac{\partial u_j}{\partial x_i} \right),$$

$$-\Omega_{ij} = \frac{1}{2} \left(\frac{\partial u_i}{\partial x_j} - \frac{\partial u_j}{\partial x_i} \right). \tag{1.4}$$

Only S_{ij} produces viscous stress. In the standard index notation used here, i and j are dummy subscripts, for which any of the numbers $(1, 2, 3)$ corresponding to the directions (x, y, z) can be substituted. For instance, with $i = 1$ and $j = 2$, Eq. (1.3) says

$$\frac{\partial u}{\partial y} = \frac{1}{2} \left(\frac{\partial u}{\partial y} + \frac{\partial v}{\partial x} \right) + \frac{1}{2} \left(\frac{\partial u}{\partial y} - \frac{\partial v}{\partial x} \right).$$

In solid body rotation, $u = -\Omega y$ and $v = \Omega x$. The first term vanishes and the second equals $-\Omega$; the rate of strain is zero under solid body rotation. An irrotational flow is one for which the second term vanishes: $\partial u/\partial y = \partial v/\partial x$. For instance, $u = \alpha y$, $v = \alpha x$ is an irrotational straining field.

The other element in the description of viscous forces is that they are character-ized as a *stress* rather than as a *force* per se. This should not be unfamiliar: pressure also is a stress. That is, it is a force per unit area, acting in the direction normal to a surface. Viscous stress is similar: it is a force per unit area, but it need not act normal to the surface; it can have both normal and tangential components. If the fluid motion is just a shearing tangential to a surface, then it is clear that the viscous stress will cause a force in the tangential direction. It is less obvious, but true, that if the fluid motion is toward the surface there will also be a component of force in the normal direction. That is best described mathematically, as will be done in the next section.

Given the stress, the corresponding force on an object is obtained by integrat-ing it over the entire area of the surface. This can be accomplished inside a CFD code, so one need only understand the origin of the force that is being computed. The reason that the force originates as a stress is that fluids are deformable ma-terial, so their dynamical properties must be defined for the fluid particles. Stress is the force per area acting on a fluid particle. It is independent of the size of the infinitesimal fluid particle. The force, by contrast, is proportional to the size. In other words, *force = stress · area*. As the area becomes tiny, the stress remains finite and the force becomes tiny. Just like pressure, viscous stress is the quantity that is defined pointwise throughout the fluid. Unlike pressure, it exists only in a flowing fluid.

Viscous stress is incorporated into the governing, Navier–Stokes, momentum equations that are solved by CFD software. The gist of those equations is the next topic.

1.3 Navier–Stokes Equations

It is assumed that the reader has studied elementary fluid mechanics and has been exposed to the basic notions of fluid flow. These include the role of pressure in momentum transport, conservation of mass in an incompressible, deformable fluid medium, and the origin of viscous, frictional forces. The last have just been discussed. This section provides an informal description of the Navier–Stokes momentum equa-tion for constant density, incompressible flow. Because we rely on CFD software for solutions to the equations, we will abbreviate the treatment that can be found more fully in standard texts on viscous flow (White, 1991).

The Navier–Stokes equations were named for the French engineer and scientist Claude Louis Marie Henri Navier and the English mathematical physicist George Gabriel Stokes. The essential form of these equations was set forth by Navier in 1822; however, he did not properly treat the origin of viscous stress. The latter was addressed by others, in particular Poisson and Saint-Venant, but independently de-veloped by Stokes in 1845. Stokes constructed a number of solutions to the equations of viscous flow, which confirmed their ability to describe fluid dynamical phenom-ena. An example is creeping flow, also called "Stokes flow," which we discuss in

Chapter 3. Navier is properly credited for the seminal formulation of the Navier–Stokes equations and Stokes for ushering their entry into theoretical physics.

Essentially, the Navier–Stokes momentum equation is an expression of Newton's law $ma = F$ applied to an infinitesimal fluid volume. Here we use the convention that bold letters denote vectors. On a volumetric basis, the mass becomes mass per unit volume or density ρ. The acceleration becomes that following the fluid element, or the convective derivative of velocity, Du/Dt, and the force per unit volume includes both pressure and viscous contributions, as follows:

$$\rho \frac{Du}{Dt} = \mathcal{F}_{\text{press}} + \mathcal{F}_{\text{viscous}}. \tag{1.5}$$

We must flesh out the meaning of these various terms.

The equations of motion referred to fluid particles is called the *Lagrangian* description. The fluid particle occupies a position $X(t; x_0)$ that changes with time. The particle is labeled here by its initial location x_0. It is more convenient to describe the flow in terms of the velocity at fixed points. We think of a flow field, $u(x)$, rather than of the dynamics of particles. The only complication in applying Newton's law to the field is transforming the acceleration of the fluid particle into velocity changes at a fixed position.

To derive the requisite expression, first consider a material that is carried with the fluid element. The material has a concentration c. An observer at a fixed point, x, will see the concentration change as different particles arrive. At any given time, the concentration is that of the particle currently at x, that is, of the particle with $X(t) = x$. At time δt later, a particle that was at $X - \delta X$, say, will have moved to x. The observer then sees its concentration $c(X - \delta X)$. Thus the observer sees the change

$$\frac{\partial c}{\partial t} = \frac{c(X - \delta X) - c(X)}{\delta t} \approx \frac{-\delta X \cdot \nabla c}{\delta t}$$

as the fluid element occupying position x changes from that at time t to that at time $t + \delta t$. $\delta X / \delta t$ is the velocity u. Hence, the motion of fluid elements produces the time variation

$$\frac{\partial c}{\partial t} = -u \cdot \nabla c.$$

There is nothing special about concentration: the same result applies to any quantity convected with the flow. Putting both terms on the same side of the last equation shows that a transported quantity satisfies

$$\frac{Dc}{Dt} \equiv \frac{\partial c}{\partial t} + u \cdot \nabla c = 0.$$

This is a statement that the changes at a fixed position are simply due to different elements arriving at that position, carrying their particular concentration. If the quantity were not simply convected but also underwent some change, then the right side would be nonzero.

Thus, we arrive at the expression of Newton's law at a fixed point – which is called the *Eulerian* description. The quantity being carried is now the fluid momentum, $\rho\boldsymbol{u}$. It is carried with the particles but also changes as a consequence of forces. The flow field obeys

$$\rho\left[\frac{\partial\boldsymbol{u}}{\partial t}+(\boldsymbol{u}\cdot\nabla)\boldsymbol{u}\right]=\mathcal{F}_{\text{press}}+\mathcal{F}_{\text{viscous}}. \tag{1.6}$$

A more general concept than the right side of Eqs. (1.5) or (1.6) is to combine pressure and viscous terms into a single stress tensor. Stresses are forces acting on a surface per unit area. *Pressure* times *area* is a force acting perpendicularly to the surface; in other words, it is a *normal stress*. Viscosity produces both *normal* and *tangential*, or shearing, stresses. The tangential stress is quite intuitive: it is analogous to the force felt when rubbing one hand over the other. As mentioned in the last section, it is also the case that viscosity produces a component of normal force, parallel to pressure. On any surface, the aggregate stress is a vector, with components both normal and tangential to the surface, composed of contributions from pressure and from viscosity.

How are forces produced by stress represented? The aggregate force can be denoted $\boldsymbol{F}_s$. It is the force produced by a stress acting on a surface. Consider that surface to be a small, differential area, dA. Further, let that area be the magnitude of a vector $d\boldsymbol{A}$ that is directed normal to the surface. The stress, $\boldsymbol{\sigma}$, now can be defined: the force is the dot product of the stress with the area vector

$$\boldsymbol{F}_s=\boldsymbol{\sigma}\cdot d\boldsymbol{A}. \tag{1.7}$$

This simply defines the *stress tensor* $\boldsymbol{\sigma}$ as a matrix relating the force vector to the area vector. Purely as a matter of consistency, Eq. (1.7) shows that stress has dimensions of force per area. In component form, the matrix relation [Eq. (1.7)] between stress and surface force is stated as

$$\begin{pmatrix}F_x\\F_y\\F_z\end{pmatrix}=\begin{pmatrix}\sigma_{xx}&\sigma_{xy}&\sigma_{xz}\\\sigma_{yx}&\sigma_{yy}&\sigma_{yz}\\\sigma_{zx}&\sigma_{zy}&\sigma_{zz}\end{pmatrix}\cdot\begin{pmatrix}dA_x\\dA_y\\dA_z\end{pmatrix}. \tag{1.8}$$

The term stress *tensor* was slipped into the text, above. A tensor is a generalization of a vector. A vector has a direction. A tensor has one or more directions. A vector is a first-order tensor, having a single direction. Stress is a second-order tensor; it is associated with two directions. The two directions are that of the force vector and that of the area vector. Matrix $\boldsymbol{\sigma}$ relates the direction of a force acting on a surface to the area vector of that surface.

Equation (1.7), integrated over a solid surface, gives the net force exerted by the flowing fluid. When, in later chapters, we consider examples of fluid forces on objects, the quantity

$$\int\boldsymbol{\sigma}\cdot d\boldsymbol{A} \tag{1.9}$$

is being evaluated by integration over the entire surface in question.

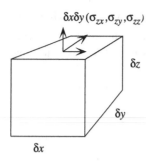

$\delta x \delta y \, (\sigma_{zx}, \sigma_{zy}, \sigma_{zz})$

Figure 1.1. Stresses on a face of a fluid element.

The force evaluation usually can be obtained with the CFD software; but it should be understood that the computer code is providing this surface integral of the stress tensor. In fact the force can be broken down into contributions from pressure and viscosity. The accuracy of the numerical evaluations depends on how finely the mesh covers the surface and on how accurately the viscous and pressure stresses are computed by the flow solution. The force evaluation is a postprocessing of that solution.

The presence of two contributions to stress can be acknowledged by writing

$$\boldsymbol{\sigma} = \boldsymbol{\sigma}_{\text{viscous}} + \boldsymbol{\sigma}_{\text{pressure}}, \tag{1.10}$$

Precise expressions for these two contributions will be given shortly.

Forces due to stress also act inside the fluid. Within the fluid, they are the force on the face of an infinitesimal fluid element (Figure 1.1). That element will move if there is an imbalance of forces. Consider two opposite sides of an element having oppositely directed normals. The force imbalance is due to a difference between the stresses acting on the opposite sides, that is, to a differential of stress. If $\sigma_L A$ is the force on the left side of the fluid element and $\sigma_R A$ is the force on its right side, then $(\sigma_R - \sigma_L)A$ is the force imbalance. If ℓ is the length of the element and $\mathcal{V} = A\ell$ its volume, then the resultant force is

$$\frac{\sigma_R - \sigma_L}{\ell} A\ell = \frac{\sigma_R - \sigma_L}{\ell} \mathcal{V} \approx \frac{d\sigma}{d\ell} \mathcal{V}.$$

That is, the force per unit volume is the directional derivative of the stress. The equation of motion is now $\mathcal{M} \, Du/Dt = \mathcal{V} \, d\sigma/d\ell$. The ratio $\mathcal{M}/\mathcal{V}$ is the density ρ. It takes only a bit of elaboration to recognize that the directional derivative $d\sigma/d\ell$ should be generalized to the divergence of the stress $\nabla \cdot \boldsymbol{\sigma}$. Essentially, the gradient operator gives the directional derivative.

Hence, the force that appears in the Navier–Stokes momentum equation is the divergence of stress, and Newton's law becomes

$$\rho \frac{D\boldsymbol{u}}{Dt} = \rho \left[\frac{\partial \boldsymbol{u}}{\partial t} + (\boldsymbol{u} \cdot \nabla)\boldsymbol{u} \right] = \nabla \cdot \boldsymbol{\sigma}. \tag{1.11}$$

This is simply Newton's law (1.5) when the force is caused by a stress gradient. As explained above, $\partial \boldsymbol{u}/\partial t + (\boldsymbol{u} \cdot \nabla)\boldsymbol{u}$ is the Eulerian form for the acceleration $D\boldsymbol{u}/Dt$.

Now to make explicit the observation that stress is composed of a part due to pressure and a part due to viscosity. The part due to pressure is just pressure times the identity matrix, with a minus sign:

$$\sigma_{\text{press}} = -p\mathbf{I}. \tag{1.12}$$

The minus sign arises because pressure applied to a surface acts inward to the surface. For instance, if the surface is the xy plane, an imposed high pressure will push down in the $-z$ direction on the surface. That is a special case of the general formula

$$\mathbf{F}_{\text{press}} = \sigma_{\text{press}} \cdot d\mathbf{A} = -p\mathbf{I} \cdot d\mathbf{A} = -pd\mathbf{A}.$$

Pressure times *area* is the inward force. The pressure contribution to stress gives rise to a pressure gradient on the right side of Eq. (1.11): $\nabla \cdot \sigma_{\text{press}} = -\nabla p$. This might be comforting to the reader who is wondering why there is no pressure gradient in Eq. (1.11).

The representation of viscous stress is less obvious. In a Newtonian fluid, it is assumed proportional to the rate of strain of the fluid motion, as described in §1.2. Viscosity is simply the coefficient of proportionality. This is stated as

$$\sigma_{\text{viscous}} = \mu(\nabla\mathbf{u} + {}^t[\nabla\mathbf{u}]), \tag{1.13}$$

where $\nabla\mathbf{u}$ is a matrix of velocity derivatives and ${}^t[\nabla\mathbf{u}]$ is its transpose. Equation (1.13) is a statement in vector form of the relation

$$\sigma_{ij} = 2\mu S_{ij}$$

in index form, with S_{ij} given by Eq. (1.4). When this formula is used for an incompressible fluid, the viscous force simplifies to the Laplacian of velocity: $\nabla \cdot \sigma_{\text{viscous}} = \mu\nabla^2\mathbf{u}$. The Navier–Stokes equations of an incompressible (Newtonian) fluid assume the form

$$\rho\left[\frac{\partial\mathbf{u}}{\partial t} + (\mathbf{u}\cdot\nabla)\mathbf{u}\right] = -\nabla p + \mu\nabla^2\mathbf{u}. \tag{1.14}$$

The pressure and viscous forces are stated explicitly here. Lengthy discussions of the derivation of this equation, and some caveats that must be made in our derivation, can be found in standard texts (Panton, 1997; White, 1991). In fact, it is an equation that is quite remarkable for its ability to describe the phenomena of fluid flow. It is equally remarkable for its mathematical intransigence. Basically, it is the momentum equation that is solved by CFD software.

That is not quite correct: (1.14) is an equation for the velocity vector $\mathbf{u}$ or for the u, v, and w components of velocity. It contains the further fluid properties ρ and p. Consider constant density flow, such as air at low speed or water without significant contaminant concentrations. Then pressure is the only additional dynamical variable. Another equation is needed to predict it.

In many situations, the interest is in essentially incompressible flow. Incompressible means that pressure changes produce negligible density changes: $d\rho/dp \approx 0$. The reader might recognize that $d\rho/dp$ is one over the squared sound speed, c^{-2}, in a gas. In that case, the approximation of incompressibility is justified if the Mach number, $M = u/c$, is small. In other words, the smallness of $d\rho/dp$ is a relative statement. It says that pressure variations go primarily into accelerating the flow rather than into changing the density. Certainly air is compressible; it can be compressed in a pump, or into a tire. But for the purpose of fluid dynamics, it can be treated as incompressible if the flowing fluid does not cause significant compression. This is normally the case when the Mach number is low. Liquids are almost always incompressible. Density variations can occur, such as those due to salt dissolved in water, but, again, it is not the fluid velocity that causes density to vary.

The condition of incompressibility is that infinitesimal fluid elements retain their volume. Their shape will deform, but the net volume associated with an element is constant. We are defining the fluid element as a fixed amount of mass, so that constant density is equivalent to constant volume. The condition of incompressibility requires the divergence of the velocity to vanish:

$$\nabla \cdot u = 0 \qquad (1.15)$$

or $\partial_x u + \partial_y v + \partial_z w = 0$. Essentially, this is saying that the volume deformations in the x, y, and z directions sum to zero. It may be justified as follows.

Consider a rectangular material element – for instruction, we work in two dimensions. The corners of the rectangle move with the fluid. Let the lower-left and upper-right corners be (X, Y) and $(X + \delta X, Y + \delta Y)$. The area is $\mathcal{A} = \delta X \delta Y$. The rectangle will deform as the fluid flows, but if it is incompressible, the area does not change. Then Eq. (1.15) follows from differentiating the area with respect to time and setting it to 0:

$$\begin{aligned}
\frac{d\mathcal{A}}{dt} = 0 &= \delta Y \frac{d}{dt}\delta X + \delta X \frac{d}{dt}\delta Y \\
&= \mathcal{A}\left(\frac{1}{\delta X}\frac{d}{dt}\delta X + \frac{1}{\delta Y}\frac{d}{dt}\delta Y\right) \\
&= \mathcal{A}\left(\frac{\delta u}{\delta X} + \frac{\delta v}{\delta Y}\right) = \mathcal{A}\nabla \cdot \boldsymbol{u},
\end{aligned}$$

where $\delta u = d\delta X/dt$ and $\delta v = d\delta Y/dt$ were substituted. In three dimensions, the same argument is applied to a fluid volume.

1.4 Reynolds Number

The boundary condition on rigid surfaces is that the fluid immediately adjacent to the surface moves with the wall velocity. This is the no-slip condition. It says that there is no discontinuity in velocity at the wall. As a stationary wall is approached, the fluid velocity tends continuously to zero. This no-slip boundary condition is not indisputable. In rarified gases, and at the intersection of a gas–fluid interface with a wall, it is violated. But at normal densities and pressures, and in homogeneous fluids,* it is correct. Certainly, if one stands in a stiff breeze, the air appears to flow freely over one's face. Nevertheless, in the immediate vicinity of a stationary body, the air velocity tends to zero. It does so across a thin boundary layer – see Chapter 5. The sensation of airflow at the surface is due to pressure and heat transfer effects produced by the flow.

No-slip is certainly not an obvious condition. Early fluid dynamicists, including Stokes, were uncertain. The book *Hydrodynamics* by Horace Lamb was called the bible of fluid dynamics in the late 19th and early 20th centuries. It contains a section entitled on the "question of slipping." The matter was resolved initially by comparing the predictions of flow through a pipe, allowing for the possibility slip, to experiment. The data implied no-slip: $u = 0$ at a stationary surface. Subsequently, high-resolution measurements of velocity proved no-slip beyond doubt.

The no-slip boundary condition is the origin of viscous drag: if the fluid away from the surface is in motion, and the velocity is brought to zero as the wall is approached, then there is a shearing stress imposed on the wall. This is illustrated by the simple examples of Couette and Poiseuille flow.

Interestingly, Poiseuille was a French physician – albeit with training in physics. His research on the flow of blood in arteries led him to investigate flow through narrow tubes. In 1838, he experimentally determined his famous law that the volume flow rate through a circular tube of radius a and length L is given by

$$Q = \frac{\Delta p}{8\mu L}\pi a^4, \tag{1.16}$$

where Δp is the pressure drop across the pipe. Stokes's mathematical derivation of this result was one piece of evidence in favor of the Navier–Stokes equations.

Couette flow is named for a French scientist who in 1890 measured the viscosity of liquids in what has come to be known as a Couette flow viscometer. Couette and Poiseuille flow are *parallel* shear flows in the sense that the flow is unidirectional. In Couette flow, the velocity is in the x direction and is a linear function of the cross-stream coordinate, y: the exact expression is simply

$$u(y) = U_w \frac{y}{H}. \tag{1.17}$$

This is the velocity between two flat walls, the lower being stationary and the upper moving with velocity U_w and being located at $y = H$. The shear is uniformly equal to $dU/dy = U_w/H$, and the viscous stress is μ times this. The no-slip condition requires fluid next to the upper wall to move at U_w and that next to the lower to remain stationary, hence producing the shear and the viscous shear stress. This is an exact, though trivially simple, solution of the Navier–Stokes Eq. (1.14).

The viscous stress is $\mu U_w/H$. Normalized by $1/2\rho U_w^2$, this becomes

$$\frac{2\mu U_w}{H\rho U_w^2} = \frac{2}{(U_w H/\nu)} = \frac{2}{Re}$$

---

* Even at interfaces, it is only in a tiny region around the three phase contact line that no-slip is violated.

Re, defined here to be $U_w H / \nu$, is the Reynolds number. It is nondimensional: the numerator has dimensions of velocity times length; the denominator has dimensions of length squared per time. The Reynolds number measures the ratio of viscous to inertial forces. In the context of Couette flow, it might not be appropriate to invoke the scaling $1/2 \rho U^2$ for inertial forces, but that is the form suggested by Bernoulli's equation [Eq. (1.38)].

The other elementary example, Poiseuille flow, is *fully developed* flow in a pipe. Fully developed means that the velocity profile is unchanged with downstream distance. Hence, it is the flow in a straight pipe, well downstream of the inlet. When the flow is not fully developed, CFD can be used to compute the velocity profile, and to examine how it approaches a fully developed condition, if the pipe is long enough for that to occur.

The Poiseuille velocity, as a function of radius from the center of the pipe, is a parabola:

$$u = 2\overline{U}\left(1 - \frac{r^2}{a^2}\right),\qquad (1.18)$$

where a is the radius and $\overline{U}$ is the cross-section-averaged velocity. The radial coordinate, r, is zero at the center and is a at the wall of the pipe. In §1.5, the parallel flow approximation will be described more fully. Parallel flow analysis shows that the cross-section-averaged velocity is related to the pressure gradient down the pipe by Poiseuille's law: $\overline{U} = -a^2(dP/dx)/8\mu$. This says that a net pressure drop, $dP/dx < 0$, is needed to overcome viscous friction and cause the fluid to flow. In a length L of pipe, this pressure drop is $8\mu\overline{U}L/a^2$.

Again, normalize the pressure drop by $1/2\rho U^2$ to compare viscous and inertial forces:

$$\frac{8\mu\overline{U}L}{1/2\rho\overline{U}^2 a^2} = \frac{16(L/a)}{\overline{U}a/\nu} = \frac{16(L/a)}{Re},\qquad (1.19)$$

where $Re \equiv \overline{U}a/\nu$ defines the Reynolds number based on bulk velocity and pipe radius. The numerator is just a nondimensional statement that the pressure drop increases linearly with the length of the pipe, in units of pipe radii.

In both Couette and Poiseuille flows, the Reynolds number emerges as a measure of the role of viscous friction. The Reynolds number is named for the British engineer Osborne Reynolds. In an article published in 1883, he addressed the question of when flow through a pipe transitioned from a laminar to a turbulent state, concluding that it was characterized by the nondimensional parameter that was subsequently named for him. His analysis was largely concerned with resolving discrepancies between the Navier–Stokes equations and observations of the rate of volumetric discharge from pipes. He showed experimentally that the discrepancy was due to unsteadiness and estimated a critical value of his parameter at which the unsteadiness could be expected. His pioneering researches earned him credit for the most commonly cited, nondimensional parameter of fluid dynamics.

A brief, entertaining history of the term *Reynolds number* was written by Rott (1990). He credits Sommerfeld with coining the term in a 1908 article at the Fourth International Congress of Mathematicians in Rome. However, he credits Prandtl with popularizing the term through his writings in the 1910s. The more expansive view of Reynolds' parameter is that it embodies the notion of dynamic similarity. His initial motive of characterizing hydrodynamic instability is just one instance. Generally, the motion of two fluids of differing velocity will be similar if they are placed in geometrically similar apparatuses and flowed with a speed adjusted so that UL/ν is the same for both fluids. For instance, the velocity field of air or water that approaches cylinders of radii a_a and a_w at speeds U_a and U_w, respectively, will be identical if these speeds and diameters are such that $U_a a_a / \nu_a = U_w a_w / \nu_w$, provided the velocities are normalized by U_a and U_w.

This broader understanding of the significance of the Reynolds number is justified by putting the Navier–Stokes equation [Eq. (1.14)] into nondimensional form. If the nondimensional velocity is defined as $\tilde{u} = u/U_r$, where U_r is a reference velocity, and the nondimensional coordinate is defined by $\tilde{x} = x/a$, where a is a reference length, then it is found that the Navier–Stokes equation takes the form

$$Re \left[\frac{\partial \tilde{u}}{\partial \tilde{t}} + (\tilde{u} \cdot \nabla)\tilde{u} \right] = -\nabla \tilde{p} + \nabla^2 \tilde{u}. \tag{1.20}$$

The fluid properties enter only through the parameter Re. The solution, in any given geometry, is a function solely of Re. That is why the chapters of this book are organized into ranges of Reynolds number. The various behaviors of fluid flow are obtained by solving the governing equations with representative values of this parameter. This is referred to as dynamic similarity: if two flows have the same geometry and boundary conditions (in dimensionless form), and if they also have the same Re, their entire flow fields will be equivalent. For instance, it does not matter whether the fluid is water or air. The kinematic viscosity of air is 15 times that of water. The product of U and a must be 15 times larger in air to obtain dynamic similarity. In a computer simulation using a fixed grid, the fluid type and inlet velocity can be selected to achieve a desired Re.

Conversely, given the geometry, fluid type, and boundary conditions, the flow is a function of Re – often quite dramatically so. Flow phenomena as a function of Reynolds number will be discussed in Chapter 4.

Actually, Eq. (1.20) is the form obtained if pressure scales with viscosity as in the above solution for Poiseuille flow: $p \sim \mu U_r / a$. But if the pressure scales as ρU_r^2, then the Navier–Stokes equation assumes the form

$$\left[\frac{\partial \tilde{u}}{\partial \tilde{t}} + (\tilde{u} \cdot \nabla)\tilde{u} \right] = -\nabla \tilde{p} + \frac{1}{Re} \nabla^2 \tilde{u}. \tag{1.21}$$

Equation (1.20) gives an insight into flows with low Reynolds number, Eq. (1.21) gives an insight into high Reynolds number flow. In the former case, letting $Re \to 0$ leaves only the viscous and pressure forces. This is the limit of inertialess flow, which is the subject of Chapter 3.

In the case of Eq. (1.21), letting $Re \to \infty$ leaves a balance between pressure forces and convective acceleration. This is the limit discussed in Chapter 5, with one important caveat: viscous forces will always come into play very close to surfaces. The implications of that caveat will be explored in Chapter 5.

1.5 Parallel Flow Approximation

1.5.1 Couette–Poiseuille Flow

The Navier–Stokes equations have a few exact solutions. They are for very simple, idealized geometries. These exact solutions can provide estimates of quantities like pressure drop or frictional drag. They can provide a framework for less simple cases, which are computed numerically. One rather useful class of exact solutions is that of parallel flows.

Mathematically, a parallel flow is one in which the velocity does not vary in the direction of the flow. For instance, if the flow is in the x direction, the derivative of $\boldsymbol{u}$ in the direction of the flow is $(\boldsymbol{u} \cdot \nabla)\boldsymbol{u} = u\partial\boldsymbol{u}/\partial x$. The operator $\boldsymbol{u} \cdot \nabla$ is the projection of the derivative into the direction of the flow. If $\boldsymbol{u}$ is only a function of y and z, this vanishes. The fully developed flow down a pipe is a velocity, u, as a function of radius, r, and hence is a parallel flow.

In steady flow, the particles move along streamlines. If the velocity does not change along the streamline, then the particle experiences no acceleration. The condition for the velocity to be constant along a streamline is that the distance to neighboring streamlines be constant: that is, that they be parallel. The reasoning is as follows: by definition, streamlines are in the direction of the fluid flow; there is no flow across them. Hence, the mass flow follows the streamlines. If the space between two streamlines were to decrease, the flow would have to accelerate to preserve the mass flux. Hence, zero acceleration requires parallel flow.

If the flow is steady and parallel, inertia vanishes, even if the Reynolds number is not small. Then the equations become linear: it is not surprising that parallel flow solutions play a large role in classical fluid mechanics.

When the acceleration vanishes, the Navier–Stokes momentum equations (1.11) become

$$0 = \nu \left(\frac{\partial^2 u}{\partial y^2} + \frac{\partial^2 u}{\partial z^2} \right) - \frac{1}{\rho}\frac{\partial p}{\partial x},$$

$$0 = \frac{\partial p}{\partial y}, \tag{1.22}$$

$$0 = \frac{\partial p}{\partial z},$$

for the case of flow in the x direction, as a function of y and z. If v and w are zero, the y and z derivatives of pressure must also be zero; hence, pressure can only be a function of x. But u is independent of x, so Eq. (1.22) implies that $\partial_x p$ cannot be a function of x and, hence, must be a constant.

Equation (1.22) describes the flow down a pipe of arbitrary cross section. When the pipe is circular, the velocity is a function of $r = \sqrt{x^2 + y^2}$. Equation (1.22) can be transformed into Eq. (E1.1) on p. 58. It will be left as an exercise for the reader to explore analyses of flow in circular cross-section pipes.

If u is a function only of y and $\partial_x p$ is constant, Eq. (1.22) is readily integrated to obtain

$$u = A + By + \frac{y^2}{2\mu}\partial_x p. \tag{1.23}$$

That is, the velocity is a second-order polynomial. This parabolic profile is sometimes used as the inlet condition for a computation. If the flow enters the domain through a channel, this might be a suitable prescription.

The integration constants in Eq. (1.23) are determined by boundary conditions. For instance, Couette flow (1.17), has zero pressure gradient and $u(0) = 0$, $u(H) = U_w$, giving $A = 0$, $B = U_w/H$. Plane Poiseuille flow has $u(0) = 0 = u(H)$ giving $A = 0$, $B = -H\partial_x p/2\mu$.

The availability of simple formulas for parallel flow makes it useful for approximate analysis. In the case of plane Poiseuille flow, the mass flux per unit width, $\dot{m}$, is

$$\dot{m} \equiv \int_0^H \rho u \, dy = \int_0^H \frac{y^2 - yH}{2\nu}\partial_x p = -\frac{H^3}{12\nu}\partial_x p \tag{1.24}$$

having substituted Eq. (1.23) with the appropriate values of A and B. This is the planar analogue to Poiseuille's law (1.18).

If flow through a duct is computed with prescribed inlet mass flow, the pressure drop can be estimated from this formula. For the purpose of estimation, let us allow H to be a function of x. Then, integrating gives

$$p_{\text{out}} - p_{\text{in}} = -12\nu \int_0^L \frac{\dot{m}}{H^3(x)} dx \tag{1.25}$$

for the pressure drop across the length from $x = 0$ to $x = L$.

Parallel flow as an approximation can be illustrated by computing the flow in a curved, contracting duct. The uppermost portion of Figure 1.2 shows the development of the velocity profile in the entrance to such a duct. Corresponding velocity vectors in the entrance region are show at the lower left. They start from the plug flow, $u = $ constant. The parabolic profile develops fairly close to the entrance in this case. The Reynolds number is the low value of 10. A standard estimate of the entrance length equates it to $0.06HRe$. In this case, the parabolic profile is established by $x \approx 0.6H$.

Pressure is assumed to depend only on x. That means that pressure contours should be vertical lines, as they are, except for a small departure in the curved region. The computed centerline pressure distribution is compared to the parallel flow approximation (1.25) at the lower right of Figure 1.2. The agreement is excellent, after the short entrance length. Consistent with Eq. (1.24), the pressure gradient

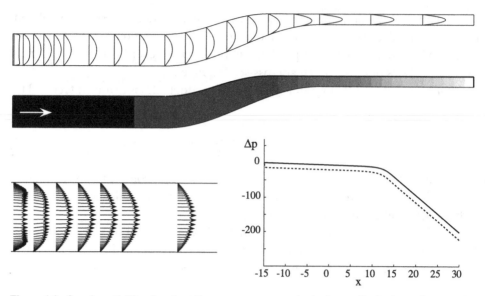

Figure 1.2. Quasiparallel flow in a duct. Pressure contours and velocity profiles in the entire domain; development of the velocity vectors in the entrance region. The computed centerline pressure distribution (– – – –) is compared to the theory (1.25) (———).

downstream of the bend is greater than the upstream pressure gradient by a factor of H^3; in this case H decreases by $1/3$ and pressure gradient increases by 27.

1.5.1.1 Branching Tubes

Rather complex examples of flow in a system of tubes are found in the human body. The cardiovascular system consists of highly branched tubes, through which blood flows, with the heart, as its pump. The pulmonary system consists of highly branched tubes, through which air flows, pumped by the lungs. Let us apply Poiseuille's law to these systems; after all, it was blood flow in arteries that motivated his studies.

At some level of abstraction, human circulatory systems can be analyzed as networks of branching pipes. A single pipe may fork into two at a junction. On average the total cross-sectional area of the human arteries and, in the pulmonary system, the bronchia, increases as the individual tubes become smaller. Let A_1 be the area before a bifurcation. Let βA_1 be the area of each of two equal tubes leaving the bifurcation. Then the area ratio is 2β. If the total area increases, $\beta > 1/2$.

Poiseuille's law (1.16), for pressure drop in a circular tube, can be restated as

$$\Delta p = \frac{8\pi\mu QL}{A^2}, \tag{1.26}$$

where $A = \pi a^2$ is the cross-sectional area and $Q = \overline{U}\pi a^2$ is the volume flux. If the flow Q splits equally into two tubes, the volume flux in each is $1/2Q$, irrespective of their radii.

The pressure drop across a length L_1 of the larger tube and L_2 of the smaller tubes is

$$\frac{8\pi\mu Q}{A_1^2}\left[L_1 + \frac{L_2}{2\beta^2}\right],$$

wherein $\beta = A_2/A_1$. The pressure drop is a factor of

$$\frac{L_1 + L_2/2\beta^2}{L_1 + L_2}$$

larger than a single tube of the same total length. Bifurcation increases the pressure drop. Even if $\beta = 1/2$, so that the total area is unchanged, the pressure drop goes up. This is because the Reynolds number decreases. The Reynolds number can be expressed in terms of volume flux and area as

$$Re = \frac{Q}{\nu\sqrt{\pi A}}.$$

Hence $Re_2 = Re_1/(2\sqrt{\beta})$. Only if $\beta = 1/4$ does Re remain constant, but then the total area is not constant.

We have asserted that $\beta > 1/2$ in human circulatory systems. (That may be violated at the first level of bifurcation, but it is satisfied on average.) After n bifurcations, the Reynolds number has fallen by $(4\beta)^{n/2}$. In the larger arteries and airways of the lung, Reynolds numbers may be in the range of a few hundred to a few thousand (Lighthill, 1975). Although the higher values may verge on transition to turbulence, after one or two levels of bifurcation, only laminar flow can be expected; human arteries branch approximately 20 times. It can also be expected that the smaller arteries will dominate the pressure drop. If tube length scales in the same way as tube radius, that is, $L_2 = \sqrt{\beta}L_1$, then the smaller tubes contribute $1/(2\beta^{3/2})$ times the pressure drop of the larger tube. Generally, this will be greater than unity, but if $\beta = 2^{-2/3}$ the pressure drop will be the same across each generation of the system.

The assumption of Poiseuille flow is not likely to be good near to the bifurcation. It will take some distance for the parabolic profile to be established. In straight pipes, this entrance length is estimated as $0.24aRe_a$ – that is, the condition for 5% accuracy of the centerline velocity, starting from plug flow. In the entrance region, the profile is steeper, and the wall stress is higher, than in fully developed flow (Figure 1.2). At a bifurcation, the stress is especially large at the junction. In the larger arteries and bronchia, where the Reynolds number can be high, the entire length of tube may be in the entrance region. Resistance will be higher than predicted from Poiseuille flow estimates. After a few levels of branching, the entrance length will be less than the length of the tube.

1.5.1.2 Circular Couette Flow

Idealized, parallel flows are commonly used for approximate analysis. To illustrate, consider flow in the annulus between two concentric circular cylinders, as in Figure 1.3 at the left. At the right, the geometry is modified to make it be a very small pump that

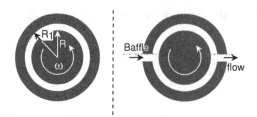

Figure 1.3. Circular Couette flow as a model.

operates by simply dragging fluid with a rotating shaft. A baffle prevents backflow. The geometry at left provides the parallel flow solution; it is used to model the situation at right.

The configuration at left is called circular Couette flow. From it, estimates of the volume flow rate (per unit length), Q, and of the torque exerted by the rotating shaft will be obtained.

The concentric circles have radii R and R_1, and the inner cylinder rotates with angular velocity ω. The flow is parallel because the velocity is in the angular direction and is only a function of radius. Let that velocity be $u_\theta(r)$. There can be no pressure gradient in the angular direction because of the circular symmetry. (If there were a gradient, it would have to be constant; but then the pressure at $\theta = 0$ would not equal that at $\theta = 2\pi$.) Because of the circular geometry, Eq. (1.22) takes the form

$$0 = \nu \frac{\partial}{\partial r}\left[\frac{1}{r}\frac{\partial(ru_\theta)}{\partial r}\right],$$

$$\frac{u_\theta^2}{r} = \frac{\partial p}{\partial r}.$$

The first is the form that the viscous force takes when expressed in cylindrical coordinates and in terms of the angular velocity. The general solution is just

$$u_\theta = \frac{A}{r} + Br$$

instead of the parabolic profile, Eq. (1.23), that obtains in planar flow. The conditions, $u_\theta = \omega R$ and $u_\theta = 0$ on the inner and outer cylinders, are met by

$$u_\theta = \frac{\omega R^2}{R^2 - R_1^2}\left(r - \frac{R_1^2}{r}\right).$$

The torque on the inner cylinder follows from Eq. (1.7). From that formula, the force in the angular direction is $2\pi R\sigma_{r\theta}$. The area of the wall is $2\pi R$ times width, which gives the first factor. The second is the viscous shear stress on the wall. The torque is R times this. In cylindrical coordinates, the formula for viscous stress is (White, 1991)

$$\sigma_{r\theta} = \mu r \frac{\partial}{\partial r}\left(\frac{u_\theta}{r}\right)$$

giving

$$4\pi\mu\omega R^2 R_1^2/(R^2 - R_1^2)$$

for the torque (per unit width) exerted *by* the cylinder *on* the fluid.

The volume flux (per unit width) that is dragged around in between the cylinders is

$$Q \equiv \int_R^{R_1} u_\theta \, dr = \frac{\omega R^2}{R^2 - R_1^2} \left[1/2(R^2 - R_1^2) - R_1^2 \log(R/R_1) \right].$$

For fixed R_1, the maximum of this formula for volume flow occurs when $R = 0.562\, R_1$, for which $Q = 0.108\omega R_1^2$. This is the maximum rate at which fluid can be pumped for a fixed shaft speed and outer radius.

Power is torque times angular velocity. With the above expression for torque, it is found that the power is

$$\text{Power} = 4\pi\mu\omega^2 \frac{R^2 R_1^2}{(R_1^2 - R^2)} = 4\pi\mu\omega^2 R_1^2 \times 0.4621,$$

when $R = 0.562 R_1$.

1.5.2 Oscillatory Boundary Layer

Another example of parallel flow is Stokes's oscillatory boundary layer. Consider a flat plate that is oscillated horizontally beneath a quiescent fluid. The plate is the plane $y = 0$, and it is oscillated in the x direction. First consider what is expected to happen. As the plate moves from left to right, it will drag the fluid immediately above it. That is because of the no-slip boundary condition. Viscosity will diffuse that movement higher into the fluid. A layer of fluid will be moving from left to right with the plate. The plate reaches its maximum displacement to the right and starts to return to the left. Again, fluid next to the wall is dragged along. Now there is a layer moving to the left with the plate. Above that, the fluid is still moving to the right, because leftward momentum has not yet diffused to that height. Following this reasoning for a few periods of oscillation leads to the conclusion that layers of alternatively left- and right-moving fluid will be created above the oscillating plate. That is indeed the case. These layers diffuse into one another, so as the height above the plate increases, the magnitude of the directional oscillations decreases.

The mathematical analysis consists in solving

$$\frac{\partial u}{\partial t} = \nu \frac{\partial^2 u}{\partial y^2}. \tag{1.27}$$

This is the unsteady parallel flow equation. At $y = 0$, the boundary condition is $u = U_w \sin(\omega t)$. As $y \to \infty$, $u \to 0$. The solution is

$$u = U_w e^{-y\sqrt{\omega/2\nu}} \sin(\omega t - y\sqrt{\omega/2\nu}),$$

or, if the layer thickness $\delta = \sqrt{2\nu/\omega}$ is introduced as an abbreviation,

$$u = U_w e^{-y/\delta} \sin(\omega t - y/\delta). \tag{1.28}$$

This embodies the physical picture of the previous paragraph. At any given time, the direction of the flow oscillates with height. The amplitude of the oscillations

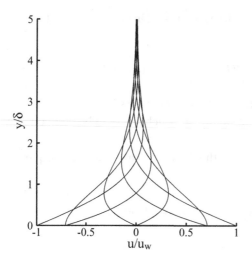

Figure 1.4. Velocity profiles at various times in Stokes's oscillatory boundary layer.

decays exponentially with height, as is seen in Figure 1.4. The exponential damping length scale is $\sqrt{2\nu/\omega}$. Lower frequencies penetrate higher above the wall than higher frequencies.

Viscous force is exerted by a moving wall. Work is done, at a rate equal to force times velocity. The viscous force, per unit area, is

$$\mu\left.\frac{\partial u}{\partial y}\right|_{y=0} = -\frac{\mu U_w}{\delta}\left[\cos(\omega t) + \sin(\omega t)\right].$$

The rate of work is this times the wall velocity:

$$F \cdot U_w = -\frac{\mu U_w^2}{\delta}\left[\sin(\omega t)\cos(\omega t) + \sin^2(\omega t)\right]$$
$$= -\frac{\mu U_w^2}{2\delta}\left[\sin(2\omega t) + 1 - \cos(2\omega t)\right].$$

The sinusoidal terms average over time to zero. Denoting the averaged rate of work by an overbar,

$$\overline{F \cdot U_w} = -\frac{\mu U_w^2}{2\delta}. \tag{1.29}$$

This, times the wall area, is the mean rate at which the wall does work on the fluid.

Where does that energy go? It is dissipated by friction. To see this, multiply Eq. (1.27) by ρu to obtain the energy equation

$$1/2\rho\frac{\partial u^2}{\partial t} = -\mu\frac{\partial u}{\partial y}\frac{\partial u}{\partial y} + \mu\frac{\partial}{\partial y}\left(u\frac{\partial u}{\partial y}\right).$$

An average of the left side over a period of oscillation vanishes because it is a time derivative: integrating is over one period gives $1/2\rho\left[u^2(t_1) - u^2(t_2)\right]$, where t_1 and t_2 are separated by one period, so that the velocities are equal and their difference is zero. The two terms on the left are the dissipation and diffusion of energy. The first is called dissipation because it is negative definite: it is viscosity times the square of the velocity gradient preceded by a minus sign. It can be only negative or zero. It will always be a sink of energy.

Integrating the two terms on the right from the wall to infinity in y and equating their time-average to zero gives

$$\overline{\mu \frac{\partial u}{\partial y} u}\bigg|_0 = \overline{F \cdot U_w} = \int_0^\infty \overline{\mu \left(\frac{\partial u}{\partial y}\right)^2} \, dy. \tag{1.30}$$

The leftmost term is recognized as the viscous force per unit of wall area times the wall velocity. The rightmost term is the total rate of energy dissipation within the fluid. The mathematics supports an understanding that work done on the fluid by the oscillating wall is balanced by dissipation interior to the flow. The dissipated energy would show up as heat.

An oscillatory flow also could be driven by an oscillating pressure gradient with a stationary wall (exercise 1.9). The solution is again Eq. (1.28) with the velocity $U_w \sin(\omega t)$ subtracted. U_w is now the velocity in the freestream. Energy dissipation depends only on the velocity gradient, so it remains unchanged; Eq. (1.29) is still valid. This is the rate at which energy must be supplied to maintain the oscillations.

An example of driven oscillations in a pipe is proposed in exercise 1.6. For Eq. (1.29) to be used in oscillating pipe flow, δ would have to be small compared to the pipe diameter. That is because it is a solution for a plane wall in an unbounded fluid. δ is small when the frequency is high. If the pipe diameter is d, the condition $\delta \ll d$ is equivalent to $\omega d^2 / \nu \gg 1$. If that condition is met, multiplying Eq. (1.29) by the total area of the wall provides an estimate of the rate of energy input that is needed to oscillate the flow against the retarding effect of viscous friction.

In §8.3.2 it will be explained how Eq. (1.29) provides an estimate of the rate at which free oscillations of a liquid sloshing in a container are damped. Dissipation is larger than might naively be supposed. It occurs in a layer of thickness δ. The naive supposition would be that in a container of radius a dissipation is spread through out the container, and hence $\overline{F \cdot U_w} = -\mu U_w^2 / 2a$. That is suitable at low frequencies, but Eq. (1.29) is correct when the frequency is high. The naive estimate is a factor of $\delta / a = \sqrt{2\nu / \omega a^2}$ smaller than the actual rate. At high frequency, the rate of energy dissipation can be very much greater than at first anticipated.

The parallel flow assumption is fairly restrictive. Without it, closed form solutions of the viscous equations are rarely possible. However, general considerations can be advanced that guide the study of fluid flow. They are the topic of §1.6.

1.6 Some Basics

Introductory fluid dynamics starts from the global conservation laws applied to a finite-sized control volume. Eventually, it leads to the differential form of the laws that apply to infinitesimal control volumes. As the differential laws have already been described, we proceed in the other direction to derive the global conservation equations from the differential form. The global conservation equations do not do

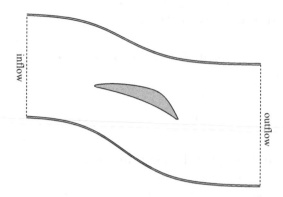

Figure 1.5. Schematic for one-dimensional conservation equations.

justice to the complexity of fluid velocity fields. However, they serve to fix ideas about the nature of fluid forces, about advection of momentum, and about the relation between fluid dynamics and forces on obstacles in fluid flow.

1.6.1 One-Dimensional Conservation Laws

The one-dimensional conservation laws, studied in introductory fluid dynamics, can be seen as the integral of the momentum and mass conservation laws (1.11) and (1.15) over a large control volume. This elementary statement of the governing laws is useful for designing computations, for making engineering estimates, and as a check on computations. They are summarized here as a reminder and for future reference; it is expected that the reader already has some familiarity with these basics.

Consider a flow domain consisting of inflow and outflow boundaries and surfaces – schematically, something like Figure 1.5. The divergence theorem of vector calculus can be used to integrate Eqs. (1.11) and (1.15) over the domain. The formal statement of this theorem is

$$\int_{\mathcal{V}} \nabla \cdot \mathbf{F} d\mathcal{V} = \int_{\mathcal{S}} \mathbf{F} \cdot \hat{n} d\mathcal{S}. \tag{1.31}$$

The integrals are over the volume and surface of any region. $\hat{n}$ is the outward normal to the surface. This basically is an elaboration of the fundamental theorem of calculus that

$$\int_{x_1}^{x_2} \frac{df}{dx} dx = f(x_2) - f(x_1).$$

The outward normal to the interval (x_1, x_2) is in the $+x$ direction at x_2 and in the $-x$ direction at x_1. Hence the right side of this equation is analogous to the right side of Eq. (1.31). The theorem applies to any volume of integration. In the case of Figure 1.5, it is the entire macroscopic region: applying the divergence theorem gives rise to the integrated, control volume equation that will be discussed shortly. In the case of a computer algorithm, the volume is a small computational cell, as will be described in Chapter 2.

First consider the continuity Eq. (1.15). Integrating over a control volume and applying the divergence theorem produces the equation of net mass conservation

$$0 = \int_{\mathcal{V}} \nabla \cdot \boldsymbol{u} d\mathcal{V} = \int_{S} \boldsymbol{u} \cdot d\boldsymbol{A},$$

where the notation $d\boldsymbol{A} = \hat{n} dS$ is introduced. Hence, in the case of Figure 1.5,

$$-\int_{\text{inflow}} \rho \boldsymbol{u} \cdot d\boldsymbol{A} = \int_{\text{outflow}} \rho \boldsymbol{u} \cdot d\boldsymbol{A}. \tag{1.32}$$

A in this equation is an area vector on the boundary, pointing outward from the flow domain. Thus Eq. (1.32) states that the mass flow out of the domain is equal to the mass flow into the domain. The minus sign arises on the left because A points out of the domain.

If u can be treated as constant on the inflow and outflow boundaries, Eq. (1.32) simplifies to the well-known, one-dimensional conservation equation

$$\rho u A|_{\text{in}} = \rho u A|_{\text{out}} \equiv \dot{m}. \tag{1.33}$$

In Eq. (1.33), u is the perpendicular component of the velocity crossing the boundary and A is the area of the boundary. $\dot{m}$ is the mass flux, which is the same across the inflow or outflow surfaces. For instance, if the outflow area is twice the inflow, and density is constant, then the velocity will fall by a factor of 2 from inlet to exit. A more correct statement is that the velocity averaged over the cross section falls by a factor of 2. The assumption that u is uniform will usually be overly crude. Nevertheless, it becomes needed when the momentum equation is analyzed.

The continuity equation allows the momentum equation (1.11) to be written in conservation form as

$$\frac{\partial \rho \boldsymbol{u}}{\partial t} + \nabla \cdot (\rho \boldsymbol{u}\boldsymbol{u}) = \nabla \cdot \boldsymbol{\sigma}.$$

If the flow is steady in time, an integral of this equation gives

$$\int_{\text{inflow}} (p d\boldsymbol{A} + \rho \boldsymbol{u}\boldsymbol{u} \cdot d\boldsymbol{A}) + \int_{\text{outflow}} (p d\boldsymbol{A} + \rho \boldsymbol{u}\boldsymbol{u} \cdot d\boldsymbol{A}) = \int_{\text{surfaces}} \boldsymbol{\sigma} \cdot d\boldsymbol{A} \tag{1.34}$$

having, again, made use of the divergence theorem. This is a statement that the sum of the forces on all surfaces internal to the domain must be balanced by the a combination of net momentum flow into the domain and pressure difference across the domain. It is assumed that viscous stresses on the inflow and outflow boundaries are negligible – so that σ contributes $-p$. The balance is stated here for time-independent flow. If the the flow is unsteady, the rate of change of momentum inside the control volume must be added in the left side of Eq. (1.34).

On the right side of Eq. (1.34), the area vector points into the surface: we have integrated over a fluid control volume; the direction outward to that volume points into the surface. In Eq. (1.7), the area vector points out of the surface because force on the surface is being computed. Hence, the right side of Eq. (1.34) is minus the surface force.

Again, we replace the local values of pressure and velocity by constant values and write the resulting quasi-one-dimensional momentum equation as

$$(pA\hat{n}|_\text{in} + pA\hat{n}|_\text{out}) + (\dot{m}u|_\text{out} - \dot{m}u|_\text{in}) = -\boldsymbol{F}_s. \tag{1.35}$$

Here $\dot{m} = \rho u A$ is the mass flux from Eq. (1.33) and Eq. (1.32) was invoked. The surface force on the right is the integral of the stress over all boundaries, except inflow and outflow, and includes both pressure and viscous forces. This represents the net force exerted *by* the fluid *on* the surface. That force is balanced by the deficit of momentum flux exiting the domain, added to the pressure drop.

For instance, in a straight duct, the sole force will be viscous drag, which is balanced by a pressure drop from inlet to exit: recall that $\hat{n}$ in Eq. (1.35) is negative at the inlet, because it points out of the domain. Hence, the momentum balance in a straight pipe is $(p_\text{in} - p_\text{out})A = F_\text{drag}$.

Or, in flow through a propellor in unbounded fluid with constant ambient pressure, there is no net pressure force on the control surface. The force exerted by the fluid on the propellor is balanced by $\dot{m}(u_\text{in} - u_\text{out})$. If the propellor is driving the fluid, the force is negative; the fluid exerts a negative force on the propellor; the propellor exerts a positive force on the fluid. It follows that $u_\text{out} > u_\text{in}$, as expected.

The integrated mass and momentum equations provide a helpful crutch for reasoning about the flow and forces. However, the real flow often is not uniform over the inlet and exit. For instance, drag-producing bodies generate a wake that must be accounted in the momentum balance. Without knowing the velocity profile, a momentum balance is not possible. In a computational analysis, rather than using the control volume momentum balance to infer the surface force, surface forces are obtained by solving for the stress tensor and integrating over the body to find the net force.

1.6.2 Bernoulli's Equation

Another basic of fluid dynamics is Bernoulli's equation. This expresses conservation of the energy of fluid elements along streamlines. It is derived by taking the dot product of the momentum equation (1.11) with the velocity. For constant density,

$$\boldsymbol{u} \cdot \nabla(p + 1/2\rho|\boldsymbol{u}|^2) = \boldsymbol{u} \cdot \left(\nabla \cdot \boldsymbol{\sigma}_\text{viscous} - \rho\frac{\partial \boldsymbol{u}}{\partial t}\right). \tag{1.36}$$

It is found that $p + 1/2\rho|\boldsymbol{u}|^2$ plays the role of potential plus kinetic energy. This quantity is not conserved: it varies along streamlines due either to unsteadiness or to viscous dissipation of kinetic energy into heat.

The operator $\boldsymbol{u} \cdot \nabla$ is the derivative in the direction of the stream. The direction of the velocity is $\hat{s} = \boldsymbol{u}/|\boldsymbol{u}|$. Hence, $\boldsymbol{u} \cdot \nabla p = |\boldsymbol{u}|\partial p/\partial s$, the derivative taken along a streamline. Equation (1.36) can be written as

$$|\boldsymbol{u}|\frac{\partial}{\partial s}\left(p + 1/2\rho|\boldsymbol{u}|^2\right) = -1/2\rho\frac{\partial|\boldsymbol{u}|^2}{\partial t} + \boldsymbol{u} \cdot (\nabla \cdot \boldsymbol{\sigma}_\text{viscous}). \tag{1.37}$$

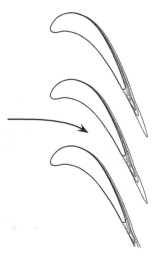

Figure 1.6. Contours of total pressure loss in flow through a cascade.

The right side can be neglected if the flow is steady and to a good approximation viscous dissipation can be ignored. Then Bernoulli's equation

$$p + 1/2\rho|\boldsymbol{u}|^2 = \text{constant} \equiv p_0 \qquad (1.38)$$

is obeyed. p_0 is the stagnation pressure or total pressure. Bernoulli's equation states that total pressure is a constant of motion; total pressure is constant along the streamlines of steady flow.

Viscous dissipation, on the right side of Eq. (1.36), would cause a loss of total pressure; some of the fluid kinetic energy would be dissipated into heat. Nonconservation of total pressure provides a measure of that loss. For example, in the application of CFD to turbomachinery, loss of total pressure is a measure of inefficiency. Contours of total pressure in constant density flow through a cascade of vanes are plotted in Figure 1.6. All of the contours are negative relative to the incident flow. The regions of pressure loss are near to the surface and in the wake. Everywhere else the total pressure is equal to that upstream of the vanes. It is not necessarily true that the viscous term on the right side of Eq. (1.36) is negative, but it almost always is. In Figure 1.6, loss of total pressure is a consequence of viscous action next to the wall.

The term *stagnation* pressure is used because if a flow with ambient pressure p_∞ and velocity u_∞ stagnates on a surface, the pressure at the stagnation point will be $p_0 = p_\infty + 1/2\rho u_\infty^2$. Stagnation pressure can be measured by stagnating the flow on a pressure transducer. At the stagnation point of a body in incident flow, the pressure coefficient, defined as

$$C_p \equiv \frac{p - p_\infty}{1/2\rho u_\infty^2}, \qquad (1.39)$$

is unity. From Bernoulli's equation (1.38), $C_p = 1 - (u/u_\infty)^2$. This is unity when $u = 0$.

The flow past a blunt, rounded leading edge accelerates away from the stagnation point to speeds greater than u_∞. This is detected by C_p becoming negative. The

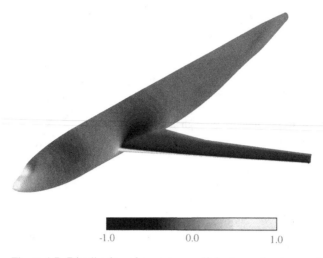

Figure 1.7. Distribution of pressure coefficient on a fuselage and wing. Figure courtesy of Edwin van der Weide.

example of Figure 1.7 shows regions where $C_p = -1$ on the upper surface of the wing. That low pressure is the origin of the upward lift force on the wing.

Bernoulli's equation provides an estimate of the pressure that is needed in integrated, control volume analysis. In the simplest cases, there are three unknowns: the exit velocity, the exit pressure, and the force exerted on surfaces inside the control volume. These are found from three conservation relations: mass conservation gives the velocity, Bernoulli's equation gives the pressure, and momentum conservation gives the force.

1.6.2.1 Expanding Ducts

A standard illustration is the gradual pipe expansion, as in the top duct of Figure 1.8. The inlet has area A_1 and prescribed velocity u_1. The exit has area A_2. By mass conservation the exit velocity is $u_2 = u_1 A_1 / A_2$. The inlet pressure is p_1. Bernoulli's equation implies the exit pressure $p_2 = p_1 + 1/2\rho u_1^2 - 1/2\rho u_2^2$. Thus

$$p_2 = p_1 + 1/2\rho u_1^2 \left[1 - (A_1/A_2)^2\right].$$

The area expands, $A_2 > A_1$, and the pressure rises.

Figure 1.8. Pressure contours in gradual, axisymmetric expansion and in step expansion.

The momentum flux and pressure force are in the x direction at both the inlet and exit plane. Hence, the momentum balance determines a surface force in the x direction. The surface forces can be both viscous and pressure; however, Bernoulli's equation is not applicable if the dominant forces are viscous. In the pipe expansion, the pressure force is that acting on the sloping side wall. It acts in the minus x direction. From the momentum equation (1.35), the net integrated force is

$$F_s = (p_1 - p_2)A_2 + \dot{m}(u_1 - u_2)$$
$$= -1/2\rho u_1^2 A_2 \left(1 - A_1/A_2\right)^2.$$

(1.40)

We have referred pressure to p_1 in the first step. Equation (1.40) gives the excess force, relative to the ambient pressure at the inlet.

Another way to derive this result is to integrate the Bernoulli pressure along the wall of the duct. The x component of the pressure force on a length dx is $-p\,dA/dx \cdot dx = -p\,dA$. Hence,

$$F_s = \int_{A_1}^{A_2} (p_1 - p)dA = -\int_{A_1}^{A_2} 1/2\rho(u_1^2 - u^2)dA$$
$$= -1/2\rho u_1^2 \int_{A_1}^{A_2} \left[1 - \left(\frac{A_1}{A}\right)^2\right] dA.$$

Evaluating the integral recovers Eq. (1.40). This shows how it is indeed the pressure that exerts the force on the sloping wall.

It is curious that the force in Eq. (1.40) does not depend on the shape of the wall. That is true of ideal flow, in which effects of viscous friction are negligible. If the duct expands rapidly, such as in the backstep in the lower part of Figure 1.8, viscosity plays an unexpected role: it injects vorticity into the flow, producing a pocket of recirculating flow next to the wall, just downstream of the step, as shown by Figure 1.9. It turns out that the pressure in the recirculating region is very close to p_1. The contours in Figure 1.8 show this.

Contrast the smooth and abrupt expansions. In the smooth duct, the pressure varies continuously along the sloping wall. In the discontinuous duct, the vertical portion of the wall is subjected to a constant pressure, which is very close to that at the inlet. The flow leaves the top of the step tangentially, following the pattern of Figure 1.9, and does not follow the wall. The streamlines do not expand at the step, and the Bernoulli pressure remains equal to p_1. Thus, there is no pressure force on the vertical face; $F_s = 0$. In fact, the pressure in the recirculation region turns out to be slightly below p_1, so the force is very slightly positive. Rather than a pressure that pushes back on the sloping wall, there is a slight suction that pulls forward on the vertical face of the step. We know that from experiments and from computer simulation. It is not something that can be deduced by simple, quasi-one-dimensional, control volume analysis. This chapter opens with the question, Why study fluid dynamics? Here is a further motive; without an understanding of fluid dynamics it would not be possible to anticipate the forces experienced by bodies in a flow.

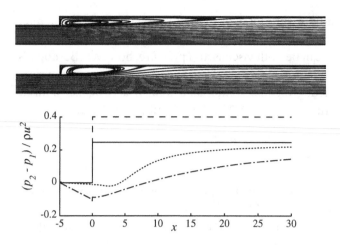

Figure 1.9. Streamlines for the step expansion, above; pressure distributions according to Bernoulli's equation (‐‐‐‐), zero force assumption (———), computation at Reynolds number of 400 (—·—), and turbulent flow (········).

Let us apply the one-dimensional conservation laws to the abrupt expansion. The three equations for mass and momentum conservation, along with Bernoulli's equation, previously provided the velocity and pressure in the downstream section, and the force on the sloping wall. Now, we replace one of these by the assumption $F_s = 0$. It is Bernoulli's equation that does not apply. The conservation of mass gives $u_2 = u_1 A_1 / A_2$. The conservation of x momentum, Eq. (1.35), with the net force set to zero, gives

$$A_1(p_1 + \rho u_1^2) = A_2(p_2 + \rho u_2^2) = A_2(p_2 + \rho u_1^2 A_1^2 / A_2^2).$$

Again, p_1 is taken as the reference pressure. After some algebra it is found that

$$p_2 - p_1 = \rho u_1^2 \left(\frac{A_1}{A_2} - \frac{A_1^2}{A_2^2} \right). \tag{1.41}$$

Because Bernoulli's equation was abandoned, total pressure is not conserved. By comparing Eq. (1.41) to the Bernoulli pressure the pressure loss can be evaluated. The Bernoulli pressure would be

$$p_2^B - p_1 = 1/2 \rho u_1^2 \left(1 - \frac{A_1^2}{A_2^2} \right).$$

Subtracting this from Eq. (1.41) gives the loss of total pressure,

$$p_2 - p_2^B = -1/2 \rho u_1^2 \left(\frac{A_1}{A_2} - 1 \right)^2. \tag{1.42}$$

Mass conservation ensures that the exit velocity is the same in both analyses, so this pressure loss equals the total pressure loss.

The right side of Eq. (1.36) contains terms that would cause departure from conservation of total pressure. In steady flow, only viscous dissipation is present. That must be the cause of the pressure loss (1.42). The precise mechanism is clouded by the simplifications inherent in the one-dimensional analysis. That analysis has an attractive simplicity, but it requires stringent assumptions and relies heavily on a priori knowledge of the fluid mechanics. The assertion that there is no pressure force on the vertical face of the step is just that: an assertion. It must be known that the flow separates tangentially from the top of the step. It must be asserted that somewhat downstream of the step the flow spreads to fill the channel. These are not obvious assumptions. The processes by which they occur require that Bernoulli's law be violated. Those processes might involve viscous diffusion of momentum; they might involve turbulent mixing. Again the question, Why study fluid dynamics? has an answer: the phenomena of fluid flow are often unexpected.

Let us consider how the simple analyses compares to computations. The geometry in Figure 1.9 is of a cylindrical pipe, with radius ratio 2/3. The area ratio is $A_1/A_2 = 4/9$. To compute a solution, a dissipative process must be represented. Computations of the centerline pressure are shown in the lower part of the figure for two dissipative mechanisms. In the first, the flow is laminar, at Reynolds number 400. Total pressure is lost to viscous friction. In the second computation, a turbulence model is used. Formally, the Reynolds number is 4×10^6. Total pressure is lost to turbulent fluctuations. The figure shows the one-dimensional, zero-force formula equation (1.41) as the solid line. It shows Bernoulli's equation as the dashed line. The recirculating zone, just downstream of the step, causes the streamlines to expand gradually. Eventually, toward the end of the duct, they spread across the radius and the pressure approaches the one-dimensional, zero-force analysis. The Bernoulli solution overestimates the exit pressure. Clearly, the zero-force assumption is more accurate.

The one-dimensional analysis assumes that the velocity is approximately constant across the duct cross section. That is more plausible for the turbulent computation than for the laminar one. The reason involves material from Chapters 5 and 6. The turbulent flow has a steeper gradient near the wall and a more nearly uniform velocity profile across the duct radius. The recirculating zone is also smaller in turbulent flow than in laminar flow at large Reynolds number. The streamlines at the top of Figure 1.6 are for the laminar case and below that for the turbulent case. The recirculation zone is about one-half as long in the turbulent computation.

1.6.3 Unsteady Flow

The integrated conservation equation (1.34) is written for steady flow. Unsteadiness is accommodated by restoring the time derivative of velocity. For generality, one can also allow a time-dependent control volume. To the left side,

$$\int_{\mathcal{V}} \rho \frac{\partial \mathbf{u}}{\partial t} d\mathcal{V}$$

is added. This is not the rate of change of momentum within the control volume: that is

$$\frac{\partial}{\partial t} \int_{\mathcal{V}} \rho \mathbf{u} d\mathcal{V} \equiv \dot{\mathcal{I}}$$

and is denoted by $\dot{\mathcal{I}}$. If the volume is changing,

$$
\begin{aligned}
\int_{\mathcal{V}} \rho \frac{\partial \boldsymbol{u}}{\partial t} d\mathcal{V} &= \dot{\mathcal{I}} - \int_{\mathcal{V}} \rho \boldsymbol{u} \frac{d\mathcal{V}}{dt} \\
&= \dot{\mathcal{I}} - \int_{A} \rho \boldsymbol{u} \boldsymbol{u}_{\mathcal{V}} \cdot d\mathbf{A}.
\end{aligned}
\tag{1.43}
$$

Here $\boldsymbol{u}_{\mathcal{V}}$ is the velocity of the boundary of the control volume. The component of velocity perpendicular to the boundary will contribute to changing the control volume. As the boundary changes, momentum will cross the surface, increasing or decreasing that contained in the control volume. Thus, two terms must be added to Eq. (1.35): one is the rate of change of momentum within the given control volume; the other is the rate of momentum change due to alterations in the boundaries of the control volume.

The modification to the integrated conservation equation is straightforward:

$$
\dot{\mathcal{I}} + \int_{cv} \left[p d\mathbf{A} + \boldsymbol{u} \, \rho (\boldsymbol{u} - \boldsymbol{u}_{\mathcal{V}}) \cdot d\mathbf{A} \right] = \int_{\text{surfaces}} \sigma \cdot d\mathbf{A}.
\tag{1.44}
$$

The rate of change of momentum is added at the left. Inside the integral, the mass flux through the control volume is defined relative to the moving surface.

Bernoulii's equation is altered by retaining the first term on the right side of Eq. (1.37). Then, in place of Eq. (1.38),

$$
p + 1/2\rho |\boldsymbol{u}|^2 + \int \rho \frac{\partial \boldsymbol{u}}{\partial t} \cdot d\boldsymbol{s} = \text{constant},
\tag{1.45}
$$

where the integral is along a streamline. This leads to the inference that inviscid pressure forces will have a component proportional to velocity squared and another proportional to the time derivative of velocity. Thus, the integrated pressure force on a body moving with velocity $V(t)$ will have the form

$$
F_p = 1/2 C_F \rho V^2 + \mathcal{M}_A \frac{dV}{dt},
\tag{1.46}
$$

where C_F is a force coefficient and $\mathcal{M}_A$ is called the added mass. We have already met the scaling of pressure force with $1/2 C_F \rho V^2$ in Eq. (1.40); the other, unsteady term represents a force that is required to accelerate the fluid.

As a simple illustration, consider a cylindrical tank with a pressure port at one end and a much smaller exit pipe at the other. The flow exits to ambient pressure. The pressure on the port is oscillated as $p = p_0 + \Delta p \sin \omega t$. We wish to find an equation for the velocity within the tank.

Let the ratio of tank cross-sectional area to exit area be A_T/A_e, which is assumed to be large. By mass conservation, $u_e = u_T A_T/A_e$. Applying Bernoulli's equation (1.45) along the axis of the tank,

$$
p + 1/2\rho u_T^2 = p_e + 1/2\rho u_e^2 + \rho \int_0^L \frac{\partial u}{\partial t} dx.
$$

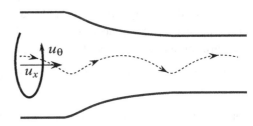

Figure 1.10. Schematic for swirling flow in a duct.

If the velocity is assumed to be uniform,

$$1/2\rho \left(\frac{A_T^2}{A_e^2} - 1\right) u_T^2 + \rho L \frac{\partial u_T}{\partial t} = p_0 - p_e + \Delta p \sin \omega t.$$

Given the oscillatory pressure, this is the dynamic equation for the velocity in the tank as a function of time. The mass of fluid in the tank is $\rho L A_T$. Hence, the last term on the left is the inertia of that fluid, so this has the form of Eq. (1.46). A more elaborate example is provided in §5.6.5.

1.7 Swirl

The simple analysis of §1.6.1 treats the velocity as a single component, directed across the surface of the control volume. Another simplistic treatment that can be informative is available for swirling flow. It also serves as an introduction to vortex dynamics.

Let the flow be axisymmetric about the x axis. Let the velocity components be u_x in the axial direction and u_θ, swirling about the axis of symmetry. This flow could be contained in a circular pipe with variable radius, $R(x)$. A particle released into such a flow will translate with velocity u_x, while simultaneously encircling the axis with velocity u_θ. The combined motion produces a helical trajectory, as illustrated in Figure 1.10.

This is referred to as swirling flow. Vorticity is concomitant to swirl — vorticity is the topic of §1.8. The discussion of swirl provides an introduction to the notion of streamwise vorticity. The direction about which the fluid rotates is the axis of the vortex. In this case, it is the x axis. x is also the direction of the primary component of velocity; hence, the terminology *streamwise vortex* refers to the fact that the axis of the vortex is in the main flow direction.

In the inviscid limit of classical fluid dynamics, it is proved that angular momentum is conserved along streamlines (Batchelor, 1967). In this case $u_\theta r$ is the angular momentum per unit mass. Its constant value will be denoted $L(r)$. r is the radial location in the upstream part of the duct in Figure 1.10. In the simplest case, the angular velocity is constant, so that $u_\theta = \Omega r$ and $L = \Omega r^2$. If a fluid element that is at r in the upstream duct moves to r_1 in the downstream part of the duct, conservation of angular momentum along streamlines implies that its angular velocity is $\Omega r^2/r_1^2$ downstream. If the duct contracts, the angular velocity increases.

Before discussing swirling flow through a duct, the notion of *radial equilibrium* of the pressure is introduced. If the fluid is to follow a curved trajectory, an inward pressure force must balance the outward centrifugal acceleration:

$$\frac{\partial p}{\partial r} = \rho \frac{u_\theta^2}{r}.$$

For a vortex in an unbounded fluid, with pressure referred to that at infinity, the radial distribution of pressure is

$$p = -\int_r^\infty \rho \frac{u_\theta^2}{r} dr. \tag{1.47}$$

The pressure inside the vortex is less than ambient to balance the centrifugal acceleration.

The Rankine vortex provides an example. It is defined by the prescription $u_\theta = \Omega r$ for $r < R_c$ and $u_\theta = \Omega R_c^2/r$ for $r > R_c$. R_c is the radius of the vortex core. The motive for this distribution will become clear in §1.8.1. From Eq. (1.47), the pressure distribution is

$$p - p_\infty = 1/2\rho\Omega^2(r^2 - 2R_c^2) \tag{1.48}$$

for $r < R_c$, inside the core.

The pressure (1.48) is subambient, dropping to $-\rho\Omega^2 R_c^2$ at the center. Low pressure at the center of a vortex is the cause of condensation that is sometimes seen near the tip of an airplane wing. If the humidity is near to 100%, the reduction in pressure can cause moisture to condense into water vapor at the core of the tip vortex, making it visible.

Suppose the swirling flow is inside a pipe of variable radius. If the cross section is reduced, the constancy of angular momentum, $u_\theta r$, requires that u_θ increases. At the same time, conservation of mass flow, $u_x \pi r^2$, implies that u_x increases. Hence, there is an association between streamwise acceleration and intensification of swirl. Conversely, streamwise deceleration is associated with reduction of swirl.

Let us consider an approximate, one-dimensional analysis. It is desired to know the pressure on the wall of a pipe containing flow and swirl. This is broken into two parts. First the pressure is determined on the axis. Then radial equilibrium is used to determine the pressure rise from the axis to the wall.

Let the upstream radius be R_u, and suppose the flow is swirling with uniform angular velocity Ω_u. The circumferential velocity at the wall of the pipe is $\Omega_u R_u$. The axial velocity is u_{xu}. Let the pipe contract downstream, to radius R_d. By mass conservation $u_{xd} = u_{xu} R_u^2/R_d^2$.

Along the axis, the circumferential velocity is zero. Applying Bernoulli's equation along the axis gives

$$p_d^c - p_u^c = 1/2\rho u_{xu}^2 \left(1 - \frac{R_u^4}{R_d^4}\right), \tag{1.49}$$

where the superscript c refers to the centerline.

From the axis to the outer wall, radial equilibrium applies, and a uniform angular velocity is assumed. Conservation of angular momentum gives $\Omega_d = \Omega_u R_u^2/R_d^2$ for the downstream angular velocity. The ratio of radii squared equals the area ratio. Hence, the angular velocity increases inversely with the area ratio: for instance, halving the area of the duct will double the angular velocity.

Now, from Eq. (1.47),

$$
\begin{aligned}
p(R_d) &= p_d^c + 1/2\rho\Omega_d^2 R_d^2 \\
&= p_d^c + 1/2\rho\Omega_u^2 \frac{R_u^4}{R_d^2}.
\end{aligned}
\tag{1.50}
$$

We wish to relate this to the pressure on the upstream wall. This is accomplished by invoking Eq. (1.49) to relate the centerline pressure to its upstream level and then invoking radial equilibrium to get to the pressure on the wall:

$$
\begin{aligned}
p(R_d) &= p_u^c + 1/2\rho u_{xu}^2 \left(1 - \frac{R_u^4}{R_d^4}\right) + 1/2\rho\Omega_u^2 \frac{R_u^4}{R_d^2} \\
&= p(R_u) + 1/2\rho u_{xu}^2 \left(1 - \frac{R_u^4}{R_d^4}\right) + 1/2\rho\Omega_u^2 R_u^2 \left(\frac{R_u^2}{R_d^2} - 1\right).
\end{aligned}
$$

This shows how swirl and area ratio relate the downstream pressure to that upstream. A nondimensional swirl parameter can be defined as

$$
S = \frac{\Omega_u R_u}{u_{xu}}.
$$

Then the normalized pressure difference can be rewritten as

$$
\frac{p(R_d) - p(R_u)}{1/2\rho u_{xu}^2} = \left(1 - \frac{R_u^4}{R_d^4}\right) + S^2 \left(\frac{R_u^2}{R_d^2} - 1\right).
\tag{1.51}
$$

For an area contraction, $R_u > R_d$. Then the first term on the right is negative, so the pressure falls as a consequence of the flow acceleration in the absence of swirl. Swirl counteracts that; the second term on the right is positive. As the area contracts, Ω increases and so does the centrifugal acceleration. To balance it by an inward pressure gradient, the pressure on the wall of the pipe must rise.

For instance, for an area ratio $R_u^2/R_d^2 = 2$, Eq. (1.51) gives a pressure drop of $S^2 - 3$. It might seem that a swirl parameter of $\sqrt{3}$ will cancel the Bernoulli pressure drop. Indeed, a large swirl would seem able to increase the pressure above its upstream level, which would cause the direction of the flow to reverse. Strong swirl can indeed cause the flow to reverse. That effect is used in swirl stabilized combustors to hold the flame in place. However, the inference of such effects from the present considerations is not safe because of an approximation made by the one-dimensional analysis. A more rigorous analysis shows the one-dimensional approximation to be valid only if S is small (Batchelor, 1967).

Here is the reason that this qualification is necessary. Because the swirl varies as the flow passes through the contracting duct, the angular velocity of fluid elements is a function of x. Consider a point at radius r_1 upstream that moves to radius r_2 downstream. It will rotate with angular velocity L/r_1^2 upstream, where L is its angular momentum, and with the greater angular velocity L/r_2^2 downstream. Consider a line drawn from r_1 upstream to r_2 downstream. The two ends of that line rotate with different angular velocities. If it is initially straight, it will subsequently twist into a helix. It might not be obvious at this point, but the one-dimensional analysis assumes that this line does not twist.

Why this is so provides an introduction to vortex dynamics. The line that twists is analogous to a vortex line. The vortex line shows the direction of the vorticity. In passing downstream that direction will be rotated from x for the same reason that an initially straight material line twists: the angular rate of rotation varies with x. After passing through the contraction, vortex lines will be twisted into helices. Correspondingly, the swirl in the downstream duct cannot be solely in the streamwise direction. It will develop a component in the θ direction too. One could say that a ring component of swirl is added to the axial component to produce the helix. The ring component alters the profile of u_x versus r from uniform. In a contraction the ring component is such that the flow on the axis is decreased and near the outer wall is increased.

1.7.1 Computation of Swirling Flow in an Expansion

Let us see how computations compare to the simple analysis. We will return to the expanding pipe of Figure 1.8, so that $R_u < R_d$ in Eq. (1.51). Introduce swirl in the upstream duct.

An inviscid flow computation was performed. The computed centerline pressure is provided in Figure 1.11 at the top left. The solid curve is a nonswirling solution, which agrees very closely with the one-dimensional analysis (1.49). In that analysis the centerline pressure is not affected by swirl. That seems reasonable, because $u_\theta = 0$ on the axis. However, it is seen in Figure 1.11 that the pressure in the downstream duct rises with swirl. The simple analysis predicted no effect at the centerline. Bernoulli's equation is valid, but the assumption that u_x is not a function of r is only correct for small swirl. The ring component of vorticity produces the radial profiles of velocity shown at the right of Figure 1.11. The low velocity on the axis causes the increase of centerline pressure with increasing swirl.

The pressure on the outer wall is just a bit lower than the nonswirling case but considerably higher than what the one-dimensional analysis predicts. With $R_u/R_d = 2/3$, $u_{xu} = 1$ and $\rho = 1$, Eq. (1.51) predicts $p(R_d) - p(R_u) = 0.4 - 0.28S^2$. That is plotted as a dotted line at the at the lower left of Figure 1.11. The computed effect of swirl is less than predicted by the analysis: it is shown by the dashed-dot line. The trend is consistent with the simple analysis, but the analysis is not accurate quantitatively. Radial equilibrium is a good approximation from the centerline to the wall. However, the centerline pressure rises more than was anticipated by the one-dimensional analysis. Ring vorticity is produced in consequence of the expansion, which is not included in the analysis.

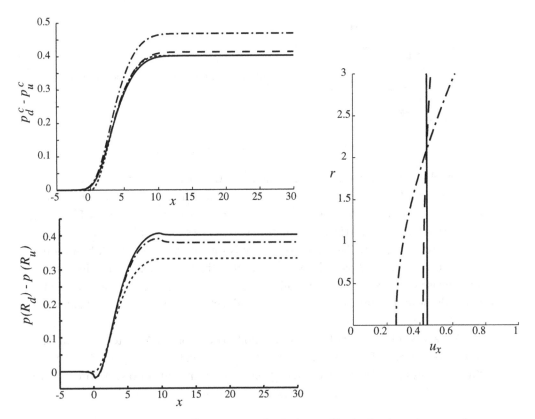

Figure 1.11. Centerline pressure, wall pressure, and velocity profiles in the downstream section of an expanding pipe with swirl. (———) $S = 0$, (– – – –) $S = 0.2$, (— · —) $S = 0.5$ (········). The one-dimensional analysis for wall pressure with $S = 0.5$.

This example is a good motivation to study the role of vorticity. That is our next focus.

1.8 Vorticity and Its Dynamics

Insights into fluid dynamics come from many sources. The integrated, control volume analysis of §1.6.1 gives a coarse-grained view of the conservation laws. The parallel flow approximation of §1.5 gives some flavor of viscous effects. Vorticity is another important notion.

Many aspects of fluid motion and convective mixing can be understood in terms of vorticity, its dynamics, and its associated velocity field. For instance, vortex dynamics might provide conceptual insights into features seen in CFD simulations. These might take the form of interpreting flow patterns in terms of vortices and their motion; it might take the form of computing vorticity, to provide a concrete structure to visualize.

The flow field associated with vortical structures can loosely be described as velocity that whirls around the vortex, in the locally perpendicular plane. This is

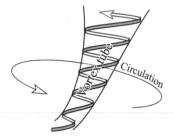

Figure 1.12. Schematic of a vortex and flow round it.

shown schematically in Figure 1.12. Sometimes we can understand convective mixing thus: vortices form in the flow; they convect fluid round themselves; they thereby transport material from one place to another. A striking example is Figure 1.13. Smoke has been released into the wing-tip vortex of a small airplane. It is swept round by the circulatory flow. The cloud of smoke is spread by advection. A by-product of the advective transport is a nice visualization of the tip vortex. As another example, streamwise vortices parallel to a wall have a tendency to increase heat transfer to, or from, the wall. In that case, they increase advective transport perpendicular to the surface.

We also study vorticity for its connection to fluid forces. A vortex in an oncoming flow experiences a force perpendicular to the approach velocity. In aerodynamics, this is the lift force. Vorticity conservation provides insights into such forces. A wing with circulation can be thought of as a vortex with strength equal to the circulation

Figure 1.13. Mixing in a wing-tip vortex. Photograph EL-1996-00130, NASA Langley.

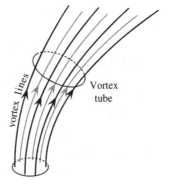

Figure 1.14. A vortex tube is constructed from vortex lines.

round the wing. Vorticity conservation requires that an equal and opposite vortex be shed into the flow to establish circulation. This type of force is understood via the production and dynamics of vortices.

There is ample motive to survey basic notions of vorticity and its dynamics, which is the objective of the following sections.

1.8.1 Vortex Kinematics and Dynamics

Mathematically, vorticity is defined simply as the curl of the velocity: $\omega = \nabla \wedge u$. It equals twice the local rotation rate of a fluid element. This is seen readily for solid body rotation with angular velocity Ω. The corresponding velocity is $u = \Omega \wedge r$; hence, $\omega = \nabla \wedge (\Omega \wedge r) = 2\Omega$. Thus the vorticity equals twice the angular velocity.

At times, that interpretation can be misleading. Fluid particles in vortical flow do not necessarily move in rotating paths. Shear flows – for instance, a jet – are long slender regions of vorticity. The velocity is unidirectional, without whorls, but it is a vortical flow – vortical shear layers will be discussed in Chapters 4 and 5. Only when it is concentrated in a closed region is vorticity associated with rotating flow or with whorls. Then one can refer to a perceptible "vortex." Such a region of vorticity has an associated circulating motion, as in Figure 1.12. The rest of this section is addressed to concentrated vortices. The notion of circulating motion is quite appropriate.

That is the motion, not the vortex. The motion produced by the vortex whirls around it, as in Figure 1.13. The vortex, itself, can be thought of as a material tube in the fluid, forming the axis of the whirling motion. Of course, vorticity is not a material, it is simply a nonzero value of $\nabla \wedge u$. Nevertheless, the equations of vortex motion are the same as those of a material tube: a mathematical analogy permits them to be thought of in a similar manner. Just as a material tube is carried by the fluid, so is a vortex tube – at least, in the inviscid limit. Viscosity adds the effect that vorticity diffuses out of the tube. In the absence of diffusion, the tube is convected, rotated, and stretched as it is carried by the flowing fluid, just as a material tube would be.

The tube is defined as a bundle of vortex lines, as illustrated by Figure 1.14. The lines are constructed by drawing arrows in the direction of the vorticity and connecting them head to tail. It is a consequence of the definition of vorticity that a vortex line cannot end in the fluid.* The lines in Figure 1.14 extend through the fluid until they hit a boundary, or until they close on themselves, to form a ring. The walls of the vortex tube are such as to contain a fixed set of vortex lines. The cross section expands, contracts, and rotates to contain the given set of lines. The arrows in Figure 1.14 point in the direction of the vorticity. The corresponding velocity is in the perpendicular plane and whirls around the vortex line, as illustrated by Figure 1.13.

A vortex is characterized by its total vorticity or circulation. Consider a circuit, encircling a cross section of the vortex tube as in Figure 1.14. The circulation, Γ, is equal to the integral of the velocity tangent to the circuit:

$$\Gamma = \oint \boldsymbol{u} \cdot d\boldsymbol{s}. \tag{1.52}$$

Gauss's theorem shows that

$$\oint \boldsymbol{u} \cdot d\boldsymbol{s} = \iint \nabla \wedge \boldsymbol{u} \cdot d\boldsymbol{A} = \iint \boldsymbol{\omega} \cdot d\boldsymbol{A},$$

so the circulation is, indeed, the vorticity contained in the tube. The area integral can be thought of as a summation of the vortex lines contained in the tube.

The definition (1.52) of circulation suggests a connection between the velocity swirling around the vortex and its circulation. Consider a circular path of integration, and let $\bar{u}$ be the average velocity tangent to that path. We average by integrating around a circle and dividing by its circumference:

$$\bar{u} = \frac{1}{2\pi r} \int_0^{2\pi} u_\theta r \, d\theta.$$

Then the right side of Eq. (1.52) is $2\pi r \bar{u}$. Hence that equation can be stated as

$$\bar{u} = \frac{\Gamma}{2\pi r} \hat{\boldsymbol{\theta}}, \tag{1.53}$$

where $\hat{\boldsymbol{\theta}}$ indicates that the velocity is in the angular direction, counterclockwise around the vortex if $\Gamma > 0$. The velocity falls off like $1/r$ outside the vortex. We refer to this as the velocity induced by the vortex. This relation between circulation and induced velocity is called the Biot–Savart law. Equation (1.53) is a kinematic relation; it states the velocity that is concomitant of vorticity. In terms of the Cartesian velocity components, Eq. (1.53) is

$$(u, v) = \Gamma \frac{(-y, x)}{2\pi (x^2 + y^2)}. \tag{1.54}$$

On the positive x axis, $y = 0$ and the velocity is in the positive y direction; on the positive y axis, $x = 0$ and the velocity is in the negative x direction.

---

* Mathematically, this follows from $\nabla \cdot \boldsymbol{\omega} = 0$, which follows from the identity $\nabla \cdot (\nabla \wedge \boldsymbol{u}) = 0$.

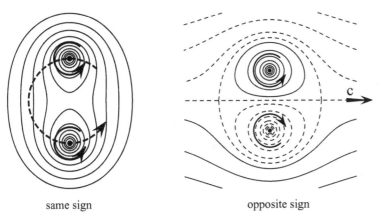

same sign   opposite sign

Figure 1.15. Interaction of two vortices of like and of opposite sign.

Given a number of vortices, the associated velocity is just a sum of expressions like Eq. (1.56), one for each vortex. Thus

$$u(x) = \sum_i \frac{\Gamma_i \hat{\theta}_i}{2\pi |x - x_i|},$$ (1.55)

where $|x - x_i|$ is the distance from a point in the fluid to the center of the ith vortex. Actually, this is an approximation, in which the vortex is represented as being concentrated at a point, x_i, at its center. It gives a qualitative perspective on vortex interactions. For example, vortices having Γ's of equal magnitude and opposite sign convect each other forward at speed $\Gamma/2\pi d$, where d is their separation. That is illustrated by Figure 1.15 at right. The reader can arrive at this result by recognizing that the lower vortex convects the upper vortex with the velocity (1.53). It does not convect itself because its own induced velocity circles around it. However, the upper vortex convects the lower one forward with the same speed, $\Gamma/2\pi d$, with which the lower convects the upper one. Hence, they move forward as a pair.

If the vortices have the same sign, they circle round each other, and hence around their midpoint, as in Figure 1.15 at left. The dashed line in that figure indicates the induced velocity. At this instant, the lower vortex induces a leftward velocity at the position of the upper vortex, and the upper induces a rightward velocity at the position of the lower. Vorticity will be advected with the fluid velocity; hence, the vortices circle about the midpoint between them. Actually, that is the idealized behavior of point vortices; finite size vortices both rotate and wrap into each other, if they are same signed. So we are primarily indicating the nature of the induced velocity rather than addressing the dynamics of a real distribution of vorticity. Those dynamics can be explored by CFD.

The right half of Figure 1.15 includes streamlines in a frame of reference moving with the vortices. A zone of fluid inside an elliptical region moves with the pair of vortices. Material in this region is carried by the vortices. This is a sort of mixing process: when the vortices eventually dissipate, the material they carry will have been transported some distance and deposited. The reason vortices dissipate is that

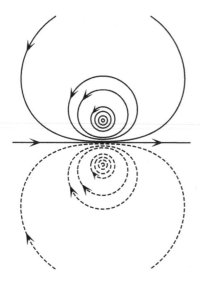

Figure 1.16. Streamlines round two equal and opposite vortices in a stationary frame of reference.

the positive and negative vorticity diffuse into one another and cancel. The diffusion coefficient of vorticity is molecular viscosity. In the presence of any finite level of viscosity, the vortex pair will eventually dissipate.

In a stationary frame of reference, the streamlines of a pair of opposite signed vortices are as in Figure 1.16. This pattern moves to the right with the speed of the pair.

The summation (1.55) implies a relation between the net vorticity in the flow and the far-field behavior, as $r \to \infty$. Assume that all vortices are located in a finite region. Let that region lie inside a circle of radius R. Now consider a point x which is very far outside that circle. When $|x - R| \gg R$ a good approximation to Eq. (1.55) is

$$u(x) \approx \frac{\hat{\theta}}{2\pi r} \sum_i \Gamma_i, \tag{1.56}$$

where $r = |x|$. This says that the set of vortices produces a velocity that corresponds to a vortex having their net circulation, located in their middle. Generally, far from the vortical region the velocity falls off like $1/r$ times the vorticity integrated over the region.

If the net circulation is zero, the previous far-field velocity vanishes. In that case, it can be shown that the velocity falls off like $1/r^2$ at large distances. Formally, that is the next term in an expansion in powers of $1/r$. As a concrete example, two vortices of equal and opposite circulation, separated by a distance d in the y direction, produce a velocity that becomes approximately

$$(u, v) \approx \frac{\Gamma d}{2\pi r^2} \left(1 - \frac{2y^2}{r^2}, \frac{2xy}{r^2} \right)$$

far from the vortices – that is, when $r \gg d$. Here r equals $x^2 + y^2$.

In addition to their relevance to vortex motion, these estimates of the rate of velocity fall off are used in designing computational domains (see Chapter 2).

The velocity will fall more rapidly when there is no net circulation than when is it nonzero. Consequently, a larger computational domain is needed in the latter case.

If the vortex is a curvilinear tube in three dimensions, the induced velocity still can be understood via Eq. (1.55). One imagines the tube to be made of segments, each with a small circulation $d\Gamma_i$ and a corresponding induced velocity. It is not hard to imagine that the small circulation of a line segment of length $d\ell$ can be defined in terms of a function $\gamma(\boldsymbol{x})$ that is the circulation per unit length. That is, $d\Gamma_i = \gamma d\ell$.

A thin ring is a tube of constant curvature. The segments that make up a circular ring are short arcs of a circle. Their induced velocity will add vectorially to produce a uniform translation of the ring. This is the proverbial "smoke ring," which is a vortex ring in nearly uniform translation. Another example of a self-propelled ring is the mushroom cloud that rises from a large explosion. The heat release causes rising air currents that roll into a vortex ring. The induced motion is upward, away from the ground.

When ideas about vortices are used to understand a flow field, that understanding often involves summing contributions of induced velocity from a distribution of vortices. This is simply an addition, as in Eq. (1.55). When these ideas are applied to understanding a computed flow field, more often than not, the "summation" is qualitative. One simply cites the distribution of vortices as an explanation of features that can be seen in the velocity field. Despite being qualitative, such notions provide a good conceptual framework.

Summing the induced velocities of a given distribution of vorticity is a kinematical exercise. Reasoning how the vorticity will be advected and rearranged by the flow, gives the exercise a dynamical component. The previous allusion to smoke rings is an example. They propagate forward in consequence of their own, self-induced velocity. In his classic 1884 monograph on vortex rings, the English physicist J. J. Thompson reasoned that two vortex rings lined up along their axes could leapfrog through and through one another. The reader might reconstruct the reasoning by thinking of the simpler case of two vortex pairs of opposite sign. Both pairs span a common axis. The pair to the rear will be brought together and accelerate through the front pair. After they pass through the front pair, they will move apart and their induced velocity will bring the other pair toward one another to repeat the process. This entertaining behavior can be simulated on a computer and has been produced in the lab. Of course, real vortices do not leapfrog indefinitely.

In general, self-induced velocities will distort the shape of a vortex. The local, self-induced velocity of a curved filament is inversely proportional to the radius of curvature. Portions of a filament that have small radius of curvature will move faster than straighter portions, thereby distorting its shape. For instance, say that a filament starts as a parabola lying in a horizontal plane. The radius of curvature is smallest at the apex of the parabola. Hence, the apex will lift out of the plane. We show this schematically by Figure 1.17. The parabolic filament will curve up or down, depending on the sign of its circulation from the horizontal. Parabolic filaments, curved up from the wall, are seen in turbulent boundary layers.

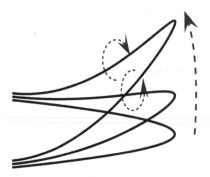

Figure 1.17. A curved filament moves with a speed inversely proportional to its radius of curvature. A planar loop lifts up unevenly. The position of the filament is shown schematically at three instants.

1.8.2 Stretching and Rotation

When a vortex is stretched, the magnitude of vorticity increases; that is, the core spins more rapidly. Stretching is illustrated at the top of Figure 1.18. As the core elongates and thins, it spins more rapidly because its angular momentum is conserved. One can reason as follows.

Circulation is conserved in an evolving vortex. That can be described as conservation of average vorticity times cross-section area

$$\Gamma = \overline{\omega}A = \text{constant}. \tag{1.57}$$

Conservation of mass says that a fluid cylinder of length $d\ell$ conserves its volume times density:

$$M = \rho d\ell A = \text{constant}. \tag{1.58}$$

In combination, these equations give

$$\overline{\omega} = \frac{\rho\Gamma}{M}d\ell. \tag{1.59}$$

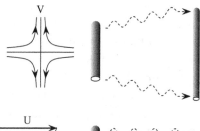

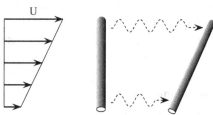

Figure 1.18. Basic distortions: stretching and rotation of vortex tubes.

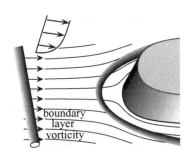

Figure 1.19. Schematic of a vortex tube being deformed in flow round a blunt body. A necklace vortex forms where boundary layer vorticity wraps around the junction between a bluff body and a wall.

Given Γ, M, and ρ, vorticity grows in proportion to increases in the length of a material line. This is a consequence of mass and circulation conservation. It represents the well-known phenomenon of spin up. For instance, as water is sucked down a drain it is often seen to form a whorl, as the vorticity is stretched and intensifies.

The relation between vorticity and material lines also applies in vector form:

$$\bar{\omega} \propto d\boldsymbol{\ell}. \tag{1.60}$$

The proportionality means that vorticity is carried in the same way as a material line is. Material lines translate and rotate as they are carried in the flow. They also elongate or contract as velocity gradients stretch or compress their length. Basic distortions in straining and shearing flow are illustrated in Figure 1.18. Straining by flow gradients parallel to the axis will stretch or compress the vortex. Shearing will both stretch and rotate the tube. Stretching decreases the cross section and increases the magnitude of vorticity inside the tube.

These deformations occur locally within a flow field. Figure 1.19 suggests distortions that occur in flow around a blunt body. The vorticity within an oncoming boundary layer lies in the y direction. It will stretch and wrap around an obstruction. In doing so, the vorticity intensifies, producing strong swirling in its core. This produces what is sometimes called a necklace or horseshoe vortex. Necklace vortices can occur at the base of a pier in a river. The intense whirling flow that forms leaves its mark by scooping out a pocket of sand.

To make the notions of stretching and rotation more concrete, it is instructive to consider simple, linear convection. For instance, the velocity of a linear straining flow is $u = \alpha x$, $v = -\alpha y$. The rate of straining along the x axis is $\alpha = \partial u/\partial x$. If $\alpha > 0$ this will cause stretching along the x axis; similarly, the negative straining along the y axis will cause compression. To see this mathematically, consider how a material line vector evolves. Convection of a material point is described by

$$\dot{X} = u = \alpha X; \quad \dot{Y} = v = -\alpha Y. \tag{1.61}$$

The capital letters denote the position of the moving material point. Now consider a short material line. Let its initial components of length be (dX_0, dY_0). It follows from the trajectory of material points (1.61) that the line element evolves as

$$dX = dX_0 e^{\alpha t}, \quad dY = dY_0 e^{-\alpha t}.$$

The analogy (1.60) implies that components of vorticity evolve in exactly the same way:

$$\omega_x = \omega_{x0}e^{\alpha t}, \quad \omega_y = \omega_{y0}e^{-\alpha t}.$$

If $\alpha > 0$, then the x component of vorticity grows exponentially, whereas the y component decreases. Vortices are stretched along the direction of positive mean rate of strain (see Figure 1.18) and compressed in directions of negative rate of strain.

Or, as another instance, if $u = \alpha y$, $v = -\alpha x$ then

$$\dot{X} = u = \alpha Y; \quad \dot{Y} = v = -\alpha X. \tag{1.62}$$

For a vortex initially aligned with the x axis, the initial condition is $dY_0 = 0$ and the solution is $dX = dX_0 \cos(\alpha t)$; $dY = -dX_0 \sin(\alpha t)$. Hence, the vortex tube rotates with angular velocity $-\alpha$.

Generally, flow gradients will stretch, compress, and rotate vortex tubes. In fact, each short segment $d\ell$ of the tube will be rotated and stretched by the local flow gradient. An initially straight vortex will be bent by mean velocity gradients, as illustrated by Figure 1.19. The portion of the vortex approaching the stagnation point, at the nose of the obstacle, moves more slowly than the portions above and below the stagnation streamline. Hence, the vortex is stretched and rotates toward the body as it approaches the surface. Along the stagnation streamline $\partial_y V > 0$; there is a positive rate of strain in the y direction. By the same token, there is a negative rate of strain in the x direction. A vortex tube lying along the stagnation streamline would become short and squat as it approached the stagnation point. The tube shown schematically in Figure 1.19 is perpendicular to the stagnation stream line, so it is stretched, becoming long and thin.

1.8.3 Origin of Vorticity

The exact equations of vorticity evolution in constant density flow show that it is carried by the fluid, being stretched and rotated. In a viscous fluid, it is also diffused. Whether viscous or inviscid, the total vorticity is conserved, and an isolated vortex conserves its circulation. In classical fluid dynamics, this is referred to as Kelvin's circulation theory, named after the physicist Lord Kelvin – whose name prior to peerage was William Thomson. Kelvin's theorem says that vorticity is neither created nor destroyed within the fluid. A similar observation is that a vortex tube cannot begin or end in the fluid. That is a matter of continuity: if the tube suddenly ended its vorticity would go discontinuously to zero. That cannot occur within the fluid. These observations are perplexing: if circulation is conserved within the fluid, how, then, is vorticity produced? If it is zero initially, why does it not stay so?

The answer is that it must be created at the boundaries. In a constant density fluid, vorticity is produced only at surfaces, through viscous action. The no-slip condition at a fixed surface requires that a fluid particle (call it particle 1) immediately next to the surface have vanishing velocity. Consider an infinitely thin plate, placed edge-on in an approaching, uniform stream. Let the flow be from left to right, with fluid entering from the left. The fluid element (1) starts upstream of the plate, very slightly above it. It moves along a streamline infinitesimally above the plate. This element will be brought nearly to rest when it reaches the plate. Now consider another element (particle 2) placed slightly higher above the plate. It will continue its motion over the surface, being far less impeded by the no-slip condition. A line joining the two fluid elements will begin to rotate in the clockwise direction as soon as the lower element reaches the plate. Hence, the fluid becomes rotational when it reaches the solid boundary. Vorticity is acquired at the leading edge of the plate. Clockwise rotation means that the sign of the vorticity is negative.

In high Reynolds number flow, this vorticity occurs in a thin boundary layer, next to the surface. If y is the direction normal to the plate, then the vorticity is $-\partial u/\partial y \approx -(u_2 - u_1)/(y_2 - y_1)$ in this case. The retarding action of viscous friction causes u_1 to be less than u_2. Friction imparts rotational velocity to the fluid.

There is a more purely mathematical view of the origin of vorticity through viscous action at surfaces. At a stationary surface $u = 0$ for all time, so $Du/Dt = 0$ in the momentum equation (1.14). If the wall is the plane $y = 0$ and the flow is in the x direction, the vorticity at the wall is in the z direction and equals $-\partial u/\partial y$. It follows that Eq. (1.14) becomes

$$\frac{\partial p}{\partial x} = \mu \frac{\partial^2 u}{\partial y^2} = -\mu \frac{\partial \omega_z}{\partial y} \tag{1.63}$$

on the surface. The rightmost term is of the form of a transport coefficient times a gradient; it describes diffusion of vorticity from high values to low. Viscosity is the diffusion coefficient of vorticity. Hence, vorticity diffuses out of (or into) the wall at a rate determined by the pressure gradient. For the infinitely thin plate, mentioned above, the pressure gradient is negligible, except at the leading edge.

Once it is created, vorticity diffuses from the surface out into the nearby fluid. Further downstream along the flat plate, the initial vorticity continues to diffuse away from the surface, but no more vorticity enters from the surface ($\nu\partial\omega/\partial y = 0$). Other than this simple case of a flat plate, the surface will continue to be a source or sink of vorticity, dependent on the sign of the pressure gradient, per Eq. (1.63).

In computational analysis, it is necessary to provide adequate grid resolution in vortical regions – as will be discussed in Chapter 2. They frequently are regions of concentrated shear along, or emanating from, a surface. That is reasonable: the wall is the source of vorticity. At high Reynolds numbers, vorticity is confined in a thin layer next to the wall and does not diffuse far from it. If the layer leaves the wall and is shed into the flow, a thin, detached layer of vorticity occurs. The computational mesh must be designed to resolve these regions.

Vorticity can also be produced by density gradients. Essentially, this is the phenomenon of hot air rising and cold air sinking. If a given pressure gradient – say, in the x direction – acts on adjacent fluid elements of different density – say, side by side in the y direction – the less dense fluid will accelerate more than the heavier fluid, producing velocity shear and, hence, vorticity. This is called baroclinic production of

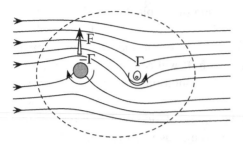

Figure 1.20. Circulation, shed vortex, and lift force on a cylinder.

vorticity. The rate of vorticity production is $\nabla p \wedge \nabla \rho / \rho^2$. In the instance just cited, this produces the z component of vorticity at the rate $\partial p / \partial x \, \partial \rho / \partial y / \rho^2$.

1.8.4 Circulation and Lift

Section 1.8.3 started with reference to Kelvin's circulation theorem. It says that the circulation round a closed path in the fluid is a constant of motion, in the absence of viscous diffusion, and of baroclinicity. The circulation is a measure of the vorticity contained within the circuit, so this is a law of vorticity conservation. A circuit with initially zero circulation will continue to have zero circulation unless vorticity diffuses into it. Any vorticity within the fluid must have diffused from the surface, as discussed after Eq. (1.63).

When the surface is closed, as in the case of a circular cylinder, a circuit can be chosen that encloses the surface – the dashed curve in Figure 1.20. Let the circulation round that circuit initially be zero. Suppose that vorticity with a total circulation Γ diffuses from the surface. Let it subsequently roll up into a concentrated vortex of circulation Γ. A net circulation has been produced in the fluid. However, as long as that vortex remains inside the circuit, Kelvin's theorem says that the total circulation round the circuit cannot change. What we have just described is not possible. If the circuit initially has zero circulation, it remains so. A nonzero circulation, Γ, cannot be produced. How is this paradox to be resolved?

It follows from Kelvin's theorem that any vortex of circulation Γ that is produced will be accompanied by an equal and opposite production of vorticity elsewhere. The sum will remain equal to zero. It is common to introduce the artifice of bound vorticity: if a vortex of strength Γ is shed from the surface, it is imagined that an equal and opposite vortex remains behind within the surface. The shed plus bound vortices sum to zero. More precisely, a circulation round the surface, equal and opposite to that of the vortex, is left behind. Actually, the circulation is contained in a thin layer of vorticity along the surface rather than literally in a vortex inside the body. The vorticity along the surface has the opposite sign to that shed. This understanding of Kelvin's theorem is shown schematically in Figure 1.20. The bound vorticity is indicated as a circulation Γ around the cylinder.

It can be shown that a closed body with net circulation develops a side force in

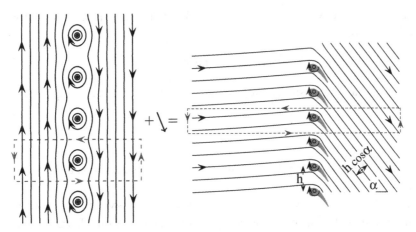

Figure 1.21. A row of vortices plus a uniform velocity represents a turning vane cascade.

the presence of a approaching flow. In fact, if the approach velocity is U_∞ in the x direction, a lift force equal to

$$-\rho \Gamma U_\infty s \qquad (1.64)$$

develops in the y direction. s is the width in the z direction. The quantities here have dimensions $\rho \sim m/\ell^3$; $\Gamma \sim \ell^2/t$, $U_\infty \sim \ell/t$; $s \sim \ell$. The product (1.64) has dimensions of mass times acceleration.

Equation (1.64) is readily explained. The circulation is modeled as a bound vortex with negative circulation – see Figure 1.20. Flow is around it, in the clockwise direction. Below the vortex, the induced velocity has the opposite direction to the approach flow; hence, the velocity decreases and, by Bernoulli's equation, the pressure rises underneath the vortex. Above it, the induced velocity is in the same direction as the approach flow: the velocity increases and the pressure falls. Hence, there is an upward pressure force exerted on a body with circulation.

Formula equation (1.64) can be arrived at analytically by applying a momentum balance to an infinite row of vortices in an approach flow, as shown in Figure 1.21. First consider the row of turning vanes at the right of the figure. Far upstream the v velocity vanishes and the u velocity is u_{in}. Far downstream, mass conservation requires that the u velocity still be u_{in}, but because of the turning the v-velocity is now $-u_{in} \tan \alpha$. It follows that the circulation round the dashed path is

$$\Gamma = -h u_{in} \tan \alpha,$$

where h is the spacing betweem the vanes.

A momentum balance applied to the volume enclosed by this same dashed line gives the lift force on the turning vane. Invoking the periodicity of the vane row, the net pressure force in the y direction vanishes. The lift force is due entirely to the change in momentum flux. From Eq. (1.35), the change of y momentum is the mass flux times the v-velocity change: $\dot{m}(v_{\text{out}} - v_{\text{in}})$. In the present case, the velocity change is related to circulation:

$$\dot{m}(v_{\text{out}} - v_{\text{in}}) = (\rho u_{\text{in}} h)(-u_{\text{in}} \tan \alpha) = \rho u_{\text{in}} \Gamma.$$

The lift force is equal and opposite to the change in momentum flux per Eq. (1.35). Hence,

$$L = -\rho u_{\text{in}} \Gamma,$$

which recovers Eq. (1.64) – when it is recognized that the momentum balance is per unit width in z.

An argument has been invoked that the turning vane is equivalent to a row of vortices. Perhaps that is obvious from the computation of circulation. However, the notion that, for some purposes, vanes with lift can be regarded as vortices warrants further discussion. Based on formula (1.55), the velocity induced by a row of vortices of strength Γ is

$$(u, v) = \frac{\Gamma}{2\pi} \sum_n \frac{(-y + nh, x)}{x^2 + (y - nh)^2}. \tag{1.65}$$

The vortices are located at $x = 0$, $y = nh$, where n ranges over the positive and negative integers. It can be shown that as $|x| \to \infty$, expression (1.65) becomes $(u, v) = [0, \text{sign}(x)\Gamma/2h]$. For a negative Γ, there is upflow on the left and downflow on the right. Streamlines corresponding to formula (1.65) are plotted in Figure 1.21. The upstream limit, $u = 0$ and $v = -\Gamma/2h$, which can be found mathematically, is evident in the figure.

Consider a rectangular circuit that encloses a single vortex passing midway between its neighbors, as shown by the dashed line in Figure 1.21. The circulation round that path is the integral of the tangential component of velocity. On the horizontal legs of the rectangle, $u = 0$ by symmetry; hence, they make no contribution to the circulation; the circulation is determined only by the vertical legs. The average vertical velocity on those legs is equal and opposite. If this average velocity is $\bar{v}_s$, then the circulation is $2h\bar{v}_s$. The circulation around any path that encloses the vortex is equal to its strength, Γ. Thus,

$$\bar{v}_s = \frac{\Gamma}{2h} \text{sign}(x).$$

Far from the row of vortices, the local velocity, $v(x, y)$, becomes equal to the average velocity, $\bar{v}_s$.

To produce an approach flow in the x direction, add $(u_{\text{in}}, \bar{v}_s)$ to the row of vortices. This gives the incident velocity $(u_{\text{in}}, 0)$. Now the sum of the row of vortices plus the uniform velocity looks like the streamlines on the right side of Figure 1.21. In fact these streamlines were computed by adding $(u_{\text{in}}, \bar{v}_s)$ to Eq. (1.65). As $x \to \infty$ the sum tends to $(u_{\text{in}}, \bar{v}_s) + (0, \bar{v}_s) = (u_{\text{in}}, \Gamma/h)$. Hence the exit flow is at angle

$$\alpha = -\tan^{-1}(\Gamma/u_{\text{in}} h). \tag{1.66}$$

(Recall that $\Gamma < 0$ in the figure.) Thus the vortex row models turning through the guide vanes. Bound vortices in the figure at right represent the circulation round the turning vanes. A row of vortices turns the flow direction, just as a set of solid vanes does.

The circulation round the vanes causes an acceleration of the flow. Given the periodicity in y, the vertical spacing of the streamlines in the exit region must be h. But the flow is at angle α. Hence, the spacing perpendicular to the flow is $h \cos \alpha$. Mass conservation gives the magnitude of the exit velocity equal to $|u_{out}| = u_{in}/\cos \alpha$. This relates the speed to turning angle; Eq. (1.66) relates turning angle to vortex strength. In combination, the circulation round the turning vanes is related to the change in velocity magnitude by

$$|\boldsymbol{u}_{out}|^2 = u_{in}^2 + \left(\frac{\Gamma}{h}\right)^2.$$

By Bernoulli's equation, this shows that the pressure drops through the vane row. Correspondingly, there is a net pressure force pushing on the vanes in the x direction.

1.8.5 Diffusion of Vorticity

An equation for vorticity evolution is derived as the curl of Eq. (1.14). This vorticity equation is not commonly used in three-dimensional fluid flow computations. However, it has value for analysis and can be regarded as the physical law that lies behind the discussion of vorticity dynamics. The equation is

$$\frac{D\boldsymbol{\omega}}{Dt} = \nu\nabla^2\boldsymbol{\omega} + \boldsymbol{\omega}\cdot\nabla\boldsymbol{u} \tag{1.67}$$

for constant density, incompressible flow. The pressure is gone. Pressure gradients do not affect the rotational motion, unless the density varies – then they produce a baroclinic torque, as explained previously. The reader might rederive Eq. (1.67) for variable density as an exercise.

The second term on the right is the mathematical form of stretching and rotation. These have been described in physical terms in §1.8.2. Were the velocity two-dimensional, of the form $[u(x, y), v(x, y)]$, then the vorticity would be entirely in the z direction. The flow could not stretch or rotate the vorticity because the last term in Eq. (1.67) would be $\omega_z \partial_z \boldsymbol{u} = 0$. To see this, place a pencil perpendicular to the plane of this page and move it with a velocity that has components only in the plane and do not depend on distance along the pencil. The pencil can only translate. To rotate it, the velocity must be higher at one end than at the other. The pencil is the direction of $\boldsymbol{\omega}$ and $\boldsymbol{\omega}\cdot\nabla\boldsymbol{u}$ is the variation of velocity along the length of the pencil.

Let us make this simplification: let the flow be two dimensional. Then Eq. (1.67) becomes the convection-diffusion equation

$$\frac{D\omega}{Dt} = \nu\nabla^2\omega. \tag{1.68}$$

Three-dimensional vortex dynamics have been removed to focus on diffusion. Viscosity is the diffusion coefficient for vorticity.

Consider a circular distribution of vorticity. Its streamlines are a set of concentric circles. The convective derivative vanishes because this is parallel flow (§1.5). Equation (1.68) simplifies to

$$\frac{\partial \omega}{\partial t} = \nu \nabla^2 \omega.$$

A solution for a Gaussian vorticity distribution is

$$\omega = \frac{\Gamma}{4\pi \nu t} e^{-r^2/4\nu t}. \tag{1.69}$$

This tells us two things that may seem obvious. The radius of the vortex grows with time and the maximum vorticity decreases with time. These are expected consequences of diffusion. From the exponential factor, a radius

$$R = 2\sqrt{\nu t}$$

can be defined. The central vorticity is

$$\frac{\Gamma}{\pi R^2}.$$

Circulation is conserved under the process of diffusion. That is expressed by the vorticity falling inversely with the area, πR^2. Diffusion spreads vorticity without changing the total vorticity or, equivalently, without changing the net circulation – which is vorticity times area.

However, vorticity can be destroyed. Opposite-signed vortices can diffuse into one another. If one has circulation Γ and the other has circulation $-\Gamma$, the net is zero. Hence, they can completely cancel each other, without violating the conservation law of circulation.

A simple analysis was provided by G. I. Taylor. Among his many brilliant contributions to fluid dynamics, Taylor was one of the pioneers of turbulence theory. He emphasized the role of vorticity. In a famous article with A. E. Green (Taylor, 1971) he introduced the Taylor–Green vortex model to explain the three-dimensional dynamics of turbulence. Somewhat earlier he had devised a two-dimensional, vortex lattice model to describe simple decay by diffusion. Taylor's solution continues to be used in turbulence research. It is a standard test of the numerical accuracy of simulation codes.

Taylor's vortex lattice is

$$\omega = \omega_0 e^{-(k^2+l^2)\nu t} \sin(kx) \sin(ly).$$

This is an exact solution to the full Navier–Stokes Eq. (1.14). It is not a parallel flow. Nevertheless, the nonlinear convection of vorticity vanishes, leaving pure diffusion. The nonlinear convection of momentum does not vanish, but it is balanced by the pressure gradient; hence, it does not contribute to rotational motion.

Figure 1.22. A periodic lattice of vortices.

The parameters k and l are wave numbers. They equal 2π divided by wavelength

$$k = 2\pi/\lambda_x, \quad l = 2\pi/\lambda_y.$$

The sine functions vanish on the lines $x = n\lambda_x/2$ and $y = m\lambda_y/2$. These mark out a rectangular grid. Within each grid cell ω is either positive or negative, as shown by Figure 1.22. Every positive cell has negative neighbors on each side. The positive and negative cells diffuse into one another and the circulation within each cell decays exponentially. Integrating over a single cell

$$\Gamma(t) = \int_0^{\lambda_x/2} \int_0^{\lambda_y/2} \omega dx\, dy = \frac{\omega_0 \lambda_x \lambda_y}{\pi^2}\, e^{-(k^2+l^2)\nu t} = \Gamma_0 e^{-(k^2+l^2)\nu t}.$$

Unlike the isolated vortex (1.69), for which Γ is constant in time, circulation in each cell of the alternating signed, vortex lattice decays.

Three-dimensional vorticity dynamics can be quite complex. The basic elements are stretching, rotation, and diffusion. There is an interplay among these components. Stretching intensifies vorticity and steepens its gradients. That enhances the rate of diffusion. Diffusion decreases the magnitude of vorticity. It is possible for amplification by stretching to exactly balance decay by diffusion. That defines an idealization called Burger's vortex. Burger's vortex is another exact solution of the Navier–Stokes equations (Batchelor, 1967). If α is the rate of strain, with dimensions of 1/time, it is clear, on dimensional grounds, that a balance between stretching and diffusion can obtain only if the vortex diameter is proportional to $\sqrt{\nu/\alpha}$. Outward diffusion increases the diameter, straining shrinks the diameter (Figure 1.18, page 42). In Burger's vortex, they come into a precise balance. More generally, stretching and diffusion are not in balance but are closely interrelated.

1.8.6 Visualization of Vortices

If vortices provide an insight into flow properties, then one would like to visualize where they occur in the flow field. The vorticity simply is concomitant to a given velocity field. In other words, it is a kinematic property of the velocity field: given $u(x)$, we simply evaluate its curl, $\nabla \wedge u$. This will show the distribution of vorticity, but other quantities are more effective for qualitative visualizations of vortices.

First, we must distinguish vortices from vorticity. The latter is produced at a wall. If it separates from the surface, vorticity might accumulate into a concentrated patch. That concentration into a compact region distinguishes a vortex from vorticity. For instance, a smoke ring that can be created by pulsing air through a circular orifice originates from vorticity on the wall which leaves the surface and rolls into a concentrated ring of vorticity. Streamlines in a vorticity layer might be straight, with no evidence of swirling flow. In a vortex they commonly form closed circuits. Streamlines swirl around a vortex, not around vorticity per se. One method of visualization might simply be to plot streamlines.

Streamlines have the shortcoming that they are not Galilean invariant: that is, adding a constant velocity to the flow field changes the pattern of streamlines. This is obvious when the added velocity is very large; then the streamlines will become straight. Even though vortices may be present, they do not deflect the streamlines significantly when a very large velocity is added. Other methods to detect vortices are needed.

Streamlines visualize flow features by a family of curves. Another approach makes use of visualization by surfaces. This approach also is Galilean invariant. A vortex is a compact region of rotational flow. How can that be made a basis for visualizing it?

The method is to identify the region where the rate of rotation is greater than the rate of strain. The symmetric derivative

$$S \equiv 1/2 \left(\frac{\partial u_j}{\partial x_i} + \frac{\partial u_i}{\partial x_j} \right)$$

defines the rate of strain tensor. The antisymmetric derivative

$$\Omega \equiv 1/2 \left(\frac{\partial u_j}{\partial x_i} - \frac{\partial u_i}{\partial x_j} \right)$$

defines the rate of rotation tensor.* Subtracting the magnitude of the rate of rotation from the magnitude of the rate of strain defines a variable Q as

$$Q = |S|^2 - |\Omega|^2. \tag{1.70}$$

---

* Formally, the vorticity and rotation tensor are related by $\omega_i = \epsilon_{ijk}\Omega_{jk}$, where ϵ_{ijk} is the skew tensor, equal to $+1$ when i, j, k can be rearranged into $1, 2, 3$ in an even number of steps, and equal to -1 otherwise.

Q is negative where rate of rotation dominates over rate of strain. The absolute values here are computed as the trace of a matrix squared: in index notation

$$|\mathbf{S}|^2 = S_{ij}S_{ji}.$$

The summation convention on repeated subscripts is invoked in this formula; that is

$$S_{ij}S_{ji} \equiv \sum_{i=1}^{3}\sum_{j=1}^{3} S_{ij}S_{ji}.$$

When ever an index appears twice in a product, it is to be summed. It can be shown that Q is just the squared velocity gradient:

$$Q = \frac{\partial u_i}{\partial x_j}\frac{\partial u_i}{\partial x_j}.$$

To see how the Q detection method works, compare a vortical layer,

$$\mathbf{u} = [u(y), 0],$$

to a cylindrical vortex,

$$\mathbf{u} = (-y, x)F(r).$$

The latter states the Cartesian components of an angular velocity $u_\theta = rF(r)$. Any choice for the function $F(r)$ provides a particular example.

In the first case

$$\mathbf{S} = 1/2 \begin{pmatrix} 0 & \partial_y u \\ \partial_y u & 0 \end{pmatrix}; \quad \mathbf{\Omega} = 1/2 \begin{pmatrix} 0 & -\partial_y u \\ \partial_y u & 0 \end{pmatrix}$$

and $Q = 1/2[(\partial_y u)^2 - (\partial_y u)^2] = 0$. We define $Q < 0$ as the interior of a vortex. Hence, the vortical layer is not a vortex.

In the second case

$$\mathbf{S} = \begin{pmatrix} -2xy & x^2 - y^2 \\ x^2 - y^2 & 2xy \end{pmatrix}\frac{F'}{2r}; \quad \mathbf{\Omega} = \begin{pmatrix} 0 & 1 \\ -1 & 0 \end{pmatrix}\frac{2F + rF'}{2}.$$

Substituting into the definition (1.70),

$$Q = -2F(F + rF').$$

The region of $Q < 0$ can be identified for any particular F.

The Rankine vortex is defined as a uniform circular patch of vorticity inside a radius R. The fluid inside the circle is in solid body rotation. Outside the circle the velocity is irrotational and given by formula (1.54). That is,

$$F = \frac{\Gamma}{2\pi} \begin{cases} 1/R^2; & r < R \\ 1/r^2; & r > R \end{cases}. \tag{1.71}$$

For this function $Q < 0$ for $r < R$ and $Q > 0$ for $r > R$. The criterion $Q < 0$ identifies the core of the vortex exactly.

Figure 1.23. Vortex visualization in a turbine passage by a constant $\mathcal{Q}$ surface.

As another example, the Oseen vortex,

$$F = \frac{\Gamma}{2\pi r^2}(1 - e^{-r^2/R^2})$$

is a smoother version of the Rankine vortex. There is no sharp boundary to the vortical region. Now

$$\mathcal{Q} \propto R^2 - e^{-r^2/R^2}(R^2 + 2r^2).$$

This is less than zero when $r < 1.121R$. That is the $\mathcal{Q}$ definition of the edge of the vortex.

A contour of $\mathcal{Q}$ slightly less than zero can be used to visualize vortices in complex flows. Figure 1.23 is an illustration in which a necklace vortex is visualized by surfaces of constant $\mathcal{Q}$. The necklace is wrapped around the leading edge of a turbine blade, much like Figure 1.19. The vortex is seen at the bottom of the picture. Its vorticity originates on the lower wall. It becomes a concentrated vortex as it wraps around the turbine blade and is stretched. In this figure, the blades themselves are not represented. Their presence is manifested by $\mathcal{Q}$ surfaces. They provide an inadvertent visualization of the geometry. This also serves as a caution: $\mathcal{Q}$ sometimes might not identify vortices per se here it has picked up the boundary layers on the blade surface.

Another way to identify vortices is via pressure contours. The pressure at the center of a vortex is locally a minimum; see Eq. (1.47). A zone around the minimum pressure can be identified with the vortex, but it is not clear how to do this. Generally, a pressure contour somewhat above the minimum will be selected.

The low pressure at its center balances the centripetal acceleration of the flow circulating around the vortex. At the center of the Rankine vortex (1.71), the pressure is given by setting $r = 0$ in Eq. (1.48)

$$p(0) - p_\infty = -\rho \left(\frac{\Gamma}{2\pi R_c}\right)^2.$$

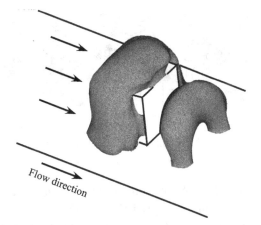

Figure 1.24. Vortex visualization by constant pressure surfaces in flow over a surface mounted cube.

Flow direction

This states that the pressure in the center is sufficiently lower than the ambient pressure to balance the centrifugal acceleration. At the edge of the rotational core, $r = R_c$, the pressure given in Eq. (1.48) has risen to half this value. In this case, the boundary of the vortex can be identified as the pressure contour midway between the minimum pressure and ambient.

In Figure 1.24, arch vortices in flow over a cube are visualized by surfaces of constant pressure. These arches reflect the presence of vorticity that separates from the surface and rolls into vortex tubes. They seem to satisfy the theorem that vorticity does not end in the fluid, because both vortices terminate on the lower wall.

1.9 Turbulence

This chapter concludes with an introduction to the notion of turbulence. This is only a bare description. More concrete material is presented in Chapter 6.

So far, we have discussed fluid and vorticity dynamics without reference to whether the flow is laminar or turbulent. From the perspective of the governing equations, there is no need to distinguish: the laws of motion apply to either. From a practical perspective, the distinction is crucial. How computations are effected is very much dependent on whether the flow is laminar or turbulent. If laminar, we simply solve the governing equations with as much accuracy as possible. If turbulent, there are choices to be made: in most cases, simply solving the equations is not practical.

Turbulence is an intriguing and immensely important fluid dynamical phenomenon. Many eminent engineers, physicists, and mathematicians have tried their hands at theories and models of turbulence in fluids. The available methods of simulation, and the theoretical understanding of turbulence, are the fruits of their labor. However, most fluid dynamicists would acknowledge that available theories and models leave much room for improvement.

Turbulence is irregular fluid flow. It is a state of fluid motion characterized by its complexity. Vorticity becomes unoriented and entangled; chaotic velocities stir the fluid. Because of the inherent complexity, full-blown numerical solution to

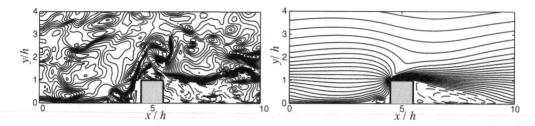

Figure 1.25. Instantaneous, turbulent contours versus averaged contours of u.

the Navier–Stokes equations – which is called direct numerical simulation (DNS) – is extremely costly. Effecting DNS requires sufficient mesh points to resolve the smallest of eddies and to integrate the discrete equations accurately in time. The panel at left in Figure 1.25 is from a DNS of flow over a rib in a channel. It provides some impression of the irregular, eddying motion. The mesh and time resolutions must suffice to represent these details.

One shortcut to full spatial and temporal resolution consists in using too few grid points to resolve the smaller of the eddies. This is called large eddy simulation (LES). Just computing the largest scales is not a solution: the small scales are not irrelevant. The usual approach is to argue that the small scales primarily dissipate turbulent fluctuations, so their role can be represented by artificially enhancing viscosity. The "large scales" are defined to be those greater than the grid spacing, so that the extra viscosity is determined by the grid spacing. Grid-dependent viscosity is added to the true viscosity to dissipate the smallest scales. LES remains an expensive prospect because of the need for time accuracy and a long enough integration time for convergence of statistics and because grid resolution must still be adequate to resolve the larger eddies. Conceptually, LES is straightforward: simply add an extra viscosity to the true viscosity and solve the Navier–Stokes equations. In practice, simulations can be dependent on the mesh resolution, especially near surfaces.

Chaotic, turbulent motion is usually described by its statistics. That is the level at which theories and models operate. Properties of turbulence are characterized by averaged values. The panel at the right of Figure 1.25 is a time-averaged velocity field in the ribbed channel. It might seem to be a different flow from that at right, but it is simply its time average. Clearly, the average is a smoother, more easily resolved, representation of the flow.

Being smoother, the averaged flow is apparently more amenable to computation. Reynolds averaged analysis (RANS) attempts to predict this averaged field directly, without first simulating the chaos, then averaging. Doing so requires the introduction of a turbulence closure model. In most practical cases, the turbulence closure model provides an eddy viscosity. Again this is an enhancement of the molecular viscosity, but in this case, it is not a function of the grid resolution. Its purpose is to represent the mixing properties of the eddies. Turbulence models predict eddy viscosity as a function of position in the flow. They do so by solving a set of transport equations for properties of the turbulence – for instance, its kinetic energy and its timescale. These transport equations are not exact laws of fluid motion; rather they contain a degree of empiricism that is designed to allow predictions of engineering accuracy in many, although not all, situations.

A decision must be made when to treat the flow as turbulent or as laminar. That decision is not always easy to make. This is the problem of *transition to turbulence*. Processes occur by which the state of turbulent motion is produced. Broadly, they are instabilities; small perturbations grow and become the irregular motions. A competition exists between destabilizing and stabilizing forces. If destabilization wins, the flow transitions to the turbulent state.

In viscous shear layers, transition to turbulence occurs above a critical value of Reynolds number. That is consistent with inertial acceleration being the destabilizing effect and viscous dissipation being the stabilizing influence. If the Reynolds number is sufficiently high, the flow can safely be treated as turbulent from the outset. But in some cases the flow is initially laminar and transitions to turbulence after some downstream distance.

The matter is complicated by the fact that eddying can occur in laminar flow without it being turbulent. This can be in the form of vortex shedding or a sinusoidal instability. Instabilities do not necessarily lead to turbulence. Despite our loose use of terminology in the previous paragraphs, *transition* to turbulence and *instability* are distinct notions. The latter implies that disturbances will amplify and change the state of motion. The former implies that the motion changes from a smooth, laminar state, to an irregular, turbulent state. Instability theory will be discussed in Chapter 6.

Free-shear flows – jets or wakes – transition at relatively low Reynolds numbers; they are inherently more unstable than wall-bounded shear layers. A jet may be unstable at Reynolds numbers, based on jet diameter, on the order of 10 and fully turbulent above a Reynolds number of a few thousand. Often transition is early enough to treat free-shear flows as turbulent from the outset.

A flat plate boundary layer becomes unstable at a Reynolds number (based on distance along the plate) of about 10^5; however, it may not transition to turbulence until a Reynolds number of about 10^6. The latter depends on the level of ambient disturbances; fluctuations of only a few percentages in the ambient velocity can reduce the transition Reynolds number to the vicinity of 10^5.

The problems of transition to turbulence and turbulence modeling are a continual source of interest and anxiety to fluid dynamicists. We will discuss them in Chapter 6.

EXERCISES

1.1 *The convective derivative.* Convection carries a contaminant concentration with the fluid velocity. Let the velocity be uniform in the x direction and equal to U. If the concentration is initially constant inside a circle, $x^2 + y^2 = 1$, and zero elsewhere, what will the concentration be after some time t?

A circular patch of dye is released into a large channel. The velocity profile is not uniform, but it can be approximated locally by $U + Sy$, where U and S are constants. Describe how the dye patch will evolve. Ignore molecular diffusion; focus on pure convection. Show that after some time t the dye will be contained in an ellipse.

1.2 *Reynolds number scaling.* Derive the nondimensional forms (1.20) and (1.21) of the Navier–Stokes equations. Write out the x and y components of these equations for two-dimensional flow.

1.3 *One-dimensional flow analysis.* Flow enters a duct from the left, with velocity u, pressure p, and constant density. It exits at right through two channels. The exit areas are $A_2 = 1/2A_1$ and $A_3 = 1/3A_1$. This is inviscid flow, and Bernoulli's equation is applicable.

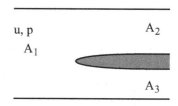

Derive a formula for the force on the divider. Explain how you chose to split the flow between the two downstream channels.

1.4 *Ejector pump.* A jet of air enters through a nozzle of cross-sectional area A_1. It has a static pressure P_1 and volumetric flow rate $\dot{m}_1$, as in the figure at right. The density is constant. Air exits the domain through a duct of area A_3 at ambient pressure, P_a. The upstream side of the duct is open to air at ambient pressure. Assume the exit velocity is uniform across A_3 and let $A_3 = 2A_1$. Frictional forces on the walls can be neglected: set the x-component of force on the walls to zero.

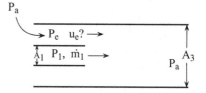

This configuration is an ejector nozzle. u_e is the air entrained by the jet. Perform a control volume analysis to determine the velocity, u_e, of air drawn into the duct. The ambient fluid is at rest. The entrance pressure is related to the ambient pressure by Bernoulli's equation.

1.5 *Poiseuille flow.* Consider flow in a straight, circular pipe. If it is fully developed, parallel flow the Navier–Stokes equations simplify dramatically to

$$0 = -\frac{\partial p}{\partial x} + \frac{1}{r}\left[\frac{\partial}{\partial r}\left(r\frac{\partial u}{\partial r}\right)\right]. \tag{E1.1}$$

If the pressure gradient is a given constant, show that Eq. (1.18) is a solution to this equation. What boundary conditions have you used?

The frictional stress on the wall is $\mu dU/dr\,|_{r=a}$. Verify that the frictional force, per unit length, on the wall of the pipe is related to the bulk velocity by $8\pi\mu\overline{U}$.

1.6 *Shock absorber.* Derive the model equation

$$\ddot{h} + \frac{32\nu}{d^2}\dot{h} + \frac{g}{L}(h - d/2) = \frac{F(t)}{M}$$

for the column of liquid in the figure to the right. d is the diameter of the tube, L is the total length of the liquid region, F is the force exerted on the piston, and M is the total mass of liquid. The top of the cylinder is open. If the piston motion is $a\sin(\omega t)$, as shown, determine the energy absorbed by the viscous liquid over one period of oscillation.

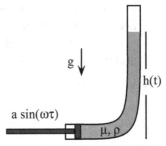

State at least two assumptions about the fluid motion that were made in deriving the model equation.

1.7 *Couette–Poiseuille flow.* A very long, solid rod is dropped exactly down the middle of a tube filled with viscous liquid. It accelerates under gravity to a terminal velocity. The tube is closed at the bottom, with enough room for the rod to drop, as in the sketch as right. Derive a formula for the terminal velocity, given the dimensions in the figure. The force of gravity is $(\rho_s - \rho)\mathcal{V}g$, in terms of the solid and fluid densities and the volume of the rod. This can be solved in either a two-dimensional or an axisymmetric geometry. The liquid has viscosity μ. The upward Couette–Poiseuille flow produces a shear force on the rod. Forces on the end of the rod can be ignored.

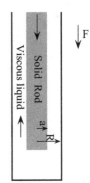

1.8 *A branching network.* Consider a network of threefold branching tubes. At each branch the tube area falls by a factor of β and the tube length falls by a factor of γ. Let the smallest tubes have length ℓ_c and area A_c. (The subscript c alludes to capillaries in the circulatory system of animals.) If there are N generations of branching, the largest tubes have area $A_c\beta^N$ and length $\ell_c\gamma^N$. Let the volume flow be Q. Mass conservation implies that the total flow is constant across the generations. After the first branching, each tube will carry a volume flow of $Q/3$.

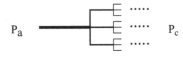

Derive a formula for the total pressure drop $P_a - P_c$, across the system of N generations. Show that if $3\beta^2/\gamma < 1$ and N is large

$$\Delta p \approx \frac{8\pi\mu\ell_c Q_c\gamma}{A_c^2(\gamma - 3\beta^2)}.$$

Hence, the pressure drop is controlled by the capillaries.

1.9 *Oscillatory boundary layer.* A pressure gradient, $\partial p/\partial x = A\sin(\omega t)$ drives oscillatory flow over a large, flat plate. Derive a formula for the net work that must be done per cycle (i.e., averaged over one period of oscillation) to overcome dissipation in the Stokes boundary layer.

1.10 *Point vortex dynamics.* Three vortices are arranged around an equilateral triangle with side a. The all have exactly the same circulation Γ. What will the individual motions be as a result of the induced velocity of the other two vortices? Explain why the triangle will rotate as a whole.

If the center of the triangle is at $(0, 0)$ and the topmost vortex is at $(0, a/\sqrt{3})$, find the angular velocity of the triangle.

1.11 *Vortex pair.* Derive the velocity

$$u = \frac{\Gamma d}{2\pi(x^2 + d^2/4)}$$

on the line $y = 0$ between the vortices in Figure 1.16. Show that in the frame moving with the vortices (Figure 1.15) the front stagnation point of the bubble is at $x = \sqrt{3}d/2$.

2 Elements of Computational Analysis

The basic laws of fluid dynamics are the Navier–Stokes momentum equations described in Chapter 1. Computational fluid dynamics (CFD) is the practice of solving those equations,* along with the mass conservation equation, by numerical algorithms. The ability of such seemingly simple governing equations to describe a wealth of complex fluid motions is quite remarkable. That remarkable capability is revealed most notably by computer simulation.

Numerical solution of Navier–Stokes equations nowadays has become almost routine. A variety of algorithms and solution methods for both incompressible and compressible flow have been developed over time and successfully implemented in a large number of computational codes (Ferziger and Peric, 2002; Fletcher, 1991; Tannehill et al., 1997). Initially, this software was primarily for research, mainly in academic institutions, government labs, and corporate research centers. But the appearance (and disappearance) of a number of general purpose, commercial CFD codes has been seen since the early 1990s. These were developed for use by nonexperts, as well as by those experienced in the practice of computation. Some of these codes have matured over time, becoming increasingly powerful as the latest techniques, methods, and analytical models were adapted to their requirements, and as high-speed computing power became increasingly available. Computational capabilities, previously mastered only in the research environment (higher-order numerical schemes, multigrid methods, advanced modeling capabilities, parallel processing) are now being used widely, through the medium of software packages. The engineer, student, or scientist no longer needs to have an intimate familiarity with computational methods to make productive use of CFD. Other technologies that have facilitated CFD include the graphic user interface, software for geometry and mesh creation, and techniques for plotting and visualization.

Advances in computer speed, memory, and communication bandwidth have led to larger and larger simulations being accomplished in a modest amount of time; simulations lasting not more than a few hours provide solutions to complex engineering problems.

This widespread availability of fluid dynamics software creates a new impetus to learning fluid dynamics. In this chapter, we describe some basic elements of

---

* The equation of energy conservation, equation of state, constitutive equations, etc., would be added for effects of compressibility, non-Newtonian stress, or other additional factors

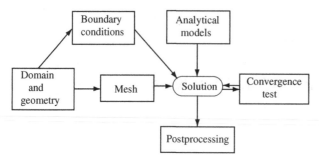

Figure 2.1. Elements of configuring a computational analysis.

computer-aided fluid flow analysis. The perspective is that of a code user. An analogy might be drawn to an experimentalist who uses existing instrumentation and an existing flow facility. We view the CFD code as a facility, a sort of *instrumented, virtual wind tunnel*. This chapter describes the basic nature of computation in general terms, including aspects of geometry and grid generation and of numerical methods. The treatment is cursory, relative to the detailed manuals that are provided by software vendors. The intention is to provide a sufficient background for the reader to understand the manner in which CFD is used throughout this text to illustrate fundamental ideas of fluid dynamics.

Although the present discussion is of common aspects of many CFD codes, a few of the exercises require that particular CFD software must be selected. It is optional whether the reader chooses to do these particular exercises. To a large extent, the steps in preparing a computer simulation of flow phenomena are the same irrespective of the choice of software: geometry definition, grid generation, boundary condition specification, selection of flow model and of a numerical method are required, in that order. That sets the stage for numerical solution, convergence and accuracy checks, and, finally, postprocessing of numerical results. The elements of computational analysis are diagrammed in Figure 2.1.

Those who write codes must fill the oval *solution* box of Figure 2.1. The elements of a program to perform computer analysis of a physical system are diagrammed in Figure 2.2. Although this is not a text on computational methods, the topics of this chapter are intended to provide some familiarity with these elements. The starting point is the governing laws, which have been reviewed in the previous chapter. These laws are stated in differential form. Their solution is a continuous function of

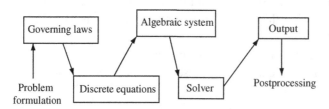

Figure 2.2. Rudiments of a computational fluid dynamics code.

position. The equations solved by computation are in discrete form, and their solution is a set of values at a finite number of points. The discrete equations are derived from the governing laws by restricting them to the computational points. This replaces the differential equations by a system of algebraic equations. Following the flow of Figure 2.2, the algebraic equations are solved to provide the desired output.

The rest of this chapter follows the outline of Figure 2.1, delving a bit into Figure 2.2. The treatment is informal.

2.1 Geometry and Grid Generation

Setting up a computation might at times seem a work of art. The starting point might be a sketch of the geometry, the boundaries to the domain, and a schematic of the mesh. The final product, after the computational model has been created, might be described as beautiful or ugly depending on the appearance of the mesh and how it fills the domain. These are the aesthetics of computational modeling. (In this context, the term *modeling* refers to the numerical representation of the surfaces and the computational mesh. The numerical representation models the physical reality.)

The notion of mesh quality is meant to be a more objective assessment of beauty than the aesthetic judgment. The quality of the mesh might be characterized by its smoothness, by the fractional change in mesh spacing, or cell size, from one element to the next, and by the acuteness of the angles of mesh cells. The ultimate judgment is the accuracy of the computational solution. The objective of accurate solution underlies any exposition on methods for computational modeling.

2.1.1 Computational Domain

The region in which the flow is computed is the *computational domain*. It is bounded by geometrical surfaces and by hypothetical surfaces drawn in the fluid. The latter are an inevitable artifice: the region being computed is an isolated part of a real apparatus. For instance, if fluid is pumped through a channel, the flow domain might be the channel, without the pump. The computational zone begins with a prescribed flow of fluid at its inlet. The inlet is an artificial surface that is required to conduct the computation. Similarly, the domain ends somewhere, and outflow conditions must be specified; look ahead to Figure 2.25, page 91. The outflow conditions can be perplexing: the real flow continues beyond the end of the domain; an exit condition must either emulate this unknown downstream flow, or, at least, be innocuous in the region of interest. This is a bit like the experimentalist's concern to smoothly diffuse the flow downstream of a test section, so as to have minimal upstream effect. Often the computationalist's concern is to situate the downstream boundary in a region of pure outflow and far enough away from the zone of interest.

The types of surface that define the computational domain can be listed: solid bodies, inflow and outflow surfaces, far-field boundaries, symmetry planes. Others can occur, such as fluid–fluid interfaces in two-phase flow. Usually there is freedom in how some of the boundaries are selected.

Figure 2.3. The passages through guide vanes of a combustor illustrate periodicity. The geometry is shown in cutaway, with the periodic sector darkened. Figure courtesy of Jorg Schluter.

It might seem that the solid boundaries are set unambiguously. They represent objects in the flow or walls that surround the flow. Sometimes the objects are simplified versions of complex shapes. We are not concerned here with that aspect of computer modeling. The geometry might have symmetries or periodicities. Those degrees of freedom may allow a significant reduction in computational cost.

A geometry is periodic if it is unchanged by a finite translation or rotation. The distances of translation, and angles of rotation are determined by a repeating unit of the geometry. If an object lies at (x, y, z), the same object must lie at $(x + L_x, y, z)$ if L_x is the period in the x direction.

A complex example is shown in Figure 2.3: 18 guide vanes direct air into a combustor. They are arranged symmetrically within a cylindrical geometry. The figure shows a cutaway view, with a periodic cell shown darkened. In the cutaway, some vanes at the top are not shown. The computational domain is just the darkened sector, with periodic conditions.

As another instance, consider a fan with six identical blades. Rotating the fan by $60°$ about its axis will leave the geometry unchanged. A 1/6 domain sector is the unit cell that is repeated six times in the full fan. By modeling one-sixth of the domain and making six copies, each rotated from the previous by $60°$, the entire geometry and mesh can be constructed. The unit cell must contain one entire fan blade, but the blade can lie anywhere in the domain.

In fact, the sector can lie between the top of one blade and the bottom of the next, so that the blade surfaces form part of the boundary of the domain (dashed lines in Figure 2.4), or the sector can contain a single, contiguous blade (chain-dotted lines in Figure 2.4). These two choices – blades along the boundaries or contiguous blade – produce singly and doubly connected computational domains. That has an effect on meshing.

A singly, or simply, connected domain is one in which any sphere drawn in the fluid can be contracted to a point without encountering a boundary. If some spheres can be contracted, but others cannot, then the domain is multiply connected. Empty space is simply connected. Unbounded space with an airplane in it is doubly connected. Any sphere that does not contain the airplane can be contracted to a point; a sphere that contains it cannot. The mesh for a doubly connected region

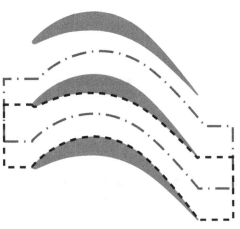

Figure 2.4. Two possible domains for flow between fan blades. The dashed line defined a simply connected domain; the chain-dot is doubly connected.

must encircle the object. When we discuss structured grids in two dimensions (see Figure 2.10), this will be called an O-topology. The choice in which the fan blades bound a simply connected region allows the grid to have an H-topology; the mesh can span from inlet to exit, without encircling an object. The choice of domain containing a full blade is doubly connected and might be gridded with an O-topology.

Another simplifying property is symmetry. Symmetries are, again, operations that leave the geometry unchanged. The operations are reflections in a plane, or rotations about an axis. Periodicity is a type of symmetry, but now we are referring to symmetries of an object rather than of the entire domain. For instance, a square is unchanged after rotation by any multiple of 90° about its center. It is also unchanged by reflection across either of its diagonals or across either of its bisectors. Symmetries might be exploited when generating a mesh; however, they are especially helpful when they permit a reduction in the problem size. For that to be possible, the entire flow must have the symmetry.

As an example, consider a flow approaching a square, side on, from the left. The square is symmetric with respect to reflection in either one of the two diagonals. However, reflection of the flow in the diagonal will cause the incident direction to switch from left-to-right to top-to-bottom. So the flow problem does not have symmetry with respect to reflection in a diagonal, as is seen in Figure 2.5. There is a viable symmetry: reflection in the bisector parallel to the flow will leave both the geometry and the incident velocity unchanged. Hence, the domain can be cut in half, solved in the top, and simply reflected in the symmetry axis to generate the entire flow field. Invoking symmetry cuts the problem size in half.

The remaining types of surface – inflow, outflow, and far field – have more obvious degrees of freedom. Where they are situated relative to the solid surfaces is often a matter of choice. When they are artifices, the guiding principle is to minimize harm and computational expense. A far-field boundary must be placed at considerable distance from the surface but not so far that the number of grid points in the domain becomes excessive. The far-field condition arises commonly in aerodynamics. Flow might be computed around an airplane, or a wing section, in unbounded space. The

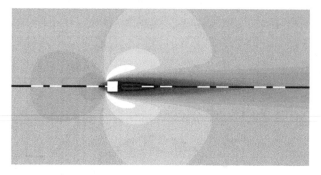

Figure 2.5. Contours of velocity magnitude in flow round a square. Symmetry across the midplane allows the domain to be halved.

computational domain is not unbounded. The artificial boundary could be a box, a sphere, or another closed surface that surrounds the body. The radius of that surface must be sufficient for the velocity to closely approach its incident value, u_∞.

An understanding of fluid dynamics helps to design the domain. Asymptotic behavior of solutions provides one guideline, as has been mentioned in connection with Eq. (1.56). According to that equation, the velocity around a two-dimensional, lifting wing decreases at large radius as one over distance. Thus, far from the surface $u \to u_\infty + C/r$, where C is a constant, that depends on the lift. This follows from the relation (1.64) between lift force and circulation. For the flow to become uniform within 10%, the radius must be on the order of 10 times the airfoil chord. Figure 2.6 is an example. The outer boundary is 15 or more chord lengths from the wing. In the case of this figure, the specific shape of the far-field boundary is determined by the method of grid generation.

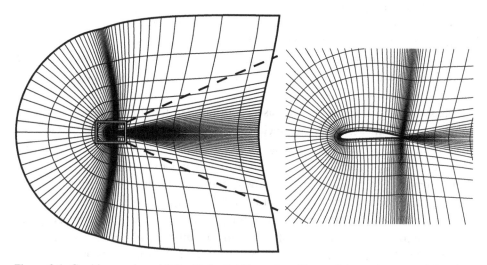

Figure 2.6. C-grid around an airfoil with far-field boundary. Zoom of the region around the wing.

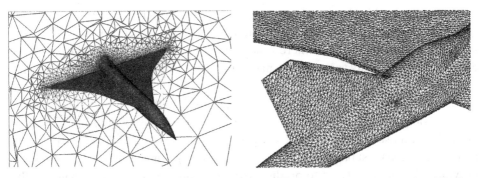

Figure 2.7. Volume and surface meshes for an aircraft. The geometry is represented by a distribution of triangles; the domain is filled with tetrahedra. Figures courtesy of project Gamma, INRIA, Rocquencourt.

2.1.2 Discrete Representation of the Geometry

Computational analysis is done on a discrete approximation to the real geometry; this approximation is usually referred to as a computational grid or mesh. Geometrical discretization means, for example, that curved lines are replaced by a connected set of points. The connection is usually by a straight line. Similarly, curved surfaces are faceted by flat segments; for example, the surface of a sphere might be covered by triangles. Figure 2.7 is a more complex, triangulated body. To the extent that the discrete surface is not the actual geometry, this is an approximation. The difference between the discrete approximation and the continuous geometry is a form of *discretization error*.

Consider a very simple example of geometry discretization. Let the curve $y = x^2$, $0 < x < 2$, be approximated by the line $y = 2x$. The maximum error occurs at $x = 1$ and equals $|1^2 - 2 \cdot 1| = 1$. A finer mesh will more closely approximate the real, curved surface: approximate the curve by $y = x$ for $0 < x < 1$ and $y = 3x - 2$ for $1 < x < 2$. Now the maximum error is $1/4$: see Figure 2.8.

The total integrated error for $0 < x < 2$ – the hatched area in Figure 2.8 – is also reduced from $4/3$ to $1/3$. If the length of the x intervals is h, the error decreases

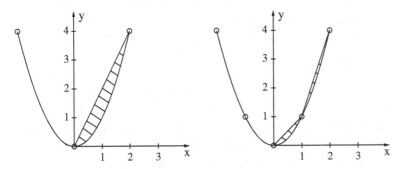

Figure 2.8. Approximation of a parabola by line segments.

in proportion to h^2. Halving the size of h cuts the error by four: this is called a second-order approximation. The definition of the order of accuracy is the power of h, which here is 2. Obviously there is advantage to high-order approximation: the error decreases rapidly as the grid is made finer. But in numerical algorithms second order is usually the most that is practical.

Geometry modeling will not be discussed at any length in this text. It is a central issue in computer-aided design. For fluid dynamical analysis, two distinct possibilities exist. Almost all commercial CFD packages provide a basic level of geometry modeling. In the most basic form, generic objects and geometric transformations are provided. These permit one to produce two- and three-dimensional geometric objects that have a limited degree of complexity: basic surfaces like channels, cylinders, spheres, ellipsoids, and so forth, can be created, or curves defining surfaces can be read in. This may suffice for many examples a student wishes to solve.

However, engineering applications usually require simulation of fluid flow around objects with geometries too complicated to be generated by the simple, generic objects. Then more elaborate solid modeling software may be needed; in particular, software for computer-aided design (CAD) provides highly sophisticated surface modeling capability. Its output, CAD geometry files, can be imported into most mesh generation packages, meant for fluid dynamics simulation. The reader should consult references on computer-aided design for further information on this type of geometry modeling.

2.1.3 Structured Grids

The computational domain delimits the fluid region. To effect a numerical solution, this region must be filled with a mesh. The mesh underlies the discrete representation of the governing equations (see Figure 2.2). The governing partial differential equations are developed into an algebraic system of equations for the particular discretization of space. A good mesh is critical to obtaining accurate solutions. It also affects the speed of numerical convergence.

A computational grid must not only resolve the geometry in sufficient detail but also *anticipate* where in the computational domain small features of the flow will appear and resolve them adequately too. Figure 2.9 suggests the idea that some features of the flow are not obvious from the geometry alone. This is a lid-driven cavity. The upper wall is moving from left to right. The other walls are stationary. The moving wall drags fluid with it, forming a large circulation in the center of the cavity. In the lower corners, smaller circulations are observed, rotating oppositely to the main vortex. Some knowledge of fluid dynamics helps to anticipate types of flow-field variation. In the case of this figure, the corner vortices at the bottom of the cavity should be expected (e.g., see §3.1.5). However, they are quite weak and will not be seen without sufficient grid resolution. This is a simple, two-dimensional example; in three dimensions, in complex geometries, at higher speeds, the location and nature of flow patterns becomes harder to anticipate. The topics and examples

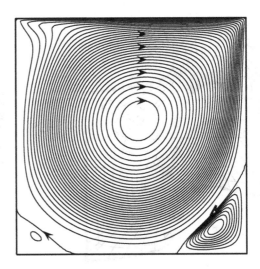

Figure 2.9. Example of flow features that must be resolved. If the grid is not sufficiently fine, the eddies at the bottom of the cavity will not be seen.

encountered in later chapters of this book serve to develop a familiarity with patterns of fluid flow.

Various techniques that have been developed for grid generation are surveyed in Thompson et al. (1999). They might be divided into two broad categories: structured and unstructured. Structured grids consist of identifiable grid lines. They can be straight, Cartesian coordinate lines, curved lines, or a family of closed curves. Examples are shown in Figure 2.10. The upper-left and lower-central grids are structured, the upper-right is unstructured. The former fill the domain with curved lines; the latter fills it with triangles.

The grid at top right in Figure 2.10 encircles a turbine blade; it is referred to as an O-mesh. At the bottom of the figure is a simple channel grid; it is referred to as an H-mesh. Grid type is often termed grid topology. Any grid with O-topology can be deformed into a nested set of circles; any with H-topology can be deformed into a set of Cartesian grid lines. The C-topology, shown in Figure 2.6, can be unwrapped into the H-topology. The C-cut extends downstream from the airfoil trailing edge. The grid is split along this line. Opening it up to separate the top and bottom halves shows that C-grids have H-topology. The C is produced by identifying portions of the boundary that border the cut.

The lines of a structured grid can be numbered sequentially, say by integers (i, j, k). Computationally, that is the defining property of a structured grid. The grid is a set of points in the form $x(i, j, k)$, $y(i, j, k)$, $z(i, j, k)$, where i runs from 1 to i_{max} and similarly for j and k. Neighboring points are separated by one increment of an index, i, j, or k. That underlying structure allows for efficient numerical methods.

The notion of computational space arises quite naturally from this description. It is illustrated in Figure 2.11. In computational space, each grid point is located by a set of integers: in two dimensions (i, j). Computational space simply corresponds to this numbering of the cells. The grid, or mesh, is a mapping from computational space to physical space.

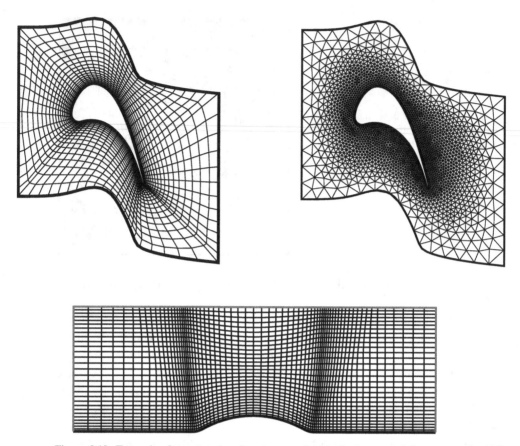

Figure 2.10. Example of structured and unstructured grids. In the upper left, one set of grid lines is a family of closed curves. At right, a triangulated grid fills the same domain.

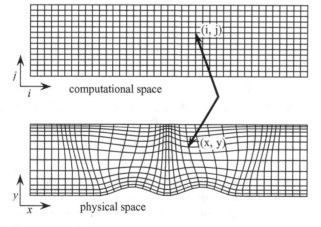

Figure 2.11. Computational and physical space.

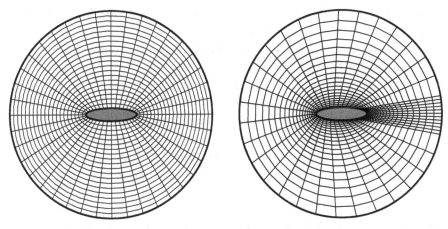

Figure 2.12. Grids that fill in between an ellipse and a circle.

The mapping that defines a structured grid might be generated by solving differential equations or it might be generated by algebraic procedures. A simple example of algebraic generation helps to solidify some relevant concepts.

Consider the domain bounded by an ellipse and a circle, as in Figure 2.12. The ellipse is the solid boundary and the circle is the far-field border of the computational domain. Gridding starts by distributing points along the boundaries. In this case, they are separated by a fixed angle: 41 points are placed at an angular spacing of $2\pi/40$. They lie at $\theta_i = (0, \pi/20, 2\pi/20, \ldots, 2\pi)$. The ellipse has principal axes of a and b. On the ellipse the surface is discretized as $(x_i^e, y_i^e) = (a\cos\theta_i, b\sin\theta_i)$. The circle has radius R. In the case of Figure 2.12 at left, the surface points are $(x_i^c, y_i^c) = (R\cos\theta_i, R\sin\theta_i)$.

The domain is filled by connecting corresponding points of the two surface grids. A line between them is given by

$$x(s) = x_i^e + s(x_i^c - x_i^e),$$
$$y(s) = y_i^e + s(y_i^c - y_i^e),$$
(2.1)

where s is a variable that ranges from 0 to 1 as the line is traversed. The radial discretization is accomplished by choosing points along s. For instance, 21 points are placed at an interval of $1/20$: $s_j = (0, .05, 0.1, \ldots, 1)$. Then the grid locations are $x(i, j) = x_i^e + s_j(x_i^c - x_i^e)$. That produces the grid at left in Figure 2.12.

Another important concept of meshing can be illustrated. The grid just described provides a good resolution of the geometry. Points are clustered in the region of the ellipse having the smallest curvature. If the surface has radius of curvature R_c, and the angular separation is a fixed $\Delta\theta$, then the spacing of points is $\Delta\ell = R_c\Delta\theta$. The spacing becomes small where the radius is small, thereby resolving the curvature. That is a good guideline for representing the geometry.

However, the flow field introduces further requirements that are not met just by meshing the geometry. If a fluid flows with high velocity, from left to right, over the ellipse, there will be a wake behind the surface. There will also be a boundary layer near the wall. The boundary layer has a steep velocity gradient (see Figures 5.1 and 5.3). The grid must be revised to capture such features. One might describe this as a desire to choose spacing that distributes the velocity changes evenly over the grid. That is a very loose description of the motive.

The desired distance between grid points might be something like $\Delta \ell \approx |\delta u|/|\nabla u|$, with $|\delta u|$ prescribed as equal to the desired velocity resolution. In regions of large gradient, the spacing should decrease because the denominator is large. In practice, the gradient is not known. Solutions to archetypal problems can provide estimates. Boundary layer and shear layer thickness will be discussed in Chapters 4 and 5. These give an idea of how the grid should be clustered near the wall and in the lee of the body.

The grid at the right of Figure 2.12 suggests these modifications. Grid lines have been clustered behind the body, to capture the wake, and near to the surface, to capture the boundary layer. This was accomplished by redistributing the points on the bounding circle and spacing the points unevenly in s. The revised grid gives an impression of the anticipated flow field around a streamlined body.

The subject of structured grid generation goes well beyond this introductory discussion. A compendium of grid generation methods can be found in Thompson et al. (1999). When the geometry is complex it may be impossible to cover the domain with a structured grid. In that case multiblock grids might be used. The domain is split into pieces that can be gridded individually. The pieces are then united to provide a full grid. The result is not structured; it is block structured. Nevertheless, many of the efficiencies of a structured grid are retained. A variant on the multiblock grid is the chimera grid, in which blocks overlap and holes are cut in the grid. These techniques are not particularly germane to the present book. They suggest how one's conception of gridding can be expanded. Even if the grid is unstructured, its design may well be influenced by ideas about how a structured grid would fill various regions of the domain.

When generating the grid, structured grids offer an ability to control the density of nodes in a particular region of the computational domain. That is especially important when dealing with boundary layers and separation. In the region near a boundary, long narrow cells may be required. Those arise naturally when the wall normal direction is identified with an index in computational space. Without structure, it is far harder to provide the required resolution.

2.1.4 Unstructured Meshes

The distinction between an unstructured and a structured grid could be a matter of labeling. For instance, instead of cells being labeled by the pair (i, j) they are identified by the single number $n = i + (j - 1) \cdot j_{max}$ in an unstructured format. The actual grid is unchanged, but it is now treated as unstructured for computational purposes.

Similarly, the neighbors of a given cell must identified explicitly if an unstructured formulation is invoked. At an interior point of a structured grid, we know there is a neighbor at $(i - 1, j)$. There is no formula to compute the neighbor in an unstructured

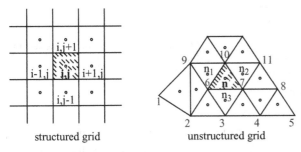

structured grid   unstructured grid

Figure 2.13. Grid structure in two dimensions.

grid; indices of the neighbors must be listed explicitly. This list of neighbors is called the connectivity list. In place of a grid, we have a list of cells and their neighbors. In a structured grid, the cells are topologically rectangles in two dimensions and rectangular hexahedra in three dimensions. An unstructured grid will usually involve other types of polygons and polyhedra.

The gist of the idea is illustrated by Figure 2.13. A structured grid is shown at left and an unstructured mesh at right. Let us define cell centers and nodes. The corners of the cell are the nodes. The center is the point in the middle of the cell, the average of the coordinates of the corners. The nodes at the right of Figure 2.13 are numbered. The cell centered at n is composed of nodes 6, 7, and 10: those three nodes are connected to form the triangular cell. Cell number n shares faces with neighbors, n_1, n_2, and n_3. These connections between cells define the connectivity of cell n. In this case connectivity is restricted to identifying cells that share faces.

The description of the unstructured mesh consists of a table of nodes, giving a number and its (x, y, z) location in space, and another table stating which nodes are joined to form the cells. The latter are identified by a cell center.

It might at first seem that the number of nodes is greater than the number of cell centers; in fact, the opposite is true. For instance, consider the rectangular grid at the left of Figure 2.13: a one-to-one correspondence can be drawn between the cell center and the node at its lower left. Away from the boundaries, all centers and nodes are thereby put into one-to-one correspondence. Hence, they are of the same number. If each cell is now split into two triangles, the number of centers doubles, whereas the number of nodes is unchanged. The number of centers in a triangular mesh is nearly twice the number of nodes – not quite twice, because of the boundaries.

In three dimensions the structured grid cell is topologically a cube. Again, a one-to-one correspondence exists between centers and nodes. If this is sliced into an unstructured mesh made of tetrahedrons, then one finds that each cube is divided into six elements: the number of centers is approximately six times the number of nodes. Elements other than tetrahedra will lead to a different ratio, but generally the number of centers will be greater than the number of nodes.

Standard elements for fluid flow computation in three dimensions are tetrahedra, hexahedra, and prisms, as illustrated in Figure 2.14. Prisms might be used near surfaces. If the surface has been discretized with triangles, then the prisms are oriented

Figure 2.14. Various types of elements in three dimensions.

prism　　　　　tetrahedron　　　　hexahedron

with their axes perpendicular to the wall so that in section they become triangles, lying in planes parallel to the wall. This is a method to cluster the mesh next to the surface: we will encounter it in Figures 5.8 and 5.9. By analogy to Figure 2.12, the axial length of the prism cells decreases as the surface is approached. Prisms of increasing height are stacked to resolve boundary layers. At some distance from the surface, they may transition to tetrahedra. This concept is suggested by the two-dimensional mesh in Figure 2.15. In three dimensions, the rectangular wall cells become prisms with triangular sections parallel to the wall. The triangles transition to space filling tetrahedrons in the central part of the channel.

Hexahedral elements also offer advantages in thin layers, where elements of high aspect ratios may be needed. The aspect ratio is defined as the longest side divided by the shortest side of the cell. High aspect ratio cells are long and thin.

Tetrahedral elements offer advantages of flexible, automatic gridding in complicated regions. Most commercial CFD packages allow for grids that include a mixture of elements types, referred to as hybrid grids. Some allow arbitrary polygons; an example of a surface discretized with polygonal facets is shown in Figure 2.16. A section through a polyhedral volume mesh is shown in Figure 2.17. In principle, higher-order polygons will produce the smoothest, best-quality meshes. However, there are no general methods to construct them, and many unstructured algorithms are not able to accept arbitrary polyhedral meshes.

2.1.5 Mesh Quality

What makes a mesh good? Its purpose is to provide a discrete solution that is a faithful approximation to the continuous solution. To that end, the grid cells must be small enough to provide smoothness. At the same time, the number of cells must be low

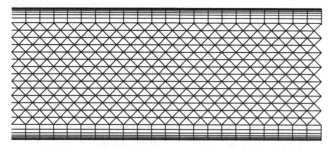

Figure 2.15. Wall and interior meshing of a plane channel.

Figure 2.16. Polygonal surface mesh. Figure courtesy of M. Peric.

enough for computations to be practical. This argues for placing more cells in regions where the velocity gradients are largest. One guideline might be to limit the change of a flow variable, like pressure or the velocity magnitude, from one cell to the next. Another might be to distribute the cells so as to limit the size of the computational error, or to distribute the error evenly over the mesh. As these objectives depend on the solution, they would have to be applied after an initial solution has been generated; in other words, the grid would have to be modified adaptively to the solution. That notion of adaptive grid refinement is indeed a capability of some mesh generators.

However, the notion of mesh quality is commonly applied on a more intuitive level. Geometrical factors are considered: the size of elements, h, which might be Δx, Δy, or Δz across the element; the shape of elements, tetrahedra versus prisms, for instance; the relative size of neighboring elements, such as $r = h_{n+1}/h_n$, which characterizes the degree of grid stretching.

A rectangular cell is characterized by its aspect ratio: the ratio of the maximum to minimum length sides. When this is an element within an unstructured mesh, a

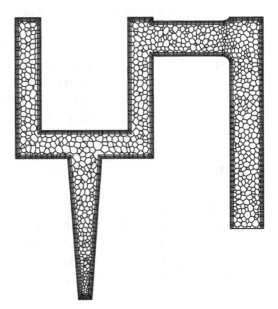

Figure 2.17. Section through a polyhedral mesh. Figure courtesy of M. Peric.

rule of thumb is that the ratio should be less than 100 or so. Structured solution techniques usually can handle significantly larger values.

An equilateral triangle has interior angles equal to $60°$ and equal side lengths. Generally, an effort is made for the interior angles in a triangular mesh to lie between $30°$ and $120°$. The faces of tetrahedral elements are similarly constrained.

These guidelines on angles for triangulated meshes can be understood by reference to Figure 2.13. Suppose that solution variables are stored at the cell centers n_i. If a value is needed on the face, midway between the centers, it must be interpolated. Face values might be needed to compute fluxes of mass or momentum in discretized conservation laws. If the triangles are very long and narrow, with highly acute angles, the line connecting two adjacent cell centers will cross the intervening face very far from its midpoint. The interpolated value will not be that of the face center, and the flux evaluation will be inaccurate.

The numerical errors will be reduced by using elements that are small compared to the length scale of the gradients of flow variables. Error control also suggests having a degree of smoothness in the variation of element size and a stretching ratio not far from unity. There are two ways of understanding the numerical role of mesh quality. One is the control of truncation error, the other is stability and convergence.

The role of stability can be explained by considering simple convection. Some solution algorithms are subject to the Courant–Freidrichs–Lewy (CFL) stability condition. This states that in computation of an unsteady flow, disturbances should move less than one grid spacing per time-step. If the disturbance is convected with velocity u, it moves a distance $u\Delta t$ in one time-step. The CFL criterion is that this should be less than the grid spacing, Δx. If the disturbance is convected in the y direction, the criterion is $v\Delta t < \Delta y$. Hence, $\Delta t < \min(\Delta x/u, \Delta y/v)$. If the u and v velocities are comparable and the cell aspect ratio is large, this can be overly restrictive. For instance, if $\Delta x/\Delta y = 100$ and $u = v$, the time-step restriction in the y direction will cause the disturbance to take 100 time-steps to cross the x extent of the cell. Hence the computation will be rather slow. Of course, if the cell is aligned with the flow, so that $u \gg v$, the high aspect ratio could be beneficial. Generally, it is only possible to ensure such alignment near to walls and in the free stream.

This CFL criterion only applies to certain basic solution algorithms. In fact, if the flow is steady, the time-step per se, is not relevant. But, some other stability condition will inevitably come into play. The process of generating a computer solution begins with an initial guess and proceeds through successive improvement, until a solution of acceptable accuracy is obtained – at least, that is the case if the algorithm is stable. However, the number of iterations required to reach convergence can be quite large. Mesh quality influences it. One can consider that the successive improvements are obtained by propagating and dissipating errors. If the true solution to the discrete equations is $u(x)$ and the approximation after n iterations is $u^n(x)$, the error is $||u^n(x) - u(x)||$. The "$||$"s in this expression are a norm: it might be the maximum difference at any grid point, or it could be the root-mean-square over the mesh. The norm of the error is called the residual. The objective of the solution algorithm is to drive the residual toward zero.

A large error at one point will cause the solution in its neighborhood to adjust, driving down the residual. If we consider the error to diffuse across the faces of a cell, then errors will adjust more quickly in directions where the cell face is largest. Again, cell aspect ratio plays a role in the rate of convergence. Long, narrow cells can produce slow convergence because the error diffuses more slowly across the smaller faces.

The other consideration in assessing mesh quality is solution accuracy. The discrete equations approximate the continuous equations. As the mesh element size decreases, the discrete solution should converge to the continuous solution. That will be the case if the discretization is consistent with the exact equations and if the algorithm is stable. However, the size of the error for any finite-sized mesh element depends on mesh quality. If the error is second order, as defined in §2.1.2, the error will decrease as h^2, where h is a measure of the cell size. Simply making the mesh fine will improve its quality. However, making the mesh fine everywhere can lead to a very expensive computation. It is not necessary.

The factor of h^2, in a second-order method, is multiplied by a coefficient that depends on the solution gradient. This is simply a matter of Taylor series: a function is expanded as

$$f(x + \Delta x) = f(x) + \Delta x f'(x) + 1/2 \Delta x^2 f''(x) \dots. \tag{2.2}$$

If it is approximated by the first two terms, then, by Taylor's theorem, the error is $1/2 \Delta x^2 f''(x + s\Delta x)$, for some $0 \le s \le 1$. The error can be large if f'' is large, unless it is compensated by small Δx. Hence, the mesh must be finest in the locations where f'' is largest. This is a formal perspective on the notion of refining the mesh to capture steep gradients.

In the final analysis, mesh quality is measured by the accuracy of the solution. If the resolution is inadequate, the mesh must be revised. That might involve regenerating it. It may be necessary to decrease the distance of grid points from the wall, to increase the number of cells in regions of steep gradient or to reduce the aspect ratio of poor-quality cells. The whole grid might be regenerated or parts of the mesh might be revised; blocks might be added to multiblock grids; control parameters might be reset in mesh generators. All of these require that judgment be exercised. The mesh is improved where the solution looks to be poorly resolved.

Can this be approached more objectively? What does it mean to say that the solution looks to be poorly resolved? By comparing the solution on two grids, say, one twice finer than the other, the points where the solution changes the most from the coarser to the finer grid can be identified to be where the resolution is poorest. Suppose u_1 is the solution with grid spacing Δ and u_2 is the solution with spacing 2Δ. Suppose the numerical discretization has a second-order error. Then, if u_0 is the solution on an infinitely fine grid, when Δ is small, $u_1 \approx u_0 + E\Delta^2$ and $u_2 \approx u_0 + E(2\Delta)^2$, where E is the coefficient of the second-order error. u_0 can be eliminated to find the error. Let $\epsilon_n \equiv |E\Delta^2|$ denote the error at node n. We find

$$\epsilon_n = 1/3|u_2 - u_1|.$$

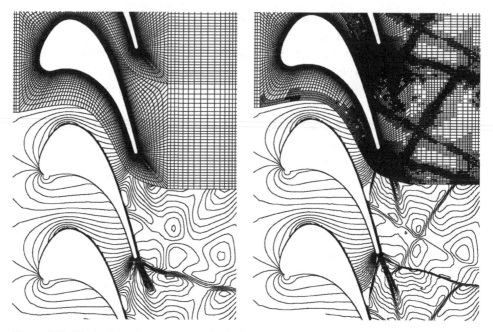

Figure 2.18. Mesh adaptation to capture shocks in a transonic turbine cascade. At left is the initial grid; at right is the solution-adapted grid. Upper portion shows the mesh; lower portion shows Mach number contours. Figure courtesy of G. Iaccarino and G. Kalitzin.

This can be normalized by its average over the entire grid, $\bar{\epsilon}$, to identify where errors are largest. Ideally $\epsilon_n/\bar{\epsilon}$ should be near to unity.

Taking this a step further, a mesh generator can automatically refine the mesh where $\epsilon_n/\bar{\epsilon}$ is large – or coarsen it where $\epsilon_n/\bar{\epsilon}$ is small. The simplest version of refinement is bisection. Every side of an element is divided in two. That is also called isotropic refinement. If refinement is limited to one direction, say, that of the gradient of a variable, it is called anisotropic. Isotropic refinement is illustrated by Figure 2.18. In that example, shock waves (see Chapter 7) are barely visible on the initial grid. After three levels of refinement, the pattern of shocks emerges clearly in the lower part of the figure. In this case, the mesh size has increased from 11,008 to 71,326 cells as a consequence of isotropic cell splitting.

Adaptation to reduce error is the basic concept. However, in practice halving grid spacing to estimate the error can be difficult to effect. A less formal approach is to examine the variation of a solution variable from point to point. The variable could be pressure or magnitude of pressure gradient. The mesh is refined in regions of large variation.

2.2 Computation of Fluid Flow

The Navier–Stokes equations are discretized on the given mesh, in the given domain, subject to the given boundary conditions. These all are particular to the problem being solved. The algorithms to convert them into the algebraic system (see Figure 2.2) and obtain a solution are generic. Descriptions of mathematical methods of computational fluid dynamics can be found in Ferziger and Peric (2002), Tannehill et al. (1997), and many other texts. In the following sections we provide a terse description, for its relevance to the use of computational fluid dynamics software.

2.2.1 Discrete Equations

It seems remarkable that a solution can be effected on a mesh like that on the top right of Figure 2.10 on page 70. If this is thought of as a sort of coordinate system, how can equations be solved in such a convoluted way? The answer is cell by cell. This is not so different from the development of the Navier–Stokes equations in the first place. Navier and Stokes did not think in terms of the complexities of fluid motion; they concentrated on the forces acting on fluid elements. Having deduced the simple laws, it took about a century to fully realize their potential. The fact that such elementary considerations as those leading to the Navier–Stokes equations produce the governing laws for a phenomenon as intricate as turbulent flow is a great triumph of scientific method.

By identifying the essential forces at an elemental level, the endeavor to compute fluid flow is made comprehensible. The integrated conservation equations of §1.6.1 require specification of fluxes and forces on the walls of the control volume. When integrated conservation is applied to the small volumes of a computational mesh, the elemental laws require cells to communicate fluxes and forces to their neighbors; that is how the global flow field is established. The fine-grained momentum balances of the mesh elements are assembled to solve the global momentum equation. A great deal of computer arithmetic is required before that solution emerges from the local conservation laws. However, the computer is just integrating the exchanges between the elemental cells.

It must be acknowledged that the assertion that computational fluid dynamics is a matter of applying the integrated conservation laws cell by cell is only true for finite volume methods. Other methods exist, such as finite difference, finite element, or spectral. All of these involve decomposition of the full laws of motion into elemental interactions. However, only the finite volume approach will be described here – in fact, only the basic version, with cell-centered, collocated storage of variables.

The majority of Navier–Stokes codes use finite volume discretization. It has the flexibility to deal with either structured or with fully unstructured meshes. The mesh decomposes the computational domain into discrete cells. Values of flow variables are associated with each cell. The velocity and pressure are considered to be stored at the central point of each elemental volume; point c, of Figure 2.19. In an unsteady calculation, the numerical procedure is to integrate the solution forward in time by solving the conservation equations at these central points. That is accomplished by forming the integrated conservation equation of the cell, then solving it by a method of successively improved approximations.

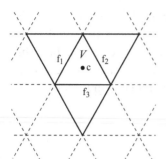

Figure 2.19. Finite volume computational cell, V and neighbors.

Momentum conservation for the discrete cell looks like the macroscopic, integrated conservation Eq. (1.34):

$$\frac{\partial}{\partial t} \int_V \rho \mathbf{u} dv + \sum_{f_i} \int_S \rho \mathbf{u}\mathbf{u} \cdot \hat{\mathbf{n}}_s ds = -\sum_{f_i} \int_S p\hat{\mathbf{n}}_s ds + \sum_{f_i} \int_S \mu(\nabla \mathbf{u} + {}^t[\nabla \mathbf{u}]) \cdot \hat{\mathbf{n}}_s ds,$$

where V is the volume of the cell. The first term is the rate of change of momentum within the cell. The remaining terms are sums over the faces of the cell – f_1, f_2, and f_3 in the case of Figure 2.19. These represent the momentum flux in, or out, of the element, and the pressure and viscous forces acting on its faces.

The face contributions are area integrals. In the second-order, finite volume approximation, the integrands are replaced by their values at the centers of the faces and volumes. The integral becomes the central value times the face area:

$$V\frac{\partial}{\partial t}\rho \mathbf{u}_c + \sum_{f_i}(\rho \mathbf{u}\mathbf{u} \cdot \hat{\mathbf{n}}_s S)_i = -\sum_{f_i}(p\hat{\mathbf{n}}S)_i + \sum_{f_i}(\mu(\nabla \mathbf{u} + {}^t[\nabla \mathbf{u}]) \cdot \hat{\mathbf{n}}_s S)_i, \qquad (2.3)$$

The values that are needed at the face centers are interpolated between the cell centers of the given cell and of its neighbors. The velocity gradients that are needed in the viscous term are also constructed from neighboring values. Thus, the equations of any particular cell become coupled to those of its neighbors. The various cells assemble into a system of equations for the velocity and pressure. The equation of mass conservation is treated similarly to complete the system for incompressible flow.

There are intricacies in writing a computer code to solve the discrete equations. Most of them are not germane to our purpose. However, a few issues of numerical analysis must be cited to understand the proper use of computational fluid dynamics.

Consider a simple illustration: solve

$$\frac{d^2\phi}{dx^2} - k^2\phi = 0 \qquad (2.4)$$

on a uniformly spaced grid in x. The discretization is embodied in Figure 2.19 or, more particularly, in the one-dimensional perspective of Figure 2.20. The cell faces are the points $\{f_i\} = (0, \Delta x, 2\Delta x, \dots, N\Delta x)$. The length of the domain is $L = N\Delta x$. The cell centers are midway between the faces: $\{c_i\} = (1/2\Delta x, 1\,1/2\Delta x, \dots, (N-1/2)\Delta x)$. Starting from the left side of the domain, the points are face at $x = 0$, center at $1/2\Delta x$, face at Δx, center at $2/3\Delta x$, and so on, until the face at $N\Delta x$.

Figure 2.20. Convective flux discretization. The shaded region is a piecewise constant approximation to the flux, the solid lines are piecewise linear, and the dotted curve is quadratic.

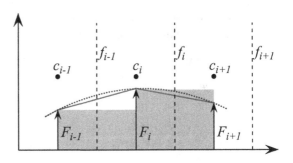

Let the boundary conditions be $\phi = 1$ at $x = 0$ and $d\phi/dx = 0$ at $x = L$. The exact solution is

$$\phi = \cosh kx - \sinh kx \tanh kL.$$

Equation (2.4) is discretized similarly to Eq. (2.3). Integrating from face to face and using a midpoint approximation gives

$$\frac{d\phi_{f_i}}{dx} - \frac{d\phi_{f_{i-1}}}{dx} - k^2 \Delta x \phi_{c_i} = 0, \tag{2.5}$$

where $f_i = (i - 1)\Delta x$ and $c_i = (i - 1/2)\Delta x$. The derivatives can be approximated by centered differences to obtain a set of equations at the cell centers:

$$\frac{\phi_{c_{i+1}} - \phi_{c_i}}{\Delta x} - \frac{\phi_{c_i} - \phi_{c_{i-1}}}{\Delta x} - k^2 \Delta x \phi_{c_i} = 0. \tag{2.6}$$

Let us go further and solve this equation in a way that is relevant to upcoming material in §2.2.3. Boundary conditions must be stated in discrete form. The face at $x = 0$ is midway between the center at $1/2\Delta x$ and a ghost point at $c_0 = -1/2\Delta x$. The boundary condition $\phi = 1$ at $x = 0$ is imposed by linear extrapolation, which corresponds to averaging between values at c_0 and c_1:

$$1/2(\phi_{c_0} + \phi_{c_1}) = 1 \tag{2.7}$$

or $\phi_{c_0} = 2 - \phi_{c_1}$. ϕ_{c_0} is required when Eq. (2.6) is evaluated at $i = 1$.

At the other end of the domain, we return to Eq. (2.5), setting $d\phi/dx = 0$ at f_N. Thus, at the last cell center, located at $i = N$

$$\frac{d\phi_{f_{N-1}}}{dx} + k^2 \Delta x \phi_{c_N} = 0$$

or

$$\phi_{c_N} - \phi_{c_{N-1}} + k^2 \Delta x^2 \phi_{c_N} = 0.$$

Then, at the final cell center,

$$\phi_{c_N} = \frac{\phi_{c_{N-1}}}{1 + k^2 \Delta x^2}. \tag{2.8}$$

The system of Eqs. (2.6), (2.7), and (2.8) could be solved directly by matrix inversion; however, for generality, we describe the Gauss–Seidel iterative method. The method starts from an initial guess, say $\phi = 1$ everywhere. New approximations are obtained by looping through Eqs. (2.6), (2.7), and (2.8), with ϕ_{c_i} on the left side and the remaining terms on the right:

```
START
```
$$\phi_{c_0} = 2 - \phi_{c_1}$$
```
DO   1=1,  N-1
```
$$\phi_{c_i} = \frac{\phi_{c_{i+1}} + \phi_{c_{i-1}}}{2 + k^2 \Delta x^2}$$
```
ENDDO
```
$$\phi_{c_N} = \frac{\phi_{c_{N-1}}}{1 + k^2 \Delta x^2}$$

(2.9)

```
Converged?
If not go to START
```

How does one check whether the solution is converged? Replace the equation inside the loop by

$$\Delta \phi_i = \frac{\phi_{c_{i+1}} + \phi_{c_{i-1}}}{2 + k^2 \Delta x^2} - \phi_{c_i}$$

and

$$\phi_{c_i} = \phi_{c_i} + \Delta \phi_i.$$

Then an error can be defined as the square root of the sum of the squares of all the $\Delta \phi_i$. Convergence is proclaimed when this drops below a preset tolerance (say, 10^{-4}). Convergence can be accelerated by replacing the last equation by

$$\phi_{c_i} = \phi_{c_i} + \lambda \Delta \phi_i,$$

where λ is a relaxation parameter. If λ is greater than unity, this is called over-relaxation; if it is less than unity is it called underrelaxation. For the present equation overrelaxation speeds convergence; the SIMPLE algorithm, described on page 89, is accelerated by underrelaxation.

In Figure 2.21, λ has been set to 1.9. Convergence to the exact solution is demonstrated by stopping the loop (2.9) after 100, 200, 500, and 1,000 iterations. The last is obscured by the exact solution, shown as a solid curve.

2.2.2 Centered and Upwind Differencing

An important concept arises in interpolating the convective flux,

$$F \equiv \rho \boldsymbol{u} \boldsymbol{u} \cdot \hat{\boldsymbol{n}}_s$$

onto the cell face, as is needed in Eq. (2.3). To simplify matters, consider interpolating in one dimension. The cell centers are at $c_i = 1, 2, \ldots$; the cell faces are midway, at $f_i = 1\,1/2, 2\,1/2, \ldots$ as in Figure 2.20. The flux can be evaluated at c_i, where the velocity is stored, and then interpolated to f_i. Approximating the face value by

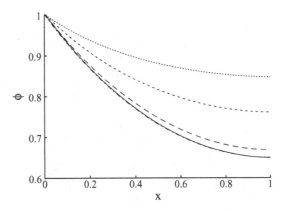

Figure 2.21. Numerical solution for 100 ($\cdots\cdots$), 200 (– – – –), 500 (— — —), and 1,000 iterations; the last overlays the exact solution (——).

$F(f_i) = 1/2[F(c_i) + F(c_{i+1})]$ might seem natural. That is called the centered value. Centered differencing is sometimes used, but more often the convective term is biased in the upwind direction.

For instance, if $u > 0$, when $F(f_i)$ is needed, the value at the upwind cell center, $F(c_i)$, might be used. This is called the first-order upwind approximation. As the term *first order* suggests, it is less accurate than central differencing, which has second-order accuracy. Why use a less accurate approximation? Upwinding adds stability to the numerics. Often upwind approximations will converge when central will not. First-order upwinding is very stable. It converges quite robustly. Unfortunately, it does so by sacrificing accuracy.

As a flux approximation, first-order upwind is piecewise constant. It is illustrated by the shaded regions in Figure 2.20. If $u > 0$, then between $i - 1$ and i the flux is equal to its value at c_{i-1}; between i and $i + 1$ it is equal to its value at c_i and so on. If $u < 0$, then between $i - 1$ and i the flux is equal to its value at c_i; fluxes are obtained by looking to the right instead of to the left. Central differencing corresponds to assuming that the flux varies linearly within each interval, with a discontinuous slope. This is illustrated by the solid line in Figure 2.20. The flux is linear between c_{i-1} and c_i and between c_i and c_{i+1} but with a different slope in each interval.

Upwinding need not represent a loss of accuracy. It can be made to have second- or higher-order accuracy. This can be done by fitting a parabola through three values of F, as in the dotted curve of Figure 2.20. The upwinding is a consequence of using $F(c_{i-1})$, $F(c_i)$ and $F(c_{i+1})$ to evaluate the flux on face f_i. Two out of three points are upwind of the face, if $u > 0$. This is call upwind biasing. The quadratic approximation is called QUICK for quadratic, upwind interpolation of convection kinematics. (The terminology was adjusted to make the acronym catchy.) In QUICK, the interpolated flux is

$$F(f_i) = 1/8[3F(c_{i+1}) + 6F(c_i) - F(c_{i-1})],$$

as can be verified by fitting a parabola and evaluating F on the face.

Accuracy increases as the order of the curve fit to the flux function increases. The upwind biased approximation is best of the three possibilities in that respect.

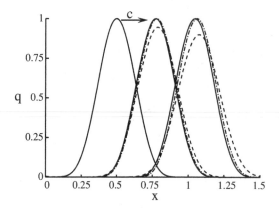

Figure 2.22. Convection schemes. Solid lines are the exact solution at three times; dashed lines are a first-order upwind approximation; chaindash lines are either central or QUICK.

However, there is another basis for comparison, additional to this formal accuracy. The nature of the error is different for these approximations. The two upwind approximations cause a *dissipative* error. The central approximation produces a *dispersive* error at leading order. Those attributes are a useful qualitative distinction between centered and biased discretizations.

Figure 2.22 is a standard illustration of the effect of convection numerics. A pulse is carried at a constant speed, c. The solid curve shows its position at three times. The exact solution is that it simply moves along, undistorted. The dashed line shows what happens when first-order upwinding is used. The pulse decays in amplitude and spreads as it propagates. This can be understood to be a consequence of an extra viscosity added by the numerics. The numerical viscosity is $c\Delta x$, the convection speed times the grid spacing. The central and QUICK approximations are second order and do not cause noticeable numerical diffusion in this example. Either one produces the chaindashed curves in the figure.

In less simple examples, the QUICK scheme will cause some degree of diffusion. The central scheme has a tendency to produce spurious oscillations and might become unstable. Upwind biased methods are favored for the latter reason. Figure 2.23 shows the example of a wave that is steepening as it propagates. The dashed curve shows how central differencing tends to produce spurious oscillations.

A simple analysis provides a nice insight. The particular equation solved in Figure 2.22 is

$$\frac{\partial q}{\partial t} + c\frac{\partial q}{\partial x} = 0, \tag{2.10}$$

Figure 2.23. Nonlinear wave propagation. Solid line is upwind solution; dashed line is central.

with c being a positive constant. This has the exact solution $q = q_0(x - ct)$, where q_0 is any function; it is equal to the initial distribution $q(x, 0)$. That initial shape translates to the right with speed c.

The different flux approximations correspond to different numerical representations of $\partial q/\partial x$. If a grid of cell centers, $x_j = (j - 1)\Delta x$, $j = 1 \ldots N$, is introduced and the approximation to the x-derivative is denoted $\delta q/\delta x$, then the model equation becomes

$$\frac{\partial q_j}{\partial t} + c\frac{\delta q_j}{\delta x} = 0. \tag{2.11}$$

In the upwind scheme

$$\delta q_j/\delta x = (q_j - q_{j-1})/\Delta x;$$

in the centered scheme

$$\delta q_j/\delta x = (q_{j+1} - q_{j-1})/2\Delta x.$$

Let us examine the distortion of a sine wave. Let the initial condition be $q_0 = \sin(2\pi x/\lambda)$. On the discrete grid $q_0 = \sin(2\pi j \Delta x/\lambda)$, where λ is the wavelength of the sine wave. Consider the centered scheme. It can be verified by substitution that the semidiscrete equation

$$\frac{\partial q_j}{\partial t} + c\frac{(q_{j+1} - q_{j-1})}{2\Delta x} = 0 \tag{2.12}$$

has the solution $q_j(t) = \sin[2\pi(j\Delta x - at)/\lambda]$ with the wave speed being

$$a = \frac{\sin(2\pi\Delta x/\lambda)}{2\pi\Delta x/\lambda} c. \tag{2.13}$$

The terminology *dispersive error* refers to the fact that this speed depends on λ and is not equal to c. Each wavelength moves with a different speed. That will cause a packet of sine waves to spread out rather than all move together. The smaller wavelengths move more slowly. The tendency of dispersive errors is to cause spurious spatial oscillations of the solution. That tendency is increased by coarseness of the grid. Some degree of numerical diffusion is usually required to prevent the oscillations.

The first-order upwind method also gives the speed of propagation (2.13), but it also gives an exponential decay rate of

$$\sigma = \frac{1 - \cos(2\pi\Delta x/\lambda)}{\Delta x} c;$$

that is, in time, the wave amplitude decays as $e^{-\sigma t}$. The damping is numerical, not physical. Because errors also damp at this rate, upwinding can improve the stability and convergence of the computer solution.

In the Navier–Stokes equations, convection is nonlinear: the convection velocity is the solution variable u. Convection is the most difficult aspect of numerical solution of incompressible flow. The nature of the discretization errors is far harder to elucidate than for the simple, linear equation (2.11). Nevertheless, the basic notions of dispersive and dissipative error remain pertinent to an understanding the numerical treatment of convection. Issues of accuracy, stability, and convergence guide the selection of a flux discretization in any particular application. Upwinding is diffusive. It enhances stability, but first-order upwinding can be quite inaccurate. Central differencing is dispersive. It can lead to slow convergence and cause spurious wiggles in the solution. Higher-order upwinding, such as QUICK, is often the preferred route.

2.2.3 Solvers

Returning to Figure 2.2: the discrete system system of equations is based on Eq. (2.3), along with choices of upwinding, or similar considerations. The box following "discrete equations" is labeled "algebraic system." The algebraic system derives from selecting a method to solve the discrete system. Choices include *implicit versus explicit* treatments and *segregated versus coupled* formulations. There are other decisions to make, but we will describe only these two aspects of the algebraic system.

2.2.3.1 Implicit and Explicit Methods

Solution by successive approximation involves iterating to a suitable level of accuracy. The stages of iterative approximation are indexed by n. The starting guess is u^0; the next improvement is u^1; the nth improvement is u^n. In an explicit treatment, a formula is provided to obtain u^n from previous evaluations of u^{n-1}, u^{n-2} .... The new approximation is obtained directly by substituting previous values into the formula.

To make the concept of explicit iteration concrete, consider solving $x = \cos(x)$. Try the simple method

$$x^n = \cos(x^{n-1}). \tag{2.14}$$

For instance, start from $x^0 = 1$. Substituting that into the right side gives $x^1 = 0.5403$; substituting that gives $x^2 = 0.8576$; then $x^3 = 0.6543$, and so on. This converges to $x \approx 0.739085$ in about 35 iterations.

An implicit statement is $x^n = \cos(x^n)$. Barring an exact solution, this equation still must be solved iteratively. But now both sides of the equation are at the same level of iteration. Newton's method is a good approach. To solve $f(x) = 0$, this method iterates on $f(x^{n-1}) + (x^n - x^{n-1})f'(x^{n-1}) = 0$ or,

$$x^n = x^{n-1} - \frac{f(x^{n-1})}{f'(x^{n-1})}.$$

In the present case, $f = x - \cos(x)$

$$x^n = x^{n-1} - \frac{x^{n-1} - \cos(x^{n-1})}{1 + \sin(x^{n-1})}. \tag{2.15}$$

Starting from $x^0 = 1$ gives $x^1 = 0.75036$ and $x^2 = 0.73911$. By the third iteration, it has converted to $x \approx 0.739085$. This example illustrates that implicit methods often converge more rapidly than explicit methods, but they involve more work per iteration.

An important distinction between implicit and explicit treatments is explained by considering the ordinary differential equation

$$\frac{dq}{dt} = Aq. \tag{2.16}$$

This would be an analogy to CFD, were q a large vector and A a large matrix. However, for explanatory purposes, let both be scalars. If the right side of Eq. (2.16) is evaluated with q^{n-1}, the method is explicit. For instance,

$$\frac{q^n - q^{n-1}}{\Delta t} = -Aq^{n-1}.$$

This has the solution $q^n = q^0(1 - A\Delta t)^n$. In an implicit treatment, the right side is evaluated with q^n

$$\frac{q^n - q^{n-1}}{\Delta t} = -Aq^n.$$

This has the solution $q^n = q^0/(1 + A\Delta t)^n$.

The exact solution is $q^n = q^0 e^{-At}$, where $t = n\Delta t$. The explicit treatment is quite wrong if $\Delta t > 1/A$; it would oscillate with n from positive to negative values. Explicit treatments involve a restriction on time-step; if Δt is small compared to $1/A$ the accuracy will be acceptable. The implicit treatment is not so obviously wrong when $\Delta t > 1/A$. It has the right qualitative behavior – a damped disturbance. This is a sense in which implicit methods are more robust.

The comparison of implicit and explicit methods in full CFD is not quite so simple. In an implicit treatment, the formula to update u^n involves u^n itself, in addition to previous values. The updated value is not given explicitly; an equation must be solved to find u^n. This usually involves inverting an extremely large matrix. If the number of grid points is N, the matrix will be on the order of $N \times N$ in size. The cost of exact inversion grows as N^3. In the application to CFD, most of the entries in the matrix will be 0, and the cost might be closer to N^2, but if N is on the order of 10^6 this is too expensive. Iterative methods are invoked to reduce cost. They effect an approximate solution in order N operations. Examples are the Gauss–Seidel (see page 82) and incomplete LU factorization methods.

The effectiveness of iterative solutions to large systems hinges, in part, on techniques that accelerate convergence. The iterative solution process can be characterized as an effort to damp out the error. As has been mentioned previously, if the true solution is $u(x_i)$, the error after n iterations is $E(x_i) = u^n(x_i) - u(x_i)$ at the ith mesh point. The error varies from point to point. It might vary greatly from one point to the next or it might vary slowly, with the variation being spread over a large number of points. The rapid variations are the short-wavelength component of the error, the slow variations are the long-wavelength components. Many solution methods damp the short-wavelength components efficiently but are less effective on the long-wavelength components.

For instance, simple three-point smoothing damps the error according to

$$E_j^n = 1/4\big(E_{j+1}^{n-1} + 2E_j^{n-1} + E_{j-1}^{n-1}\big).$$

Substitute a sinusoidal error $E^{n-1} = \cos(kj\,\Delta x)$ on the right. It is damped to

$$E^n = G\cos(kj\,\Delta x) = E^{n-1}G,$$

where

$$G = 1/2\cos(k\Delta x) + 1/2 = 1/2\cos(2\pi\Delta x/\lambda) + 1/2.$$

G is called the damping factor. If $k\Delta x \gtrsim \pi/2$ the error damps by a factor of $G \lesssim 1/2$. After n iterations, the error damps by a factor of G^n. In 10 iterations, the error drops by a factor of more than $1/2^{10} = 0.00098$, which is very good damping.

The definition of wave number is $k = 2\pi/\lambda$; inserting into $k\Delta x \gtrsim \pi/2$ gives $\lambda \lesssim 4\Delta x$. Errors with wavelengths $\lambda \lesssim 4\Delta x$ are damped nicely. But for long wavelength error components, $\lambda \gg \Delta x$ or $k\Delta x \ll 1$, the damping factor G is close to 1. For instance, if $k\Delta x = 0.01\pi$, then $G = 0.99975$. The error damps very slowly: in 10 iterations it damps by only $G^{10} = 0.9975$. It would take over 18,000 iterations to damp by a factor of 0.01. Long-wavelength errors, those with $\lambda \gg \Delta x$, are reduced very slowly. That can be the cause of slow convergence. How can this be overcome?

The long-wavelength errors have a scale comparable to that of the entire computational domain. For a given domain size, or given λ, a smaller G would be obtained by increasing Δx. That means decreasing the number of grid points in the domain. But coarsening the grid would also increase the size of the discretization error. Convergence would be faster, but the solution would be less accurate. The *multigrid* procedure is a method to accelerate the error damping, without decreasing accuracy of the solution. A set of increasingly coarse grids is constructed. The solution is obtained on the finest. To accelerate convergence, the error is interpolated from the finest to the next coarsest, thence to the next coarsest, and so on. On each grid the error is damped by the solver. Thus, the long-wavelength errors are damped efficiently on coarse grids, whereas shorter scales are damped on finer grids. In this way, both long- and short-wavelength errors are dissipated efficiently.

The solver evaluates contributions to the solution from each of the coarse grids and then adds them to the solution on the finest grid. The test of convergence is only applied on the fine grid. Hence, the multiple grids are only an intermediary. The solution is obtained with the highest resolution and assessed by the error on the finest grid.

The performance of multigrid acceleration is illustrated by Figure 2.24. Global numerical error is plotted against iteration number. The unaccelerated method, shown by the line without symbols, starts to converge, then stalls. Large wavelength residuals are preventing convergence. Cycling through coarser grids restarts the convergence. In this particular case, cycling through five grid levels provides a uniform, exponential rate of convergence.

Multigrid is a methodology rather than a single method. Unstructured codes often invoke a variant called algebraic multigrid; the original version is called geometric multigrid. In the algebraic case, the coarse grids are not actually needed, but the philosophy is the same: propagate errors on successively coarse discretizations to accelerate convergence.

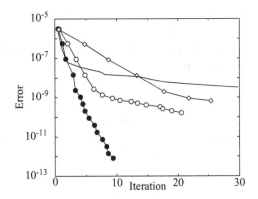

Figure 2.24. Performance of a multigrid Poisson solver. Symbols indicate the number count of V cycles (•) with five multigrid levels; (○) with three multigrid levels; (◇) with two multigrid levels; (——) SLOR without multigrid.

2.2.3.2 Coupled and Segregated Methods

The two topics cited at the outset of this section are implicit versus explicit treatments, which has just been described, and segregated versus coupled formulations. Fluid flow is governed by a system of equations for the dependent variables u, v, w, and p. Four equations, three for momentum and one for mass conservation, are solved for these four variables. The equations can be solved successively or simultaneously. That is what the terms *segregated* and *coupled* mean. In the segregated approach, the velocity components are solved first, say, u and then v and then w. The pressure is then computed such that continuity is satisfied.

This is the pressure correction scheme. It has its origins in incompressible flow computation, although it can be modified for compressible flow. In pressure-correction methods, the velocity components as first computed do not satisfy incompressibility, $\nabla \cdot \boldsymbol{u} = 0$. The pressure computation step provides a correction to the velocity such that continuity is met.

The pressure correction method is also called the projection method and, in some contexts, the fractional step method. Variants of pressure correction go under the names of SIMPLE, SIMPLEC, PISO, and so forth [Fletcher (2002)]. SIMPLE is the archetype for this class of methods.

SIMPLE is an acronym for semi-implicit, pressure-linked equations. The method is first to solve the momentum equations for each component, with the pressure prescribed, and then to solve for a correction to the pressure. Then to correct the velocity to make it divergence free. The method for advancing from the nth iteration to the $n + 1$st iteration is summarized by

    1. Given $\boldsymbol{u}^n$, p^n compute an intermediate velocity $\boldsymbol{u}^*$.

    2. Use $\boldsymbol{u}^n$ and $\boldsymbol{u}^*$ to compute the pressure increment Δp.

    3. Update $p^{n+1} = p^n + \Delta p$ and correct the velocity to   (2.17)

       $\boldsymbol{u}^{n+1} = \boldsymbol{u}^* + \Delta \boldsymbol{u}$.

    4. Check if converged; go to step 1 if not.

In the third step, the velocity update is a correction to u^* that adjusts $\boldsymbol{u}^{n+1}$ to be divergence free. It has the form $\Delta \boldsymbol{u} = \nabla(\Delta p)/A_p$, where A_p is a coefficient that depends on the particular method. The velocity correction is obtained from the gradient of the pressure increment. For our purposes, it is sufficient to note that in a segregated solver, the velocity and pressure are advanced separately. The solution is a sequential process, that is repeated until convergence.

The alternative to the sequential, segregated strategy is the coupled solver. Segregated solvers are primarily developed for incompressible flow. When extended to compressible flow, especially at supersonic speeds, such methods tend to be inefficient. Codes primarily meant for compressible computations invariably invoke a coupled formulation. That is partly because compressible flow phenomena are related to sound propagation (Chapter 7). Sound is a consequence of pressure-density coupling. At appreciable Mach numbers the pressure variations are caused by the flow. Hence pressure and flow also are intimately coupled to one another.

In place of "sound propagation," substitute "error propagation": error propagation involves the coupled interaction of pressure and velocity. As the errors propagate, the approximate solution adjusts toward the converged solution. Compressible flow solvers respect the intimate coupling between variables in the interest of efficient convergence. Methods that couple the solution variables are also used for incompressible flow, although they are less common.

The basic idea in a coupled strategy is to regard the dependent variables at each grid point as a vector. If the variables are u, v, w, and p, the vector is $\boldsymbol{q} = (p, u, v, w)$. So the first component of $\boldsymbol{q}$ is the pressure, and the rest are the velocity components. In compressible flow the temperature, or enthalpy would be included in the vector, and the density might replace the pressure.

Conservation laws, like the momentum equation (1.14), take the form

$$\frac{\partial \boldsymbol{q}}{\partial t} + \nabla \cdot (\text{convective flux}) = \nabla \cdot (\text{stresses}) + \text{external forces}. \tag{2.18}$$

This equation contains evolution, divergence of flux, forcing by divergence of stresses, and possibly external forces, like the force of gravity. In a coupled solver, all equations – momentum, mass, and energy conservation – are put into this same form. Then they are solved for the vector $\boldsymbol{q}$ of dependent variables. In that sense, it is a unified treatment. One reason this is less common as an incompressible method is because the continuity equation $\nabla \cdot \boldsymbol{u} = 0$ is not in the form of Eq. (2.18). The compressible flow mass conservation equation $\partial \rho / \partial t + \nabla \cdot (\boldsymbol{u}\rho) = 0$ does have the form of Eq. (2.18). We will return to coupled solvers briefly at the end of Chapter 7.

2.2.4 Boundary Conditions

The laws of motion are incomplete without boundary conditions. They are a mathematical necessity. The conditions that are required, and the conditions that are tenable, are determined by mathematics. In the context of a course on differential equations that would be the tack to take; in the present context, boundary conditions have a physical rationale. That will be our emphasis.

Section 2.1.1 listed some types of surface that bound a computational domain: solid bodies, inflow and outflow surfaces, far-field boundaries, symmetry planes. A

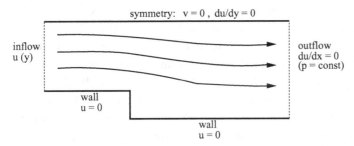

symmetry: $v = 0$, $du/dy = 0$

inflow
$u(y)$

outflow
$du/dx = 0$
(p = const)

wall
$u = 0$

wall
$u = 0$

Figure 2.25. Generic flow domain and boundary conditions.

selection is illustrated in Figure 2.25. Before the equations can be solved, appropriate conditions must be specified on these boundaries.

The condition on solid surfaces is no-slip; see §1.4. The velocity components are equal to the velocity of the surface; if it is stationary, they all vanish: $u = v = w = 0$. The normal derivative of pressure is usually equated to zero. This is derived from the inviscid momentum equation: $\rho Du_n/Dt = -\partial p/\partial x_n$. If $\boldsymbol{u} = 0$, this reduces to $\partial p/\partial x_n = 0$. In viscous flow, this is not correct; the derivative of the total stress, including viscous stresses, should vanish. If the normal direction is y, then the correct condition is $\partial p/\partial y = \mu \partial^2 v/\partial y^2$. However, the right side is often very small; vanishing normal pressure gradient is usually quite accurate.

Symmetry conditions apply only to planes. If the plane is $y = 0$, say, then the condition on velocity is $v = 0$, $\partial u/\partial y = \partial w/\partial y = 0$. Symmetry means that the velocity field is unchanged by reflection in the plane. Upon reflection v is inverted: $v \to -v$. For v to be unchanged $v(y) = -v(-y)$ must be satisfied on $y = 0$; that is, $v(0)$ must vanish. Upon reflection in the plane $y = 0$, $u(y) \to u(-y)$. Differentiating with respect to y: $\partial u/\partial y \to -\partial u/\partial y$, showing that on $y = 0$ this derivative must vanish. The same condition applies to the pressure, for the same reason.

Solid wall and symmetry conditions leave no options and can be coded without choices. Identifying the boundary type implies the boundary condition. The remaining types, inflow, outflow, and far field, do present choices. We discuss only the first two. The far-field condition might be specified as inflow, outflow, or a mixture of them.

Despite its name, inflow can occur at an outflow boundary. Often that is not desirable. Especially in an unsteady simulation, it is preferable to situate the boundary such that the flow is directed out of the domain. The outflow is an artifice that is needed so that the domain can be terminated. Its position and shape are matters of choice. It should be situated so as to have limited adverse effect on the simulation. An example is provided by the flow over a backward-facing step, presented in Figure 2.26. Placing the outflow in plane A of the figure is undesirable because

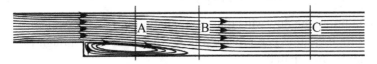

Figure 2.26. Locating the outflow boundary.

flow enters the domain over a portion of plane. Position B is also unsatisfactory, in this case because the pressure would show significant variation over the plane. A constant pressure on the outflow surface is preferable. Either choice, A or B, would have a significant effect on the flow behind the step. Placing the outflow downstream of plane C will cause minimal adverse effect on the flow near the step. By that point the velocity is unidirectional and the pressure will be nearly a function of x only.

In an unsteady simulation, the motive for wanting pure outflow is so that disturbances are carried out of the domain, not convected in. If that is the case, a convective outflow condition can be applied. This takes the form

$$\frac{\partial \phi}{\partial t} + u_c \frac{\partial \phi}{\partial x} = 0, \tag{2.19}$$

where ϕ is a flow variable – like velocity or temperature – and x is the direction normal to the boundary, as in Figure 2.26. This condition says that disturbances are convected out of the domain at the speed u_c; the latter might be the local fluid velocity or an averaged velocity. This boundary condition is not perfect, and some unnatural disturbance will occur in the region before the outflow surface. The distance of upstream influence must be short enough not to affect the region of interest.

In a steady-state computation, the time derivative is absent from Eq. (2.19) and it reduces to a condition of zero normal derivative, $\partial \phi / \partial x = 0$. Then the tenet of pure outflow becomes less compelling. In fact, examples will be presented later in this book in which inflow occurs at the exit boundary.

The pressure at exit can either be constant or have zero normal derivative. In a duct flow, the exit pressure might be a constant which is adjusted to achieve a desired flow rate. In a compressible flow, it might be adjusted to achieve a selected exit Mach number. Either of these require some iteration to reach the desired condition. That is quite like adjusting the pressure in an experimental apparatus.

At inflow boundaries there are more choices. In the incompressible case, most often all the components of inflow velocity vector are specified. These can be constant values, or profiles across the inlet plane,

$$\mathbf{u}_{\text{inflow}} = \mathbf{u}(\mathbf{x}, t). \tag{2.20}$$

For instance, the parabolic profile of Poiseuille flow might be specified at the inlet to a channel.

Another option is to specify the pressure together with the flow direction but not the magnitude of velocity. This might seem nonintuitive. How can specifying pressure replace specifying velocity magnitude? Consider flow from a pressurized tank, through an orifice. The velocity is determined by the pressure drop across the orifice. The flow rate will adjust in accord with the discharge coefficient, C_I, according to

$$1/2 \rho u^2 = \Delta p / C_I.$$

Given the pressure condition, the flow will adjust to that admitted by the geometry. It is clear that a pressure condition can replace a flow condition, even though the precise flow rate cannot be anticipated.

In compressible flow, boundary conditions must be consistent with the direction of characteristic propagation. That topic will be left to Chapter 7.

2.3 Solution (In-)Accuracy

2.3.1 Residuals and Convergence

Numerical solution by successive improvement is an iterative process. As the iterations proceed, a stable solution method will converge to an accurate approximate of the exact solution. Because the Navier–Stokes equations are nonlinear, the iterations may become unstable in some cases while remaining stable in others. The instability often can be overcome by adjustment in numerical parameters, such as relaxation factors or time-steps. Grid quality also can have an influence on convergence.

A perspective on convergence rate can be had by describing the successive approximations as an exact solution plus an error. The objective of successive approximation is to reduce the error through an iterative process. After n iterations, suppose that the solution is

$$u_n = u_e + Er^n, \tag{2.21}$$

where u_e is the exact solution, E is a constant and $r < 1$ is the damping rate of the error. Er^n is a model of the error; in practice, neither it nor the exact solution u_e is known. Equation (2.21) is one conception of the solution process.

A residual is a quantity that is introduced as a measure of the unknown error. It monitors the rate that the iterations progress toward convergence. One definition of the residual is the change of u_n from one step to the next, normalized by its initial value. According to the above model

$$res_n \equiv \frac{||u_{n+1} - u_n||}{||u_1 - u_0||} = r^n. \tag{2.22}$$

A log-log plot of residual versus n has slope equal to the log of the damping ratio: $\log res = n \log r$. Hence, if $r < 1$ the residual will decrease as the iteration proceeds and the solution will converge toward its exact value. If $r > 1$ errors grow, the iterations are unstable, and the solution moves toward blow-up.

A better model is to represent the error as a sum of components:

$$\sum_\alpha E_\alpha r_\alpha^n.$$

For a stable method all r_α must be less than unity. The components with the smallest r will damp relatively quickly. After a large number of iterations, the components with the largest r will make the dominant contribution to the sum. The rate of convergence may slow down as these become prominent.

Formulas other than (2.22) to measure the residual can be conceived, but (2.22) is common and is suited to the present exposition. Figure 2.27 illustrates the progress of an iterative solution. After an initial quick drop, the residuals continue a monotonic descent. Eventually, they level out. In this computation, further convergence is not

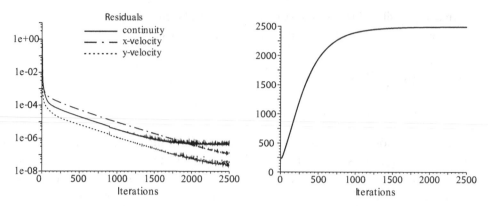

Figure 2.27. Convergence: residuals and drag coefficient in a low Reynolds number computation.

possible. However, the residuals have dropped by six orders of magnitude, which usually is sufficiently accurate. Going further may require improvements in the smoothness and resolution of the grid.

Convergence is monitored by following the reduction of the magnitude of the residuals. A criterion like a four-order-of-magnitude reduction can be imposed for stopping the computation. However, any such fixed value must depend on the particular definition of the residual. Normalization by initial value, as in Eq. (2.22), makes the connection between residual and accuracy ambiguous.

The objective is for the iterative solution to converge to the exact solution. In that spirit, in addition to following the decline of the residual, selected solution variables should be monitored directly. The right side of Figure 2.27 illustrates the evolution of drag force on an object in the flow. It corresponds to the residuals at the left of the figure. The drag becomes accurate only after the residuals have dropped by six orders of magnitude. Thus, monitoring how well the drag, or the velocity at selected points, or some similar quantity, is converged should always be a part of the assessment of solution accuracy.

In unsteady problems, the solution itself is changing in time, from one global iteration to another. For simple vortex shedding flows, such as a flow over a circular cylinder, again, the forces on the cylinder can be monitored. It is expected in this case that those forces will oscillate periodically in time. The time-dependent solution is converged when the periodic limit cycle is approached. This is illustrated by Figure 2.28. The solution goes through a transient phase, then converges to repetitive oscillations. In cases like this, it is very difficult to judge convergence by an examination of residuals, without also examining a solution variable.

2.3.2 Grid Independence

Once a converged numerical solution is obtained, we would like to have an idea how accurately it solves the fluid dynamical equations. There are several possible sources of uncertainty in the simulation: Have we chosen the right flow equations? Have we

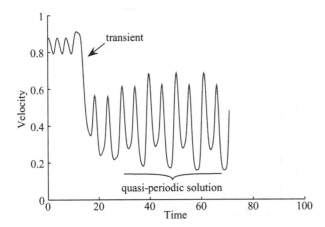

Figure 2.28. Convergence in an unsteady simulation: monitoring flow variables. An accurate solution should be very nearly periodic in time.

defined the right boundary conditions? Have we chosen the right numerical scheme? How good or bad is the grid? Is the time-step small enough?

The last two are questions about resolution in space and in time. The latter is relevant to unsteady flows. To answer it requires a study of the dependence of the solution on the size of the discrete time-step Δt. It is difficult to know in advance what is the appropriate Δt; it depends on the physical timescale T of the flow in question. For periodic flows, $\Delta t \approx T/50$ is often sufficient. A test of grid independence might consist of comparing this to a solution with $\Delta t \approx T/25$. If the solution is accurate, it should change very little when the time-step is doubled.

That notion of grid independence also applies to spatial discretization. Indeed, the choice of computational grid will always be an important issue. Again, there might be some guidelines on how fine the grid should be for certain flow problems: the fluid dynamical theory described in subsequent chapters of this book provides a framework for anticipating grid requirements. We will return many times to this question of the need to adequately capture features of the flow field on the computational mesh.

However, the only way to gain confidence in the adequacy of grid resolution is to conduct a mesh refinement study. That involves computing the numerical solution on a sequence of finer and finer meshes and determining whether the solution changes significantly. Refining the grid, and so forth, by reducing Δh, should improve the solution. At a certain point, further refinement will cause negligible change in the solution – operationally, grid independence has been obtained.

This informal definition of grid independence is too naive: obviously, a very small refinement of the grid will produce negligible change of the solution. The rule of thumb is that refinement should be by a factor of 2. Poisson's equation was solved with increasingly finer grids and the results on 25×25 and 101×101 grids are displayed in Figure 2.29. The latter is essentially grid independent; the former is underresolved.

Figure 2.29. Coarse- and fine-grid solutions are needed to test solution accuracy. Here the solution on a 25 × 25 grid is compared to a solution on a 101 × 101 grid.

In an ideal case, the rate at which the error decreases with refinement of the mesh is determined by the order of accuracy of the discretization; see §2.1.2. The error should vary as in Figure 2.30. On this log-log plot, a first-order scheme has slope equal to 1; a second-order scheme has slope 2. As the grid is refined, the solution moves to the left. This plot gives some idea of why first-order accuracy is usually insufficient: the mesh must be extremely fine to obtain an accurate numerical approximation to the continuous solution.

The ideal is not what one usually confronts. In practice an initial computation might be conducted on a coarse grid. A factor of two refinement can then be effected. However, it is rare that many refinements can be done to check for grid convergence. Indeed, the process of improvement may also involve redistributing grid cells, as well as reducing some cell sizes. For instance, it may be necessary to refine locally within wakes or boundary layers. The notion of doubling the mesh is rarely

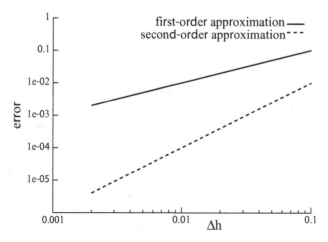

Figure 2.30. Error vs. grid size: convergence under grid refinement Δh. (———), First-order accuracy; (− − − −), second-order accuracy.

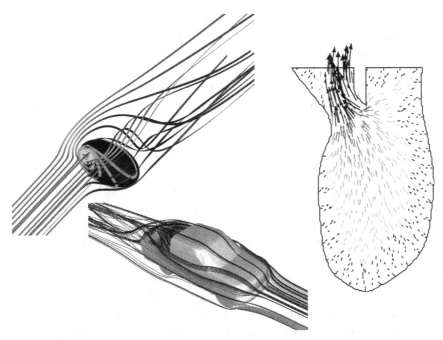

Figure 2.31. Streamlines in flow around an ellipsoid and over a pickup truck. Velocity vectors in a model of a heart.

practicable. Refinement studies tend to be less systematic. They aim to achieve a solution that is insensitive to the grid, without an excessively large mesh.

2.4 Postprocessing and Visualization

A fluid dynamical simulation might consist of many thousands, if not millions, of numerical values, distributed within an intricate geometry. Exploiting that wealth of information can be quite a challenge. Visualization is a technique to aid that endeavor. Data are presented in forms that engage the human perceptual system. The analyst probes the data, identifying essential features, possibly deciding what quantitative information to extract from the massive results file. The ability to interact with the graphical display is often an important feature. Images are rotated, zoomed, clipped, and so on, to come to an understanding of their content. The images may be contours, vector fields, or volumetric renderings of fluid dynamical variables. In unsteady flow, contours at a set of equally spaced times can be assembled into a movie of the flow evolution.

Efficient, special-purpose hardware and software have been developed for visualization. The scientific and engineering community are the unwitting beneficiaries of popular video games, which motivated much of this technology. Algorithms for scientific visualization overlay this well-developed framework of computer graphics tools.

The objective of postprocessing and visualization is to convert numerical output into intelligible information. Sometimes this takes artistic forms, such as the rendering of streamlines and surface pressures, at the left of Figure 2.31. Pressure contours are displayed on the surface of the spheriod. Stream ribbons are shaded by the vertical component of velocity; more information can be displayed by color contours. The wake and flow near the truck are rendered by particle paths. An auxiliary observation may concern some readers: the wake should be quite turbulent, but the streamlines are smooth curves. In fact, this is a computation of the ensemble averaged flow – which was discussed in §1.9 and will be described further in Chapter 6. The computed field is smooth because it is the averaged state of the unsteady, turbulent flow.

2.4.1 Streamlines

Streamlines have been cited above and in Chapter 1. What are they? Streamlines are an intuitive notion. They are simply the paths that particles follow if released into a steady flow. They are curves whose tangents are everywhere parallel to the velocity vector. At each point of a computational grid, a vector can be drawn in the direction of the flow. Connecting them, head to tail, draws the streamlines. This is equivalent to releasing particles in the given velocity field and tracing their trajectories. In two dimensions, the streamlines are readily visualized: Figure 2.9 is an instance. However, in three dimensions it is usually impossible to visualize a large family of streamlines; instead, they are chosen judiciously to extract flow patterns. Some examples will be encountered in Chapter 4.

Trajectories in a given velocity field are curves traced out by integrating

$$\frac{dX}{dt} = u(X, Y, Z); \quad \frac{dY}{dt} = v(X, Y, Z); \quad \frac{dZ}{dt} = w(X, Y, Z). \tag{2.23}$$

The capital letters indicate the position of a moving particle. Time here is an artifice. The flow is frozen and these trajectories are computed by following them artificially in time. The trajectories are integrated from an initial position (X_0, Y_0, Z_0).

Postprocessing software permits particles to be started at a number of selected locations, and their trajectories traced. Particles are introduced in selected locations to map out some portion of the flow, as in Figure 2.31. Chapter 4 contains several examples of streamlines created in this way. The initial location determines which features are seen. In the case of Figure 4.16 on page 147, vortices are isolated by careful selection of the particle tracks.

Sometimes a distinction is drawn between streamlines and streaklines. Streaklines are trajectories that a particle would follow in an unsteady flow as it evolves. They are a function of when the particle is released. In a lab experiment, these are what are visualized. However, in CFD it is more common to plot streamlines at fixed instants. The velocity field is frozen in time and particle tracks are computed. At each instant, a different set of streamlines is obtained.

Lines of surface stress can be informative, as will be discussed in Chapter 5. The surface stress is a force vector. Surface stress lines connect the vectors to form continuous curves. An example is shown in Figure 2.32. Just behind the wing, the stress lines spiral, showing a small region where the flow separates from the surface. The pattern of curves, imprinted on the body, gives a feel for how the flow proceeds over the surface.

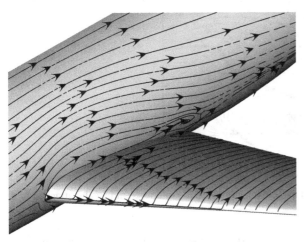

Figure 2.32. Surface stress lines on a wing-body configuration. Figure courtesy of Edwin van der Weide.

Surface stress lines are analogous to streamlines, projected onto a two-dimensional surface. The projection consists in removing the normal component of the velocity vector to leave a vector field tangent to the given surface. If the velocity is (u, v, w) and the surface is a plane $z = $ constant, then the projected velocity is simply (u, v). The streamlines in this plane are the trajectories

$$\frac{dX}{dt} = u, \quad \frac{dY}{dt} = v.$$

As a simple example, if $u = -0.5x + 0.6y$ and $v = -y - 0.6x$, the spiraling surface stress lines shown in Figure 2.33 are obtained.

Figure 2.34 is an example of a planar projection of the instantaneous streamlines in unsteady flow about a surface mounted cube (see Figure 1.24, page 55). Care must be taken when interpreting such flow visualizations. It should be recognized that the pattern in Figure 2.34 does not represent the full streamlines of the flow. The actual streamlines have a component directed out of the plane, corresponding to the w velocity. For instance, the arrows in Figure 2.34 seem to show flow swirling into a sink. Mass would seem to disappear into the sink. However, it is actually flowing

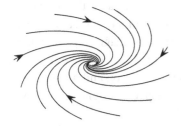

Figure 2.33. An illustration of spiraling surface stress lines.

Figure 2.34. Top view of streamlines in a cross-sectional plane in flow round a cube attached to a wall.

out of the plane, not disappearing. The three-dimensional incompressible flow is divergence free:

$$\frac{\partial u}{\partial x} + \frac{\partial v}{\partial y} + \frac{\partial w}{\partial z} = 0.$$

However, the projection onto a two-dimensional plane, is not:

$$\frac{\partial u}{\partial x} + \frac{\partial v}{\partial y} \neq 0.$$

Where this divergence of the projected field is negative, the streamlines converge, as they do into the sink.

2.4.2 Vectors and Contours

In the terminology of computer graphics, vectors are glyphs: objects drawn at each mesh point to convey information about the data contained there. In this case, the glyph is an arrow. The arrow points in the direction of the flow, its length – and possibly its color – indicate the magnitude of the velocity. Thus the set of numbers $[u(x), v(x), w(x)]$ is converted into images that produce a ready view of the flow field. Because the glyphs are rooted in the grid, they can interfere with one another – then, not producing so ready a view. Software for vector plotting usually permits the user to select a fraction of the computational nodes for display. A sparse set of vectors can provide a clear image of flow patterns. In the example of Figure 2.35 a vector is plotted at every sixth grid point.

The right side of Figure 2.31 visualizes a computation via a plot of velocity vectors in a cross-sectional plane. The wall is moving inward to pump the fluid out of the chamber, simulating the type of flow that would occur in a heart – albeit in a very simplistic form. The velocity vectors are grayscale coded by magnitude, showing a zone of high velocity where fluid is being pumped out of the heart. The field of vectors provides pointwise information. One sees the flow direction and magnitude, perhaps gaining some feel for the flow pattern. Streamlines give a better overall view of the flow field.

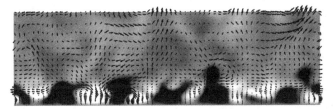

Figure 2.35. Velocity vectors in a yz plane overlaying contours of u velocity. From a computer simulation of a boundary layer undergoing transition from the laminar to the turbulent state.

Streamlines and velocity vectors identify flow patterns. Contour plots display quantitative values without information on flow direction. First, a set of constant contour levels is selected, either by subdividing the interval between the maximum and minimum or by user specification. Then the scalar field, such as pressure or velocity magnitude, is evaluated on a surface. Curves of of constant value are created by interpolating between data available at the computational nodes. The curves may then be displayed directly, or the regions between them may be filled with colors or grayscale. The figures in this book are all in grayscale. The coloring (or grayscale) is prescribed by a mapping between color and numerical value. The airplane seen in Figure 1.7, page 26, is actually a mapping of surface pressure contours. The highest pressure is mapped to white and the lowest to black.

A scientific visualization of velocity perturbations in a boundary layer undergoing laminar to turbulent transition is the subject of Figure 2.35. Velocity vectors show the pattern of eddies in the yz plane, whereas the u component is indicated by contours. The averaged flow is out of the page. In the dark region, u is larger than average; the velocity fluctuation is positive. In the light patches, the fluctuating part of u is negative, although the averaged velocity is still out of the page. The figure provides data on the full, three-dimensional velocity field by combining vector and contour plotting.

Velocity contours are easily confused with streamlines, but they should not be. Figure 2.36 contrasts the two. Contours are represented here by lines, without grayscale fill. The upper plot shows contours of the u component of velocity, the lower shows streamlines. The streamlines contain a rounded region of recirculating flow. In

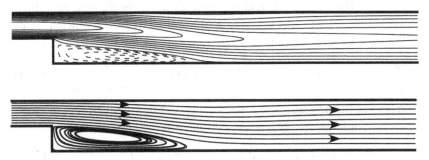

Figure 2.36. u-component velocity contours (upper) and streamlines (lower) in flow over a back-step. Dashed lines indicate negative u.

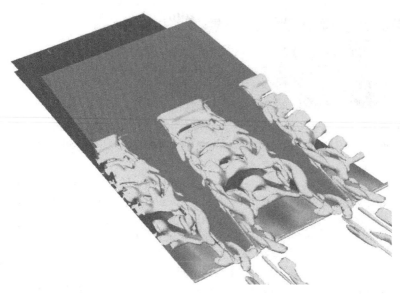

Figure 2.37. Visualization in three dimensions. Isosurfaces of $\mathcal{Q}$ identify vortices in flow out of slots.

the lower part of the recirculation zone, u is negative. Hence, the u velocity contours indicate a region of negative flow, indicated by dashed lines. In the upper part of the recirculation zone, solid lines indicate that u is positive. Hence, the contours of u velocity have a different shape from the recirculating streamlines. The streamlines are computed from both the u and v components of velocity.

Contours in three dimensions must be plotted as surfaces. In §1.8.6, the quantity $\mathcal{Q}$ was defined as an indicator of vortices. Figure 2.37 is another example in which surfaces of constant $\mathcal{Q}$ are visualized. They show unsteady jets emerging from slots along a wall. The vorticity in the jets rolls up into an irregular pattern of unsteady vortices. The figure shows them at one instant. The information conveyed is a subjective understanding of the phenomenology.

2.4.3 Quantitative Data

In addition to visual images, postprocessing involves extracting numerical values of interesting quantities. These may be raw numbers, or they may be derived quantities. Properties at the surface, such as the pressure coefficient c_p, friction coefficient c_f, or heat transfer coefficient h_t, are examples. These three are defined by

$$c_p(\boldsymbol{x}) = \frac{p(\boldsymbol{x}) - p_\infty}{\frac{1}{2}\rho_\infty u_\infty^2}, \qquad c_f(\boldsymbol{x}) = \frac{\tau_w(\boldsymbol{x})}{\frac{1}{2}\rho_\infty u_\infty^2}, \qquad h_t(\boldsymbol{x}) = \frac{q(\boldsymbol{x})}{C_p(T_{\text{wall}} - T_\infty)}. \qquad (2.24)$$

where p, τ_w, and q are the pressure, shear stress, and heat flux evaluated at the wall. The second two are computed from gradients of the primitive flow variables:

$$\tau_w = \nu\hat{n} \cdot (\nabla\mathbf{u} + \hat{\imath}[\nabla\mathbf{u}]) \cdot \hat{\imath}, \qquad q = \kappa\hat{n} \cdot \nabla T, \qquad (2.25)$$

where κ is the heat conductivity and $\hat{\imath}$ and $\hat{n}$ are tangential and normal unit vectors.

Generally, these surface properties are computed internally to the CFD code. Knowing their definition gives an idea of the grid requirements. To resolve transport to the wall, the distribution of velocity and temperature near the surface must be accurate. Computing gradients requires very accurate values of the variable. The discrete derivative is $[T(x + \Delta x) - T(x)]/\Delta x$. The numerator might be the small difference between two large values. An accurate solution is needed to evaluate it.

For engineering purposes, integrated fluid forces exerted on an object might be evaluted. Equation (1.9) is a formal statement of how they are computed. Let $\hat{n}$ be the unit vector, normal to the surface, and let $\hat{x}$ be the direction of the flow approaching an object. The force in the direction of $\hat{x}$ that is exerted on the surface, by the flow, is defined as the drag. It is evaluated as

$$D = \int_{\text{surface}} [\hat{n} \cdot \sigma \cdot \hat{x}] dA, \tag{2.26}$$

where σ is the stress tensor introduced in Chapter 1. Most commercial CFD software includes procedures for computing standard quantities of this type. Nonstandard quantities can be evaluated at the postprocessing stage, after the simulation has converged.

2.5 Packages and Codes

Commercial software packages have had a large impact on the use of CFD. It has been greatest on engineering applications, but researchers also use software packages as a convenience. These packages hide the inner workings behind a graphical user interface (GUI). Two GUIs are illustrated in Figure 2.38. The first is a step-by-step guide to setting up the analysis. The user clicks through the steps, at each stage being invited by subwindows to select methods, models, and parameters.

The second GUI is laid out as pull-down menus, again offering a selection of models and methods. The lower panel illustrates some of the options: segregated versus coupled solver, implicit or explicit treatment, steady or unsteady in time, and so on. Other menus offer further choices that configure the analysis.

An interface for geometry generation is shown in Figure 2.39. The user can select shapes and then size and configure them to create objects. This provides the surfaces. The domain that they define is subsequently filled with a mesh. The interface portrayed in Figure 2.39 also provides meshing. The number of cells that is desired can be specified. A layer of fine cells can be created around objects by first inflating them and then gridding the space between the actual surface and the inflated surface. This might be done with prisms, hexahedra, or other shapes that are conducive to resolving steep gradients near the wall.

These software packages come with tutorials to provide quick instruction on their use. Extensive manuals are also available for more advanced usage. Most packages also allow some element of user programming. They can be used as general purpose convection-diffusion equation solvers or as partial differential equation solvers. As

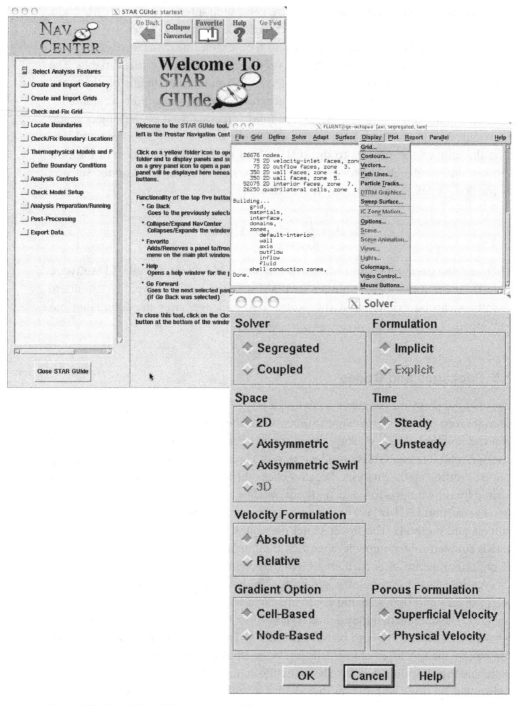

Figure 2.38. Star-CD and Fluent user interfaces.

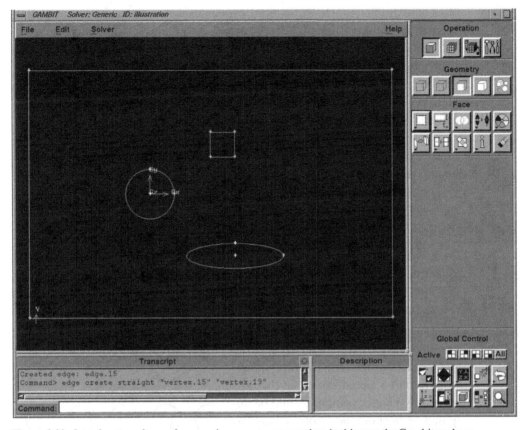

Figure 2.39. Interface to software for generic geometry generation, in this case the Gambit package.

these are well-documented, well-supported, commercial products, we do not dwell on their particulars.

Homegrown codes have not been exterminated by commercial software. The title of this section, "packages and codes," refers to software accessed through a GUI and to software in source language – Fortran or C. The research community initially expressed disdain for packages. Although that has largely dissipated, it remains true that codes are needed for many purposes that are not served well by packages.

One case in point is turbulence simulation (Chapter 6). The algorithms that are most accurate and efficient for turbulence simulation are not effective for general purpose software. Other special applications, for instance, studies of polymer rheology, require access to a code that can be extensively modified or that uses algorithms suitable for a narrow type of application.

Early packages were written for structured single or multiblock, grids. Most have evolved to unstructured formulations. Structured grids are not conducive to automated meshing of complex geometries. However, when meshing is practical, structured algorithms increase computational efficiency. Hence, many codes exist to maintain the efficiency of structured solvers.

To a fluid dynamicist, the distinction between packages and codes is immaterial: one desires an accurate numerical approximation to the exact equations, irrespective of how it is effected. Most of the examples in this book were computed with commercial packages. Many of them had, in earlier years, been computed with research codes. The packages are convenient tools for our purpose. In more complex geometries, they become indispensable. The exception, in this book, is turbulence simulation. All the examples of eddy simulation were computed with homegrown codes. These use a combination of explicit and implicit treatments for time integration and centered spatial differencing. Or, in the case of Figures 6.21–6.23, isotropic turbulence was computed with a spectral method.

EXERCISES

2.1 *Intro to modeling.* With paper and pencil, sketch grids for flow over a two-dimensional circle. Select a less simple shape and sketch a grid for high Reynolds number, laminar flow. Label all boundaries by type.

2.2 *Symmetry.* An array of circles models a bank of tubes in a flow from left to right. Discuss possible choices of flow domain for the array of circles. What are the geometrical symmetries? Which symmetries permit a simplification of the flow domain? What conditions should be imposed on the boundaries of the domains that you have selected?

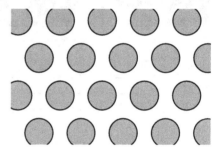

2.3 *Grid generation.* Sketch a channel with a bump on the lower wall. An analytical form for the domain is

$$h(x) < y < H \text{ with } h(x) = 0.2He^{-(x/H)^2}, \quad -5H < x < 5H.$$

Draw a structured grid such that the grid lines meet the upper and lower walls perpendicularly.

A simple algorithm for meshing propagates the grid from one surface to the other. This is an algebraic method, similar to Eq. (2.1). Write a pseudoalgorithm – that is, a set of steps that could be performed by a computer – to construct the mesh that you drew by propagating the grid lines from the lower to the upper wall. Program your algorithm and generate a grid.

2.4 *Nodes and centers.* What is the minimum number of tetrahedra into which a hexahedron can be divided? The hexahedron can be a "brick" shape and the tetrahedra need not be regular, just four-sided volumes with triangular faces. Explain how your answer provides an estimate of the ratio of cell centers to nodes in a tetrahedral mesh.

2.5 *Upwinding.* The equation

$$\frac{d^2u}{dx^2} - Re\frac{du}{dx} = 0$$

with $u(0) = 0, u(1) = 1$ has the exact solution

$$u = (e^{Rex} - 1)/(e^{Re} - 1),$$

where Re is a Reynolds number. Thus, large values correspond to dominance of convection over viscous diffusion.

A numerical solution can be effected as follows. The differential equation is discretized on the grid points $x_i = i\Delta x$, $i = 0, 1, 2, \ldots I$, where $\Delta x = 1/I$. The discrete form is

$$(u_{i+1} - 2u_i + u_{i-1}) - Re\Delta x \, \delta u = 0.$$

The first term is a finite difference approximation to the second derivative, multiplied by Δx^2. The approximation to the first derivative is either centered,

$$\delta u = (u_{j+1} - u_{j-1})/2$$

or first-order upwind

$$\delta u = u_j - u_{j-1}.$$

An iterative method to solve the discrete equation is to program two loops

```
DO n=1,N
  u_I = 1; u_0 = 0
  DO i=1,I-1
    u_i = u_i + α[(u_{i+1} - 2u_i + u_{i-1}) - ReΔ x δ u]
  ENDDO i
  print u(10)
ENDDO n
output x,u
```

This is a pseudo-time-stepping method. The value of u_i is overwritten within the inner loop. The outer loop iterates the solution until it converges – if it converges.

Try this with $I = 25$, $\alpha = 0.1$, and $N = 2,000$. The printed value $u(10)$ is a convergence check. Set $Re = 1$ and solve with the central and upwind forms of δu. Which converges fastest? Compared to the exact solution, which is most accurate?

Repeat with $Re = 10$ and with $Re = 50$. Repeat these cases with $I = 70$. Other values of I and Re can be explored. Note that the numerical method depends on the value of $Re\Delta x$, which is called the mesh Reynolds number. The central method requires this to be small.

2.6 QUICK. Derive the upwind biased flux interpolation $F(f_i) = 1/8[3F(c_{i+1}) + 6F(c_i) - F(c_{i-1})]$ cited on page 83. State some of the favorable and unfavorable properties of upwinding.

3 Creeping Flow

The terminologies *creeping flow, Stokes flow*, or *low Reynolds number hydrodynamics* are used synonymously to refer to flows in which inertia is negligible compared to viscous and pressure forces. The formal requirement is that the Reynolds number be small: $Re \ll 1$. However, in practice the low Reynolds approximation often remains satisfactory for Reynolds of order unity: $Re \sim 1$.

With inertia neglected, momentum is transported by viscous diffusion but not by convection. Some ideas about fluid dynamics must be rethought in this limit. Without convection, there is no wake on the downstream side of an object; pressure scales on viscosity not on kinetic energy; when they occur, eddies are as likely upstream as downstream of a blunt body.

Low Reynolds number can mean highly viscous; hence, one can imagine objects moving through syrupy fluid or syrupy fluid being pumped through a conduit. The dominant forces are frictional in origin. Of course, low Reynolds number can also mean very low velocity or very small scale. One application of creeping flow is to locomotion of microorganisms through a fluid. These animals are a few microns in size. They do not move by propulsion; they drag themselves through the fluid, pushing or pulling by frictional forces. In some cases, they use spiral flagella to corkscrew themselves along. On their relative scale, the fluid appears to be very viscous. One can imagine pushing against a very thick fluid to move forward. To do so, the frictional force pushing forward must be greater than the frictional force resisting motion. Swimming is possible if the organism can produce motions that create more pushing friction than impeding friction. This tends to be accomplished by propagating waves along flagella or cillia. The waves are ripples moving along the flagellum. Rotate a corkscrew and you will see a pattern propagate back along the axis; it is a pattern, not the solid surface, that translates backward. This provides forward force on the microorganism. As it moves, frictional drag on its head and on the flagellum will balance the forward force with backward resistance. That is a simple view of swimming in the low Reynolds number regime.

Another application of low Reynolds number flow is to small particles in suspension. This might be a colloid or a dusty gas. Typically, the particles might be in the size range of 10–100 μ. A fluid analysis may be required to determine settling rates of the suspension, or it may determine an effective viscous stress that is due to the presence of particles. The reason particles seem to increase the viscosity is that the flow around the them produces local velocity gradients and local dissipation of energy. Viewed on a macroscopic scale, the suspension seems to dissipate more energy than it should: the colloid appears to be a fluid with higher viscosity than that of the pure liquid. For some purposes, it can be treated as such – as a fluid with a viscosity that increases in proportion to the concentration of particles.

Flow in microscale devices is a further application of low Reynolds number hydrodynamics. Flow through tiny channels, etched in silicon is used in some types of chemical analysis. The small scale leads to low Reynolds number. Complicating factors, like electric fields or bubbles, often are involved in applications. Although these add extra phenomena, the flow remains viscous dominated.

In other applications, one is dealing with a combination of high viscosity and small channels. Hydrodynamic lubrication is an important example. There, the flow through gaps between bearings and races is governed by a balance between pressure gradient and viscous friction. Extremely large pressures can be encountered in the bearing gaps; these large pressures are the mechanism by which oil prevents surfaces from contacting each other, when hydrodynamic lubrication is in effect. Lubrication forces also may be active in the joints of animals. Synovial fluid is confined in a small gap between bone surfaces. Hydrodynamic lubrication may prevent contact and wear of the joints.

In these cases, high pressure is created by forcing viscous liquids through small gaps. As a bearing rotates, it drags oil into the gap, raising it to high pressures. This might be conceived as an analogy to stuffing clothing into a bag – but with an important qualification: the oil can exit the other side of the gap. When doing so, the pressure falls as it is released from the viscous impedance.

Thoughts about fluid dynamics in the low Reynolds regime stem from various classical analyses. The archetype is flow around a sphere; that is where we will start.

3.1 The Low Reynolds Number Limit

If the inertial term, $\rho Du/Dt$ is omitted from the Navier–Stokes momentum Eq. (1.14), it becomes

$$\nabla p = \mu \nabla^2 \boldsymbol{u}, \tag{3.1}$$

which states that the pressure force balances viscous stress. An important property of this equation is that it is linear in u. For instance, that is why a wake cannot exist: essentially, there is no nonlinear convection term to carry vorticity downstream; it simply diffuses outward from the surface.

Linearity leads to the important idea of a resistance matrix. Before discussing that, we cite a well-known, exact Stokes flow solution. It is for flow past a sphere. The streamlines of this flow are plotted in Figure 3.1. These are cross sections through axisymmetric stream surfaces, created by rotating the figure about the central axis. They show the fore–aft symmetry that is a hallmark of creeping flow. There is no wake because convection is small compared to viscous diffusion.

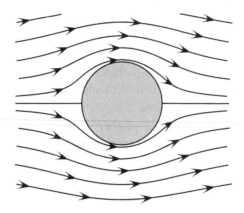

Figure 3.1. Streamlines in the vicinity of the sphere, $Re = 0.01$.

The exact solution to Eq. (3.1) for the pressure and velocity field around the sphere can be stated as

$$u = U_\infty - \frac{3a}{4}\left[\frac{U_\infty}{r} + \frac{xU_\infty \cdot x}{r^3}\right] - \frac{a^3}{4}\left[\frac{U_\infty}{r^3} - 3\frac{xU_\infty \cdot x}{r^5}\right],$$
$$p = -\frac{3\mu aU_\infty \cdot x}{2r^3}$$
(3.2)

(Batchelor, 1967). The pressure is proportional to μ. As the fluid proceeds over the sphere, the pressure falls in the direction of U_∞ due to the factor of $-U_\infty \cdot x$. Thus the pressure is lower on the rear than on the front of the sphere. The pressure force acts in concert with viscous friction to pull the sphere in the direction of the flow. To remain stationary, sphere must be held with an equal and opposite force.

In Eq. (3.2), U_∞ is the velocity approaching the sphere from the left and a is the radius of the sphere. u is the three-component velocity vector, x is position in the fluid relative to the center of the sphere, and $r^2 = x^2 + y^2 + z^2$ is distance from the center. The factors in Eq. (3.2) that depend on x can be understood to result from the viscous force on the sphere in the direction of the oncoming flow. They give rise to the famous formula

$$D = 6\pi\mu aU_\infty \tag{3.3}$$

for the drag force on a sphere in Stokes flow. This is derived by integrating the pressure and viscous stress over the surface. The exact mathematics show that 1/3 of that drag is due to the pressure force on the sphere and 2/3 is due to the viscous force per se (Batchelor, 1967).

Analysis shows that when the Reynolds number is small, the drag can be developed in a series as

$$D = 6\pi\mu aU_\infty(1 + 3/16Re + \cdots), \tag{3.4}$$

where the Reynolds number $Re = U_\infty d/v$ is based on diameter. Equation (3.3) is a first approximation; Eq. (3.4) is called the Oseen approximation. At $Re = 0.1$, the

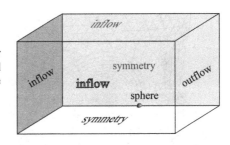

Figure 3.2. Flow domain for low Reynolds number flow round a sphere. Only one-fourth of the spherical surface is contained in the domain, in consequence of the rear and lower walls being symmetry planes.

above suggests that Stokes drag (3.3) is about a 2% under estimate. In fact, it can be proved that Stokes flow will always give a lower bound on the viscous drag.

The first bracketed factor in Eq. (3.2) gives rise to an important observation in analyzing Stokes flow: the induced velocity falls off like $1/r$ at large r. The uniform flow $\boldsymbol{u} = \boldsymbol{U}_\infty$ is approached slowly. A sufficiently large computational domain is needed to calculate flow accurately in a uniform oncoming stream.

The solution (3.2) is derived in many texts. Here Stokes flow round a sphere will be solved numerically. We can obtain low Reynolds number flow by selecting suitable parameters of a CFD simulation. A computation can be effected as follows.

3.1.1 Stokes Flow around a Sphere

Consider a small sphere with a diameter of $d = 2a = 1$ μm, which is a size representative of a typical microorganism. Its surrounding fluid is assumed to be water at room temperature ($T = 20°$ C and $p = 1$ atm) with $\mu = 10^{-3}$ Pa · s and $\rho = 1$ g/cm^3. For the purpose of numerical simulation, the flow will be treated as steady. For instance, if the sphere is translating, the computational domain moves with the sphere; this is equivalent to setting the far-field water velocity to minus the velocity of the sphere.

Initially consider a reasonable, low value of $U_\infty = 1$ cm/s for the far-field velocity. This results in a Reynolds number $Re = U_\infty d/\nu$ of 0.01. Later on, we will discuss two additional cases with $U_\infty = 0.1$ m/s ($Re = 0.1$) and $U_\infty = 1$ m/s ($Re = 1.0$).

Before preparing the grid for the numerical simulation, a flow domain must be chosen. Anticipating the flow to be left–right and top–bottom symmetric, the domain can contain only one-fourth of the sphere. Note that symmetry cannot be assumed in the direction of the flow; therefore, the domain cannot be reduced to only one-eighth of the sphere.

A second issue that arises in setting the flow domain is what shape to choose for the outer, far-field boundary. It could be a large sphere with a prescribed far-field velocity, but that is not the present choice. A more general approach is simply to select the outer boundary to be a large box, on which inflow, outflow, and symmetry conditions are imposed, as appropriate. The computational domain is shown in Figure 3.2. The rear and lower walls are symmetry planes. All other faces of the rectangular domain are treated as inflow boundaries, except the leftmost, outflow plane.

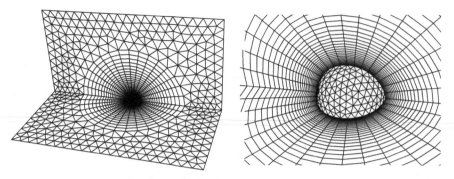

Figure 3.3. Computational grid – surface elements only. At right, zoom of region near the sphere.

The final choice is of the domain size. From Eq. (3.2), one sees that the effects of the sphere on the flow velocity will be felt at large distances from the surface. The velocity perturbation falls off like $1/r$ at large values of r. For example, at $r = 10a$, the second term in Eq. (3.2) is still large: it is $3/40$ times the far-field velocity, U_∞. At $r = 40a$, it becomes equal to $3/160\, U_\infty$; that is, U still differs from U_∞ by about 2%. We will consider that to be small enough for the purpose of this analysis; hence, the computational domain is chosen to be a $40d \times 40a \times 40a$ box.

An unstructured mesh will be used. Here we are considering creeping flow – there will be no boundary layers or separated shear layers, which arise at higher Reynolds number. Hence, a very fine grid should not be needed. The distribution of cells should provide a relatively uniform degree of angular resolution around the sphere. However, accurate computation of forces requires good resolution near the surface. To that end, tetrahedral elements are used over the bulk of the domain, with a few prism layers next to the sphere. This leads to a very simple, unstructured grid that is controlled by the node density on the sphere surface and at the outer boundaries.

The grid summarized by Figure 3.3 consists of 18,580 elements and 7,812 nodes. It has the seven boundary regions stated in Figure 3.2; the sphere is a no-slip surface.

The full incompressible Navier–Stokes equations are solved in the laminar regime, with a standard, segregated equation, implicit, steady-state solver. No concessions need be made to the Stokes flow regime. Note, however, that the choice of upwind scheme (§2.2) has practically no effect for the flow considered here because convection effects are negligible, by definition.

The standard inflow boundary conditions prescribe the velocity; they are also applied at the far-field boundary, per Figure 3.2. The outflow boundary condition is used on the remaining side. This condition is not entirely consistent with the solution given by Eq. (3.2) because it assumes $\partial U/\partial x = 0$, but it becomes reasonable if the boundary is far downstream.

On the spherical surface the standard no-slip condition is specified. That produces viscous surface forces. The convergence can be monitored through the residuals; however, it is advisable additionally to judge convergence of the solution

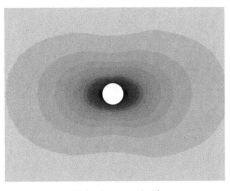
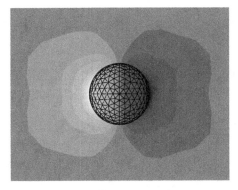

Velocity magnitude   Pressure

Figure 3.4. Velocity and pressure contours in the vicinity of the sphere, $Re = 0.01$.

by monitoring forces exerted on the sphere. It turns out that a very well converged solution is needed to get them accurately.

First, the numerical results for $Re = 0.01$. Most CFD packages integrate the pressure and viscous forces over the surface of the sphere to provide the net force. Equation (3.3) predicts the drag force D on the sphere to be $D = 0.9425 \times 10^{-10}\,N$ and the theoretical drag coefficient of $C_D \equiv D/(1/2\rho U_\infty^2 \pi a^2) = 24/Re = 2,400$. The computed drag force is equal to $0.977 \times 10^{-10}\,N$ and the corresponding drag coefficient is $C_D = 2,490$. This represents a numerical error of about 4%. Part of that error is due to the finite domain size, as discussed previously.

Streamlines in the vicinity of the sphere (Figure 3.1), and contours of the velocity magnitude (Figure 3.4) show a solution that is symmetric in the streamwise direction at $Re = 0.01$; in particular, there is no wake. Pressure-field contours are presented in Figure 3.4: the solution is antisymmetric with respect to the plane $x = 0$, x being the streamwise direction. These are hallmarks of creeping flow. At low Reynolds number, the flow obtained by reversing the direction of the velocity arrows in Figure 3.1 is still a solution. The pressure contours must reverse on changing the velocity direction, so that pressure decreases in the direction of flow.

In the vicinity of the sphere, a numerical solution for $Re = 0.1$ does not differ significantly from the solution at $Re = 0.01$. This is consistent with the fact that Eq. (3.2) does not contain Reynolds number, if the velocity is normalized by $|U_\infty|$ and the pressure is scaled by $\mu U_\infty/a$. The comparison of the computed drag force $D = 0.98 \cdot 10^{-9}\,N$ with the theoretical value $D = 0.9425 \cdot 10^{-9}\,N$ reveals discrepancies similar to those observed for $Re = 0.01$: the error is still about 4%. However, farther from the sphere, asymmetry is beginning to appear, as can be seen in Figure 3.5.

A paradox of creeping flow around a body in an unbounded region is that the low Reynolds number approximation cannot be maintained far from the surface. Viscous forces decrease more rapidly than inertia. The ratio of viscous to inertia terms decreases as $(a/r)Re$. When $r \sim a/Re$ the two become comparable; then convection plays a nonnegligible role. This is referred to as *Whitehead's paradox*. At $Re = 0.1$, we

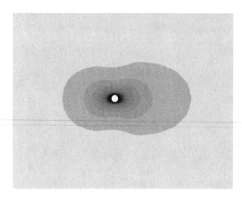

Figure 3.5. Velocity magnitude contours. $Re = 0.1$.

can expect to see asymmetry in the flow field at about 10 radii from the surface. Hence, the behavior seen in Figure 3.5 is expected. The asymmetry is not a failure of the low Reynolds number assumption; rather, it is a nonuniformity at large distances. CFD analysis nicely illustrates this: at small, but finite Re, near to a sphere the computed field is symmetric. Farther away, one sees asymmetry developing.

A rather simple understanding of Whitehead's paradox is provided by the Oseen equation. It is

$$\rho U_\infty \frac{\partial \boldsymbol{u}}{\partial x} = -\nabla p + \mu \nabla^2 \boldsymbol{u}. \tag{3.5}$$

Equation (3.5) restores the convection term, but with a uniform convecting velocity. That approximation is rational because convection is only important far from the surface, where the flow is nearly uniform, with speed U_∞ in the x direction.

The curl of Eq. (3.5) is the vorticity equation. Normalizing x by the sphere radius provides it in the nondimensional form

$$Re \frac{\partial \boldsymbol{\omega}}{\partial x} = \nabla^2 \boldsymbol{\omega},$$

where $Re = \rho U_\infty d/\mu$ is the Reynolds number. The right side of this equation describes vorticity diffusing out from the surface in all directions. The left side describes vorticity being blown downstream by a uniform velocity. The left side is small at low Reynolds number, subject to the following caveat. The vorticity decreases like $1/r^2$ at large distance from the surface. The left side decreases like Re/r^3, whereas the right side decreases like $1/r^4$. Hence, the convection term added by Oseen plays a role when $r \sim 1/Re$.

Solutions to Eq. (3.5) are described in Milne-Thomson (1968). They contain a weak wake inside the paraboloidal region $y^2 + z^2 = x/Re$ on the downstream side of the sphere. This is the cylindrical region produced by rotating a parabola about an axis through the center of the sphere. The extra vorticity in that region causes the asymmetry.

Next, we present a computed solution for $Re = 1.0$. The low Reynolds number assumption now is becoming questionable. Velocity magnitude and pressure contours in the vicinity of the sphere are presented in Figure 3.6. The velocity field

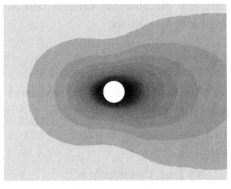
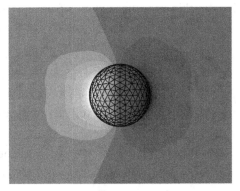

Velocity magnitude   Pressure

Figure 3.6. Velocity and pressure contours in the vicinity of the sphere. $Re = 1.0$.

has become asymmetric even near the sphere, and the pressure field deviates from antisymmetry: clearly, at $Re = 1.0$, the Stokes flow approximation is not entirely valid. The computed value for the drag coefficient is $C_D = 27.7$, compared to the Stokes flow value of $C_D = 24/Re = 24$. Oseen's approximation (3.4) gives $C_D = 28.5$; our computational value is slightly lower. In fact, it agrees quite well with the data fit proposed in White (1991),

$$C_D = \frac{24}{Re} + \frac{6}{1 + \sqrt{Re}} + 0.4 = 27.4.$$

This example shows how classical analysis provides guidance in setting up the flow domain, checking the accuracy of the numerics and in interpreting the solutions. Conversely, the CFD goes beyond the limitations of the approximations used in the analysis, allowing finite Reynolds number effects to be assessed. It also allows more complex geometries to be computed, as subsequent examples show.

It might be wondered whether we will follow the sphere with a solution for creeping flow round a cylinder. The answer is no. Whitehead's paradox becomes more profound for two-dimensional objects. It is then called Stokes's paradox. Stokes's paradox is that there is no creeping flow solution that becomes a uniform flow in the far field, for two-dimensional bodies. The paradox is resolved again by Oseen's equation (3.5). It is found that the drag, per unit width, on a cylinder is

$$D = \frac{4\pi \mu \, U_\infty}{\log(7.4/Re)}$$

(White, 1991). The dependence on Re shows that inertia can never be neglected, even at low Reynolds number.

The impossibility of a zero Reynolds number limit stems from the slow fall off of the velocity at large r. Imagine the cylinder to be composed of a row of spheres placed along its axis. Instead of a two-dimensional cylinder, let it be a finite length, thin rod, that extends from $z = -b$ to $z = b$. Because the velocity falls as $1/r = 1/\sqrt{x^2 + y^2 + z^2}$, the velocity of the sum of spheres goes as

$$A \int_{-b}^{b} \frac{dz}{r} \sim 2A \log(2b/\sigma)$$

for $b \gg \sigma$, where $\sigma = x^2 + y^2$. A is a coefficient of proportionality. On the surface of the cylinder, $\sigma = a$. If the velocity of the cylinder is U_s then

$$A \sim \frac{U_s}{2 \log(2b/a)}.$$

An accurate analysis shows that the drag force on a long, thin needle is

$$D_N = \frac{8\pi \mu b \, U_s}{2 \log(2b/a) + 1} \tag{3.6}$$

if the needle moves perpendicularly to its axis. If it moves parallel to its axis, the result is

$$D_T = \frac{4\pi \mu b \, U_s}{2 \log(2b/a) - 1}. \tag{3.7}$$

These are the normal and tangential drags, which will be discussed shortly. Note that in the two-dimensional limit, $b/a \to \infty$, these become invalid. This finite length, slender body approximation is more relevant.

3.1.2 Resistance Matrix

The drag force on a translating sphere allows an evaluation of settling velocity in the presence of a force. Under the force of gravity, a sphere of mass $m = 4/3\pi a^3 \rho_s$ will fall at a speed determined by the balance between gravitational force and viscous drag. At low Reynolds number, this balance is

$$6\pi \mu a U_s = 4/3\pi a^3 (\rho_s - \rho_f) g$$

per Eq. (3.3). The left side is the Stokes drag; the right is the negative buoyancy. Rearranging gives

$$U_s = \frac{2a^2 (\rho_s - \rho_f) g}{9\mu} \tag{3.8}$$

as the speed with which the sphere will fall. This is the terminal velocity that would be reached by a sphere dropped from rest.

Equation (3.3) is an example of a resistance formula. It is particularly simple because of the symmetry of a sphere. The ideas and computations for the sphere that were described in the last section can be extended and modified to describe nonsymmetric bodies in creeping flow. Resistance coefficients can be found numerically instead of being derived analytically. Before illustrating this, we describe the general notion of a resistance matrix.

The resistance formula (3.3) is saying that drag is proportional to velocity. That is a consequence of the linearity of Eq. (3.1). The general way to state the idea of

Figure 3.7. The drag on a spheroid is represented as a sum of forces along its centerline. They add to minus the drag on the body.

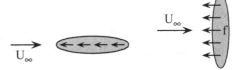

linear resistance is via a resistance matrix. The resistance matrix connects the drag force to the velocity of the object. That is,

$$D = \mu a R \cdot U_\infty. \tag{3.9}$$

R is the resistance matrix. (If the body is rotating as well as translating, a rotational resistance matrix has to be added.) Equation (3.9) is similar to Eq. (3.3), except that R is a 3×3 matrix. The essential difference between the general object and the sphere is that the drag force does not have to be in the direction of motion. If the motion is in the the x direction, then the drag force has three components

$$D_x = \mu a R_{xx} U; \quad D_y = \mu a R_{yx} U; \quad D_z = \mu a R_{zx} U.$$

For the sphere, $R_{yx} = 0 = R_{zx}$ and $R_{xx} = 6\pi$. Very few exact formulas for the resistance matrix are known; but resistance coefficients are readily evaluated by a flow computation.

As a simple, nonnumerical illustration, consider a slender, axisymmetric object falling under gravity. For instance, a cigar-shaped rod, at angle θ to the x axis, with gravity in the $-y$ direction. It will not fall in the y direction but will slip at an angle. That is because motion in y generates a component of viscous drag in x that pushes the object to the side. The matrix R is not diagonal.

This case of a long, slender rod can be analyzed in terms of normal and tangential components of the resistance coefficient [Eqs. (3.6) and (3.7), for example]. If the object moves along its long axis, the drag will be $D_T = \mu a R_T U$. R_T is the tangential resistance coefficient – the resistance to motion along the axis. If it moves perpendicularly to its axis the drag will be $D_N = \mu a R_N U$, the normal component.

The disparity between normal and tangential drag forces can be explained by decomposing the force into elemental components. A force is placed at the center of each infinitesimal slice of a spheroid, as suggested by Figure 3.7. This force and the associated flow field is called a Stokeslet. The aggregate drag is equal and opposite to the sum of these infinitesimal contributions. The Stokeslet strengths are determined by satisfaction of the no-slip boundary condition. Their induced velocity counters that of the approach flow U_∞. When the flow is along the major axis, the Stokeslets are aligned with each other. They reinforce to counter the incident velocity. When the flow is toward the major axis, the Stokeslets are parallel to one another; they must individually counter the incident flow. Thus, smaller individual strengths are needed when the flow is aligned with the major axis. Their sum produces a lower viscous resistance.

When the body is sloped at angle θ and is moving in the positive x direction, the relative fluid velocity will be $-U_\infty$. This has components $-U_\infty \cos\theta$ along and

$U_\infty \sin\theta$ normal to an axis through the body. The tangential drag force is the tangential velocity times the tangential resistance coefficient, which is now seen to be $D_T = -\mu a R_T U_\infty \cos\theta$. This force is along the axis of the body. Because the body is sloped, the tangential drag has both x and y components. They are

$$D_x = D_T \cos\theta = -\mu a R_T U_\infty \cos\theta \cos\theta,$$
$$D_y = D_T \sin\theta = -\mu a R_T U_\infty \cos\theta \sin\theta.$$

The R_{xx} and R_{xy} entries in the resistance matrix are formed by equating $D_x = -U_\infty R_{xx}$ and $D_y = -U_\infty R_{xy}$. Similarly, for the normal velocity, $D_N = \mu a R_N U_\infty \sin\theta$. This has x and y components

$$D_x = -D_N \sin\theta = -\mu a R_N U_\infty \sin\theta \sin\theta,$$
$$D_y = D_N \cos\theta = \mu a R_N U_\infty \sin\theta \cos\theta.$$

Adding the contributions to drag, component by component, gives

$$D_x = -U_\infty \mu a(R_T \cos^2\theta + R_N \sin^2\theta),$$
$$D_y = -U_\infty \mu a(R_T \sin\theta \cos\theta - R_N \sin\theta \cos\theta).$$

Similar considerations apply for motion, V_∞, in the y direction. Then the two-dimensional resistance matrix is found to be

$$\boldsymbol{R} = \begin{bmatrix} R_T \cos^2\theta + R_N \sin^2\theta & (R_T - R_N)\sin\theta \cos\theta \\ (R_T - R_N)\sin\theta \cos\theta & R_T \sin^2\theta + R_N \cos^2\theta \end{bmatrix}. \tag{3.10}$$

For a long, slender object $R_N \approx 2R_T$ is often a good approximation, see Eq. (3.6) and Eq. (3.7).

Suppose this object falls at a terminal velocity, under the force of gravity in the $-y$ direction. Drag is in balance with gravity in that direction: $D_y = mg$. But no forces are acting in the x direction, so $D_x = 0$. Hence, we find that

$$R_{xx}U + R_{xy}V = 0,$$
$$\mu a(R_{yx}U + R_{yy}V) = \text{mg}.$$

From the first

$$\frac{U}{V} = \frac{-R_{xy}}{R_{xx}} = \frac{(R_N - R_T)\sin\theta \cos\theta}{R_T \cos^2\theta + R_N \sin^2\theta} \approx \frac{\sin\theta \cos\theta}{\cos^2\theta + 2\sin^2\theta} \tag{3.11}$$

with the approximation $R_N \approx 2R_T$. For instance, a slender rod at $\theta = 45°$ will fall down and to the left with

$$\frac{U}{V} = \frac{(R_N - R_T)}{R_T + R_N} \approx \frac{1}{3}$$

or at an angle of about $18°$ to the vertical. It does not fall in the direction of gravity.

The slender body analysis finds application in the theory of swimming by microorganisms – in particular, those which use flagella to swim. Flagella are tentacles attached to the body of the organism, which can be either oscillated or rotated. At the outset to this chapter, it was described how a wave is passed along the flagellum to push against the fluid and propel the body. The slender body theory shows that the forward velocity of the animal is proportional to $(R_N - R_T)c$, where c is the speed of the wave. A normal resistance coefficient that is larger than the tangential coefficient enables the organism to push forward against the drag exerted on its head and on the flagellum, itself.

The components of the translational resistance matrix can be evaluated from three CFD computations: one, with the velocity in the x direction, gives the forces $F_x = \mu a R_{xx} U$, $F_y = \mu a R_{yx} U$, $F_z = \mu a R_{zx} U$. Dividing the forces by $\mu a U$ gives three components of R. The same, done for the other two directions, fills the matrix.

In general, there will also be a rotational resistance matrix and the body will experience a torque. For instance, Eq. (3.9) would be supplemented to

$$
\begin{aligned}
D &= \mu a R \cdot U_\infty + \mu a^2 Q \cdot \Omega_\infty, \\
T &= \mu a^2 R_T \cdot U_\infty + \mu a^3 Q_T \cdot \Omega_\infty
\end{aligned}
\tag{3.12}
$$

for the drag and torque on the body. These correspond to a body moving with velocity $-U_\infty$ and rotating with angular velocity $-\Omega_\infty$. It can be shown that the matrices R and R_T are symmetric and that Q_T is the transpose of Q.

The dynamics of Stokesian particles are determined by Eqs. (3.12). For instance, in the absence of an externally imposed torque, the particle would rotate according to the formula

$$
R_T \cdot U_\infty + a Q_T \cdot \Omega_\infty = 0.
$$

This is a bit like the idea that a car wheel rotates as it translates, because of friction against the road. Here it is the viscous force due to flow round the particle that causes it to rotate. For instance, a sphere falling at a small distance, h, from a wall will experience a moment of force in consequence of the viscous stress on the wall. That moment will cause it to rotate as it falls. An approximate analysis gives a torque on the sphere equal to

$$
T = 1/5\pi\mu a \log(a/h)\left[4U - 16a\Omega\right]
\tag{3.13}
$$

when $h \ll a$. If no external torque is applied ($T = 0$), then as the sphere falls under the force of gravity, it will rotate at an angular velocity of $U/4a$. A sphere rolling without slipping rotates at the angular velocity U/a. The viscous solution predicts slower rotation; the motion can be called rolling with slip relative to the wall.

3.1.3 Resistance Due to Relative Motion

For the previous discussion, a frame of reference attached to the object has been adopted. In that frame, the geometry stays put. Let the translational velocity of the surface be u_s and its rotational velocity be Ω_s. If the ambient fluid is at rest, then in the particle frame of reference the ambient fluid moves with velocity $U_\infty = -u_s$

and rotates with $\boldsymbol{\Omega}_\infty = -\boldsymbol{\Omega}_s$. If the fluid is not at rest but is moving with velocity, $\boldsymbol{u}_f$, then

$$U_\infty = u_f - u_s. \tag{3.14}$$

The velocity on the right side of Eq. (3.12) should be understood as that of the fluid, relative to the particle. It is this relative velocity that sets the drag. In the Stokesian regime, the resistance matrices govern motion relative to the fluid.

In the absence of externally imposed forces, $\boldsymbol{D}$ and $\boldsymbol{T}$ are zero in Eqs. (3.12). Hence, the object must translate and rotate with the fluid: $\boldsymbol{U}_\infty = 0 = \boldsymbol{\Omega}_\infty$. In other words, $\boldsymbol{u}_f = \boldsymbol{u}_s$ and $\boldsymbol{\Omega}_f = \boldsymbol{\Omega}_s$ ensure that the force and torque vanish. For instance, in a linear shear flow $u_f = u_0 + Sy$ a sphere will translate at velocity u_0 and rotate at angular velocity $-1/2\,S$. The second follows because the vorticity is $-\partial u/\partial y = S$ and the rotation rate of fluid particles is one-half of this.

Recall that the velocity gradient can be written as a sum of the rate of rotation and the rate of strain, as shown by Eq. (1.3). Although a rigid particle can follow the rotation, making the torque vanish, it cannot follow the strain. Hence, a rate of strain must create a stress on the surface of the particle. A suspension consists of a large number of microscopic particles floating in a solvent. In straining flow, each particle produces a stress perturbation. In aggregate, they add up to an additional macroscopic stress. If there is a volumetric concentration, c, of tiny particles in a volume of fluid, the average stress within that volume will be increased above that of the fluid itself. For a dilute suspension, the increment is proportional to c and to the rate of strain. The incremental, alternatively, can be characterized as an enhanced, effective viscosity (Batchelor, 1967). The formula

$$\mu_{eff} = \mu(1 + 5/2c) \tag{3.15}$$

for the effective viscosity of a dilute suspension of spheres, was derived by Einstein in his 1905 Ph.D. thesis. Einstein was not so concerned with fluid mechanics, as he was with the properties of molecules. At the time, these had to be inferred rather indirectly. By modeling large molecules in a solvent as particles in suspension, Einstein was able to estimate their concentration, and then their size, through the effect on viscosity. His result [Eq. (3.15)] has become a foundation of work on the rheology of suspensions.

3.1.4 Computation of Resistance Coefficients

The particle frame of reference is amenable to computation. If the solid particle were moving, and a fixed frame of reference were adopted, the flow would be unsteady: the body would enter the computational domain, move through it, and exit as a function of time. It is preferable to make the computation independent of time and, hence, to move with the body.

Let us illustrate resistance coefficients by a computation. Consider the small, three-dimensional, asymmetric body in Figure 3.8. This might be representative of a microorganism in an idealized sense. It is a body consisting of a sphere of a diameter

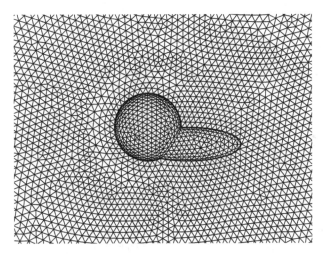

Figure 3.8. Computational grid in the vicinity of the body.

$d = 2\,a = 1$ μm interconnected with an ellipsoid with the major and minor axes of
1.3d and 0.5d. Figure 3.8 shows the surface mesh and the volume mesh in section. It
is unstructured with 12,591 nodes and 58,752 tetrahedral elements.

Two computations are needed to obtain all the coefficients in the resistance
matrix – considering a two-dimensional submatrix, because the three-dimensional
body is, in fact, symmetric in the z direction. First, the flow over the body is computed
at a certain angle and then it is computed perpendicular to that direction.

As in the example of the sphere, the surrounding fluid is water at room temperature ($T = 20°$ C and $p = 1$ ATM) with $\mu = 10^{-3}$ Pa s and $\rho = 1$ g/cm^3. Again,
for the purpose of this numerical simulation the body is not moving, and the flow
is steady. The direction of the water flow in the far field is adjusted to achieve the
various angles of incidence. The far-field velocity of $U_\infty = 1$ cm/s gives a Reynolds
number of $Re = U_\infty d/\nu = 0.01$.

The domain is chosen to be a box of dimensions $40d \times 40d \times 40a$. The velocity
is prescribed on all the far-field boundaries. $z = 0$ is a symmetry plane. The body is
a no-slip surface.

By computing flow incident first in the x direction and then in the y direction,
the data needed to evaluate the full resistance matrix are obtained. The resistance
coefficients are R_{xx}, R_{yx}, R_{xy}, and R_{yy}. The following values were computed: from flow
in the x direction, $R_{xx} = 25.38$, $R_{yx} = 1.64$; from flow in the y direction, $R_{xy} = 1.64$,
$R_{yy} = 25.68$. These have about 10% accuracy.

Streamlines in some selected planes are plotted in Figure 3.9. The velocity is
reminiscent of that round a sphere. In fact, far from the surface the velocity field
approaches that of a sphere having the same drag as the given body and having
a radius $a = D/(6\pi\mu U_\infty)$. This is the notion that the drag-producing body can be
represented by a Stokeslet. The Stokeslet is a point force. Far from the body, its
detailed shape is not as important as the force it creates; hence, the flow looks like
that which would be produced by concentrating the drag at a point.

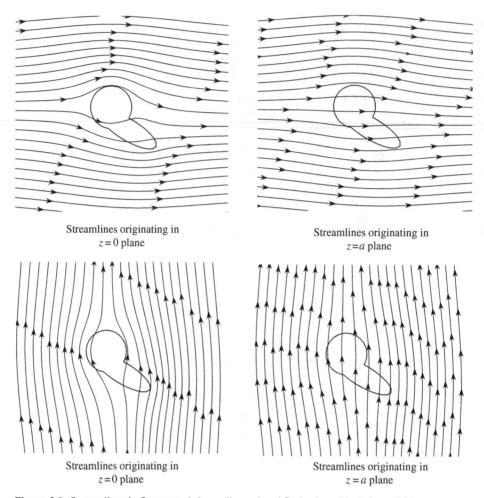

Figure 3.9. Streamlines in flow round three-dimensional Stokesian object. $Re = 0.01$.

3.1.5 Eddies in Stokes Flow

Although recirculating streamlines and eddies are commonly associated with high
Reynolds number flow downstream of bluff bodies, recirculation can occur in
creeping flow. However, it is not associated with a wake and is as likely to be
upstream as downstream of a body. Flow reversal occurs because friction prevents
an oncoming stream from flowing into corners. Thus the oncoming stream rides over
the fluid in the corner and drives a recirculating eddy – in fact a series of counter-
rotating eddies, extending into the corner. Theory suggests that eddies will form
when the included corner angle is less than 146°. If the angle is greater than this,
the flow is unidirectional, following the geometry. For smaller angles, recirculation
begins to appear. A 90° corner is shown in Figure 3.10.

An analysis of corner flow shows that the velocity varies as r^α, where r is distance
from the apex of the wedge. If the included angle is less than 146°, α is found to be
complex. Then the solution is of the form $r^\beta \cos[\gamma \log(r)]$, where β and γ are the

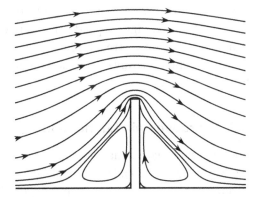

Figure 3.10. Stokes flow over a two-dimensional rib at $Re = 0.01$.

real and imaginary parts of α. The oscillations in sign of the cosine correspond to alternatively signed eddies. The size and intensity of the eddies falls in a geometric progression as $r \to 0$.

At high Reynolds number, eddying is due to vorticity advected from the surface. Hence, it tends to occur downstream. But at low Reynolds number its cause is quite different; it is due to an impedance to fluid flow into a corner or into a gap. The flow toward a corner encounters an increasing frictional resistance as the distance between the walls decreases. That deflects the flow; if large enough, the resistance prevents oncoming fluid from entering the corner. As the flow circumvents the corner, it drives a countercirculation by viscous friction. Figure 3.10 shows two large eddies driven in this way. Two very small eddies, just at the junction between the vertical rib and the lower wall, are also seen in this figure.

Given this mechanism, there is no reason for eddies to occur on the downstream side; they can occur in any sufficiently confining angle of the geometry. Theoretically, an infinite family of counterrotating eddies forms in a closed corner; typically only one or two are resolved on the computational grid. If the corner is not closed, a finite number of eddies can form. For instance, creeping flow past a cylinder close to a plane wall, but with a small gap to allow leakage, can contain a few eddies within the opening between the wall and cylinder.

The two-dimensional rib at $Re = 0.01$, shown in Figure 3.10, was selected to illustrate eddies in Stokes flow because it is a simple geometry. A simple, rectangular grid, with higher resolution near the surfaces, coarsening away from them was used for this computation. The full domain extends ± 10 rib heights upstream and down; only a portion is shown in the figure. The pressure is specified on the left, inflow boundary. The bottom wall, including the rib, is no-slip and the upper wall is a symmetry, or slip, boundary. Friction on the wall impedes the flow. The velocity profile develops into the parabolic, Poiseuille form (§1.5) before reaching the rib.

A pressure drop from inflow to exit drives the fluid in opposition to the drag force. In this computation, the pressure at outflow was adjusted to obtain a specified velocity and Reynolds number.

The computed flow shows left–right symmetry. That is evidence that inertia is truly unimportant at this Reynolds number. Corner eddies develop upstream and downstream of the rib. They illustrate the type of recirculation that develops in creeping flow. At higher speeds, the eddy on the upstream side would be smaller than that on the downstream side. That is due to convection of vorticity. In creeping flow, convection is a tiny effect; the symmetry is established by diffusion of vorticity from the surface.

3.2 Hydrodynamic Lubrication

The theory of hydrodynamic lubrication was originated by Osborne Reynolds (of the Reynolds number). His seminal 1886 article was motivated by experiments that Beauchamp Tower had conducted on the journal bearings of railroad cars. The friction in these bearings had been observed to be far lower than expected. Tower made a small hole in a bearing and found that the oil came gushing out when the axle was in rotation. This led Reynolds to develop a theory that was able to explain how hydrodynamic pressure in the lubricating oil reduced friction.

The concept becomes clear from the consideration of a rod with a hemispherical end that is immersed in oil and propelled toward a wall. As the rod translates toward the fixed surface, a film of viscous oil is squeezed out of the gap between its hemispherical end and the wall. A large pressure builds up within the gap as the hemisphere nears the surface.

There is an asymptotic solution available: if the hemisphere is at height h above the wall, and moving downward with velocity $-V$, it experiences an upward force of

$$F = 6\pi\mu a V \frac{a}{h} \qquad (3.16)$$

when $h \ll a$ (Batchelor, 1967). a is the radius of the hemisphere. This force is a consequence of the high pressure that is needed to drive the viscous fluid out of the small gap between the rod and the wall. In an idealized sense, the hemisphere can never contact the wall because the upward force tends to infinity as h tends to zero. As the film becomes increasingly thin, the shear and viscous force becomes increasingly large. This is the essential idea of hydrodynamic lubrication: when viscous fluid is squeezed out of a small gap, a large pressure will develop and hold the surfaces apart. For instance, in the joints of animals, the synovial fluid may produce large pressures when it is squeezed between bone surfaces.

By the same token, if a cylindrical shaft is set very close to a wall and rotated, a high pressure develops as the viscous fluid is dragged into the gap between the shaft and wall. For instance, the oil in between a bearing and a race develops high pressures that keep the surfaces from rubbing. For the gap between a cylindrical shaft and a wall, an approximate solution shows that a maximum pressure of order

$$P_{\max} \sim \frac{\mu U_w a}{h^2}$$

develops (see page 126). U_w is the surface velocity of the rotating shaft, a is its radius, and h is the gap height, which is assumed to be very small compared to a. The higher the viscosity, or the smaller the gap, the higher is the pressure.

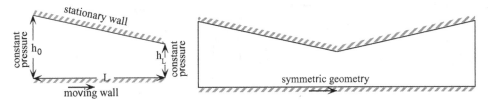

Figure 3.11. Wedge-shaped gaps provide the most elementary geometry for study of hydrodynamic lubrication.

Unfortunately, this case of the symmetric gap between cylinder and wall is not so simple. The high pressure that develops as the fluid enters the gap drives the fluid out the other side of the gap. That, in combination with the effect of fluid being dragged out of the gap by the moving wall, produces a low pressure after the narrowest portion of the gap. The upward force is the pressure, integrated over the cylinder surface. It turns out that the high and low pressures exactly cancel to produce no net upward force. This is a disconcerting result, because lubrication does occur in real bearings.

In a real bearing the low pressure can become low enough to cause cavitation. This means that gas comes out of solution, preventing the pressure from dropping lower than the vapor pressure of the gas. Cavitation bounds the pressure from below. Then the low pressure is insufficient to cancel the high pressure and a net upward force results. Experiments show that cavitation does indeed occur in bearings, confirming the theory.

Although lubrication in bearings relies on cavitation to produce an upward force, that should not cloud the issue: hydrodynamic lubrication is a consequence of the large forces that occur when viscous fluid is forced into a small gap. The mechanism of hydrodynamic lubrication can be computed by considering an asymmetric gap so that the low pressure will not cancel the high pressure. Then a net pressure force will exist. The following example illustrates a computational analysis of the lubrication pressure of oil flow through a small passage.

To illustrate a classical result of lubrication theory, consider laminar incompressible flow in a two-dimensional wedge between two bodies in near contact, with one of the two bodies moving. In particular, the slider bearing consists of a gap between a moving wall and a sloped surface, as in the left part of Figure 3.11. There is only a contracting portion to the gap, so only elevated pressure is produced.

For the computation, the wall velocity is set to $U = 10$ m/s, the width of the gap is $L = 4$ cm, the minimum height of the gap is $h_L = 0.1$ mm, the maximum height is $h_0 = 0.2$ mm. The fluid is SAE 50 lubricating oil, with $\mu = 0.625$ Pa s and $\rho = 892.86$ kg/m^3. The Reynolds number for this flow, based on the width of the gap, L, is $Re_L = 571$. Computations were performed on a structured, quadrilateral grid of 1,701 nodes and 1,600 elements.

We have previously cited approximate analyses. They invoke the lubrication approximation, which is an assumption that the wall slope squared times the Reynolds number is small. Lubrication theory requires only that $Re_L(h_0 - h_L)^2/L^2$ be small.

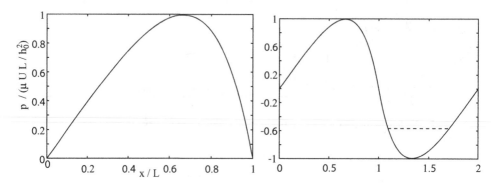

Figure 3.12. Pressure across the gap. These correspond to the geometries in Figure 3.11.

In this case it equals 0.0036. The CFD makes no use of this requirement; rather, it is a limit in which a theoretical solution is available.

The CFD predictions shown at the left in Figure 3.12 are in quite excellent agreement with analytical results. In particular, an upward pressure force of order $\mu U_w L/h^2$ is exerted on the upper wall. The pressure rises as the moving wall drags fluid into the gap, working against the frictional impedance. At the exit, an ambient pressure boundary condition is prescribed. The computed pressure passes through a maximum near $x/L = 0.65$ and then decreases toward the exit.

A somewhat surprising result of lubrication theory is that no net pressure force is exerted on a symmetric gap. That is illustrated by the second case, at the right of Figures 3.11 and 3.12. The pressure rises on entering the gap. It falls to the ambient level at the center, drops to an equal and opposite pressure after the center, and exits to the prescribed ambient pressure. This behavior is an embodiment of the flow symmetry that exists at low Reynolds number: if the flow is symmetric, so is the pressure *gradient* but, relative to ambient, the pressure itself must be antisymmetric.

The symmetric gap produces no net upward force because the pressure in the right plot of Figure 3.11 integrates to zero. However, in a real bearing the predicted low pressure can be well below the point of cavitation. Then dissolved gas comes out of solution, producing a bubble at the vapor pressure. The negative pressure drops no lower than this vapor pressure. Some of the negative portion of the pressure profile is increased to the vapor pressure, as is suggested by the dashed line in Figure 3.11. Integrating this modified pressure distribution results in a net upward force.

The nature of the flow symmetry is shown by Figure 3.13. Lubrication theory assumes that these velocity profiles are of Couette–Poiseuille form:

$$u = U_w \left(1 - \frac{y}{h}\right) - \frac{y}{h}\left(1 - \frac{y}{h}\right)\frac{h^2}{2\mu}\frac{dp}{dx}, \qquad (3.17)$$

where $h(x)$ is the local height of the gap. This profile is invoked for any shape upper wall, assuming only that it is slender, $dh/dx \ll 1$. Equation (3.17) solves the balance $\mu\partial^2 u/\partial y^2 = dp/dx$ between viscous and pressure forces in parallel flow (see §1.5). Where the pressure is rising, Eq. (3.17) provides to concave velocity profiles; where

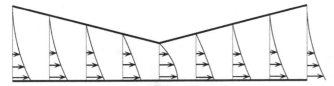

Figure 3.13. Velocity profiles in the symmetric gap.

pressure is falling they are convex. The computed velocity profiles in Figure 3.12 agree with this analytical approximation.

The volume flux through the gap is

$$Q \equiv \int_0^h u\, dy = 1/2 U_w h - \frac{h^3}{12\mu}\frac{dp}{dx}.$$

For a given volume flux, the pressure gradient is

$$\frac{dp}{dx} = \frac{6\mu U_w}{h^2} - \frac{12\mu Q}{h^3}. \tag{3.18}$$

For a given Q, the pressure gradient is most negative where h is smallest. If there is no net pressure difference between entrance and exit, the volume flux dragged into the gap by the moving wall is found by integrating (3.18) from the inlet to the exit:

$$Q = \frac{U_w}{2}\frac{\int_0^L h^{-2} dx}{\int_0^L h^{-3} dx}. \tag{3.19}$$

The smaller the gap, the larger the denominator and hence the less the flow admitted by the contraction. This helps to explain the behavior of viscous flow through narrow channels. The narrowest point exerts a controlling influence, presenting a high impedance to the flow.

Substituting (3.19) into Eq. (3.18) determines the pressure from the shape of the gap alone. That is the beauty of Reynolds's classical work. His theory has been expanded extensively over the years. The solution (3.18) is not available for three-dimensional gaps, but simplified, lubrication equations are. It is not our intent to expound on that rather extensive body of work; Hamrock et al. (2004) is a thorough reference. The following is another illustration that has been computed with a general purpose CFD code.

The gap between a spherical point and a moving wall provides this example of lubrication flow. Figure 3.13 shows a grid and pressure contours on the upper wall. This is a simple structured grid containing 85,000 hexahedral elements. The analysis is limited to the space between a square of dimensions $5 \times 5\,\mathrm{cm}$ on the lower wall, and the corresponding portion of the sphere. The radius of the sphere is 20 cm. All other dimensions and fluid properties are the same as in the previous, two-dimensional examples. Again the pressure shows left–right antisymmetry, as seen in the contour plot of Figure 3.14.

One might wonder why the inlet and exit pressures were both set to ambient. It is because this is regarded as a local solution. The three-dimensional computation

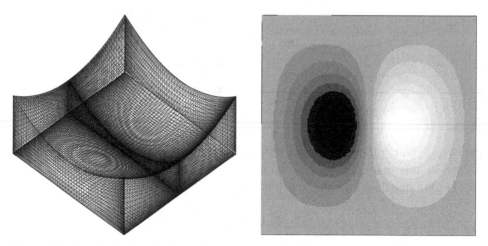

Figure 3.14. Lubrication flow in the gap between a spherical tip and a moving wall.

in Figure 3.14 could be regarded as a portion of the solution for flow between a complete sphere, or a rod with a hemispherical tip, and a wall. If the distance between the hemisphere and wall is small, the problem involves two disparate lengths: the gap height and the sphere diameter. A computation with the entire geometry could be expensive. A fine grid is needed in the tight space between the sphere and the wall, and a coarser mesh is needed around the rest of the geometry,

When lubrication theory was formulated, in the late 19th century, it would have been impossible to solve the full problem. Approximations were needed. Those approximations lead to insights that continue to be very informative. One assumption made by lubrication theory is that the slope of the surfaces is small; this justifies approximating the velocity profile by Couette–Poiseuille flow. A second assumption is that the high frictional impedance within the small conduit will cause a very high pressure. The entrance and exit pressures are far more moderate and can be set approximately to ambient.

In the mathematical analysis, the pressure boundary conditions could be improved by matching the solution in the gap to an outer solution, valid in the larger geometry. In the research literature, that has been done for low Re; if Re is not small, the outer solution would have to be numerical.

How valid the second lubrication approximation is can be illustrated by computing a larger portion of the hemisphere and comparing to the solution of Figure 3.14. The domain was extended to include the rest of the hemisphere. The wall slope times Re, which is assumed to be small in lubrication theory, is not small in the outer portion of the domain. Hence, CFD is being used to solve both within the gap and in the outer flow. The centerline pressure is plotted in Figure 3.15 for the previous, small domain, with zero pressure at the inlet and exit, along with a corresponding curve for the entire hemisphere. The maximum pressure is computed with reasonable accuracy using the smaller domain. Of course, the accuracy depends on the ratio h/a of minimum gap height to radius: here it is 0.1 mm/20 cm $= 5 \times 10^{-4}$. Nevertheless,

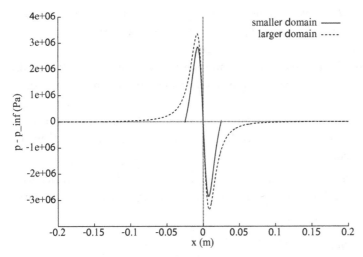

Figure 3.15. Comparison between the pressure in a small portion of the gap and a full hemisphere; (——) small and (−−−−) full.

it can be seen that an error is incurred by solving the smaller domain, with ambient pressure prescribed at the boundaries.

EXERCISES

3.1 *Basics.*
  (a) In reference to the resistance formulas (3.12): explain how translation of a body can cause a torque; explain how rotation can cause a linear force.
  (b) A solid sphere, of density 2.5 g/cm^3 and radius 1 mm is dropped in a tube fill with glycerine. What is its terminal settling velocity? Is the low Reynolds number assumption acceptable?
  (c) Explain why creeping flow about a fore–aft symmetric body must be fore–aft symmetric. Sketch the flow past two cylinders in contact, and aligned with the flow direction:

3.2 *Resistance formulas.* The resistance matrix is computed for an asymmetrical body in a particular orientation. It is found to be

$$R = \begin{pmatrix} 9 & 2 \\ 2 & 6 \end{pmatrix}.$$

Determine the angle, relative to that orientation, at which this object must be aligned if it is to fall in the y direction, under a force in the y direction.

3.3 *Stokes drag on a sphere.* Very small, spherical particles in a gas stream are removed by an electrostatic precipitator. The precipitator is simply a pair of opposed walls with an applied voltage difference.

The particles are charged electrically and then removed in the electric field. The particles have radius a; they are advected in the y direction with velocity V; and they are subjected to a force qE in the x direction (E = electric field q = charge). The particles are uniformly distributed at the entrance, $y = 0$ to the precipitator. They exit at $y = H$. If the width of the precipitator is W, what fraction of the particles is removed? Give your answer as a formula.

3.4 *Hydrodynamic lubrication.* To illustrate principles of hydrodynamic lubrication theory, G. I. Taylor developed a solution for a parabolic gap. The upper boundary is $y = h(x)$ with

$$h(x) = H[1 + e(x/L)^2], \quad -L < x < L.$$

The lower wall is flat, located at $y = 0$ and moves with velocity U_w in the x direction. Use the lubrication equations (3.19 and 3.18) to recover Taylor's solution. What is Q, the volume flow (per unit width) dragged into the gap? Apply the conditions $P = 0$ at $x = \pm L$. Plot the pressure distribution versus x for the range $-L < x < L$ for $e = 0.1, 0.4, 1.0, 2.3$. Comment on how your solution could be used to estimate the lubricating pressure.

Evaluate the maximum pressure in atmospheres above ambient for a gap length of 1 cm, gap height of 0.05 mm, velocity 1 m/s, SAE10 oil ($\mu = 1$ g/cm $\cdot$ s), and $e = 1$.

3.5 *Stokes flow round a spheroid.* A spheroid is the axisymmetric surface defined by

$$\frac{x^2}{L^2} + \frac{r^2}{R^2} = 1.$$

Its eccentricity is defined as $e = L/R$. $e = 1$ is a sphere; $e > 1$ is a stretched (prolate) sphere; $e < 1$ is a squashed (oblate) sphere. Lamb (1932) constructs an exact Stokes flow solution for a spheroid (as a special case of an ellipsoid). The drag is approximately

$$6\pi \mu U a \left(\frac{4 + e}{5} \right)$$

for flow along the x axis and

$$6\pi \mu U a \left(\frac{3 + 2e}{5} \right)$$

for flow perpendicular to the x axis when $0 < e < 5$ (Panton, 1997). When $e = 0$ the exact formulas in Lamb (1932) are

$$6\pi \mu U a \left(\frac{8}{3\pi} \right)$$

and

$$6\pi\mu Ua\left(\frac{16}{9\pi}\right).$$

The limit $e = 0$ corresponds to a flat disk.

If a disk is set with its principal axis at an angle α to the direction of gravity, at what angle will it fall?

3.6 *Spheroid continued.* Compute Stokes flow over a spheroid with e of about 2 and with it about 0.5. Select a domain size that will provide about 2% accuracy. Create the geometry and volume mesh. Select the viscosity and incident velocity to achieve a Reynolds number of 0.1. Determine the drag for flow incident along the x axis and for flow incident along the y axis.

4 Intermediate Reynolds Numbers

4.1 Separation, Vorticity, and Vortex Shedding

Most classical theory of viscous flow is based on approximations valid at low or high Reynolds number. Although there are a few exact solutions that illustrate aspects of the intermediate range, they are rather limited. The range of phenomena that fall under this heading is commonly illustrated by photographs taken in laboratories. Computational fluid dynamics makes the intermediate Reynolds number range quite accessible.

Inertia is now comparable to viscous stress, and all terms in the Navier–Stokes equations must be retained. Convection destroys the upstream–downstream symmetry of creeping flow (Chapter 3). A distinct wake can be identified leaving the downstream side of a body in an incident flow. Forces on blunt bodies become increasingly due to pressure rather than to viscous stress.

Vorticity is increasingly confined to regions near to walls on the upstream portions of a body and to wakes on the downstream side. The upstream vortical regions become boundary layers in the high Reynolds number limit. As vorticity diffuses away from the surface it is convected downstream, ultimately to form the wake.

The upstream–downstream asymmetry leads to another important idea, that of separation. For example, flows into and out of a nozzle are quite different. The flow *into* a trumpet shaped orifice, say, will follow the walls. As the opening narrows, the flow accelerates to conserve mass [loosely, $\rho UA =$ constant, per Eq. (1.33), implies U increases as A decreases]. The accelerating flow convects vorticity toward the wall, keeping it confined near the surface.

The flow *out of* such an orifice can be quite different. It can start to follow the surface but then leave it. That behavior is referred to as separation of the boundary layer or just separation. As the geometry expands, the flow slows, vorticity convected downstream accumulates in an increasingly thick layer near the surface, and ultimately it separates from the wall. Below the layer of separating vorticity, the direction of the flow reverses. In flows around bluff bodies, this effect becomes noticeable at Reynolds numbers on the order of 10; the flow over streamlined bodies can remain attached up to much higher Reynolds number, but in general wall-bounded flows subject to appreciable deceleration will separate.

The term *separation* refers to streamlines leaving the vicinity of the wall or to vorticity being convected away from the surface. Schematically, this occurs as in

Figure 4.1. Schematic showing flow separating from a surface.

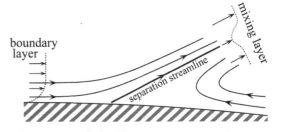

Figure 4.1. The boundary layer follows the wall up to the separation point and then flows away from the wall along the separation streamline. The boundary layer is a layer of vorticity. When it leaves the surface it forms a free-shear layer – a mixing layer in the case of Figure 4.1.

In boundary layer theory (to be discussed in Chapter 5), separation is attributed to the effect of a rising ambient pressure; in other words, to an *adverse* pressure gradient. This means that the flow is slowing down in the direction of motion. Near a stationary wall, the velocity is already low. As the decelerating pressure gradient slows the flow, that near the wall can decrease through zero to a negative value. This backflow, encountering the incident forward flow, erupts from the wall as in Figure 4.1.

The matter is not quite as simple as flow reversal by deceleration. The ambient, forward flow, well above the surface, diffuses forward momentum toward the wall. This wallward diffusion of forward momentum counters the decelerating pressure gradient. Thus, we understand separation via a competition between viscous stress and deceleration. The decelerating effect must be strong enough to overcome viscous diffusion for separation to occur. Their ratio is characterized by the Reynolds number. For diffusion to be overcome, the Reynolds number must be sufficiently high. How high? That depends on the nature of the deceleration.

Flow in an expanding orifice decelerates, but it might not separate if the Reynolds number is low, if the orifice angle is small, or if the decelerating portion is short. Behind a bluff body, separation is inevitable because of the bluntness at its rear – unless the Reynolds number is quite low.

The next CFD example illustrates the asymmetry between flow into and out of an expanding duct. This is a classical illustration of separation induced by flow deceleration. It is the first of five computational studies to illustrate the development of flow patterns in this intermediate range of Reynolds number. The first two show how separation and backflow develops. The theme of the last three studies is asymmetry between upstream and downstream regions in consequence of vorticity advection.

4.1.1 Five Examples

4.1.1.1 Flow in a Wedge-Shaped Region

Consider a two-dimensional, laminar, incompressible flow in a wedge-shaped region – a diffuser-like geometry. From the classical analysis of Jeffery and Hamel

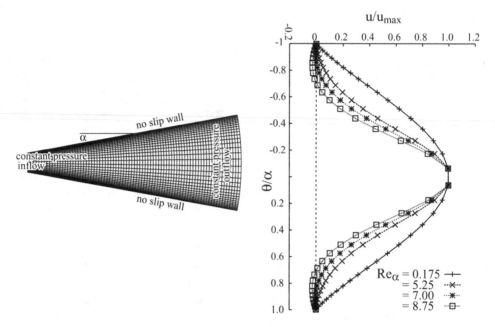

Figure 4.2. Grid for flow in expanding duct. Velocity profiles at the exit plane. The highest Reynolds number shows a region of negative velocity near the walls.

(Batchelor, 1967; White, 1991), we know that the flow in such a geometry is defined by two parameters: they are the Reynolds number, defined by $Re = u_{\max} r \alpha / \nu$, and the angle of the wedge α, measured in radians. The theoretical analysis of Jeffery and Hamel is of flow from a point source at the apex of the wedge. In that idealization $u_{\max} r$, in the Reynolds number definition, is constant with downstream distance, and proportional to the mass flux down the duct; in particular, Q/μ is an equivalent definition of the Reynolds number, where Q is the mass flux (per unit width, because this is a two-dimensional flow).

The analysis shows that when $Re \gg \alpha$, the parameter that defines whether the flow separates is αRe. In the Jeffery and Hamel analysis, the flow "separates" at $\alpha Re = 10.31$. Actually, because of the idealizations, this is not a realistic separation criterion: the solution corresponds to backflow along the entire wall, with "separation" at the apex. A real flow separates at some point along the ramp, as will be seen. Nevertheless, the classical analysis gives an understanding that a given, finite α will become critical when Re becomes high enough.

The computations are not of a point source at the apex of a wedge; they have a finite height inlet, as in Figure 4.2. For the present computations, the wedge angle is set to $\alpha = 10° = 0.175$ rad. At $Re = 30$, $\alpha Re = 5.24$; at $Re = 50$, $\alpha Re = 8.73$. The flow, as we shall see, is indeed separated at the higher, but not at the lower, Reynolds number.

The computational domain is defined to be the $\pm 10°$ wedge, with the inlet being a circular arc, radius $r_{\text{in}} = 1$ cm, and the exit being a circular arc, $r_{\text{out}} = 5$ cm. The molecular viscosity is $\nu = 0.1$ cm^2/s. The Reynolds number is defined by $Re = u_{\text{in}} r_{\text{in}} \alpha / \nu$. The inlet velocity is uniform, and radially directed, so the volumetric flow rate, Q/ρ, is simply $2 u_{\text{in}} r_{\text{in}} \alpha$.

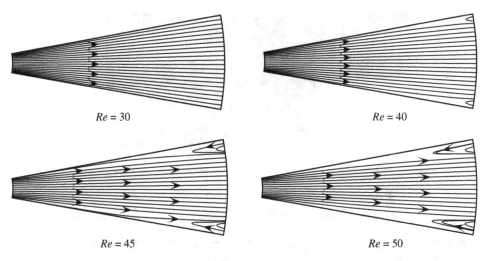

Figure 4.3. Flow in an expanding duct. Streamlines at various *Re*.

The inflow and outflow boundaries are specified as surfaces of constant pressure and the flow is assumed normal to the inflow surface. The top and bottom, 10° walls are no-slip surfaces. The inflow pressure was adjusted to achieve the desired values of Reynolds number. The simple, structured grid at the left of Figure 4.2 was used; it contains 81 × 36 points.

The streamlines plotted in Figure 4.3 illustrate that the flow remains unidirection up to a Reynolds number of about 30. After that, backflow develops near to the exit plane. These streamline plots are consistent with the radial velocity profiles in the outflow plane, shown at the right of Figure 4.2.

Although the flow stays attached to $\alpha\,Re$ just below 7, it must be noted that this value depends on the length of the flow domain: a longer domain will show separation at lower *Re*. That is because separation is the consequence of the net deceleration, which is determined by the ratio of exit to inlet radius, r_{out}/r_{in}.

Were the direction of the flow reversed, the streamlines for flow into the inlet would look quite like those at $Re = 30$ in Figure 4.3 – with the direction of the arrows reversed. That is the form of the flow at all Reynolds numbers; the flow does not separate. The inflow accelerates in the direction of the stream; the pressure gradient is favorable, which prevents separation. There is a clear distinction in the behavior of inflow and outflow. This is analogous to the difference between blowing air out of one's mouth and sucking it in, mentioned on page 1.

4.1.1.2 Flow down a Step

The backward-facing step was encountered in §1.6.2. Bernoulli's equation was abandoned in favor of an assumption that the pressure is constant on the vertical face of the step. The fluid in the lee of the step is nearly stagnant. An ideal model is for the fluid to leave the top of the step tangentially and to continue downstream as a strictly parallel flow, riding over a stagnant layer; the top part is a parallel jet, and the lower region is at rest. Indeed, this is a valid inviscid limit.

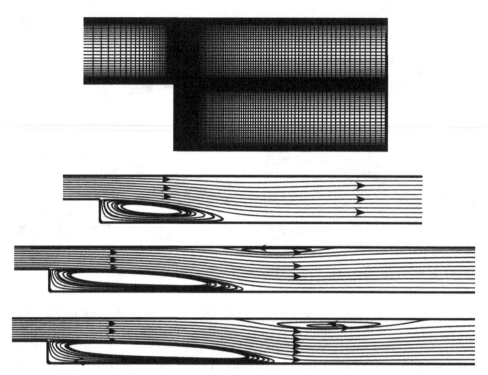

Figure 4.4. Flow over a backward-facing step at Reynolds numbers of 250, 500, and 650, from top to bottom.

However, viscosity will diffuse momentum into the lower region, driving a recirculating flow. Thus, the streamlines appear as in Figure 4.4. The streamline that leaves the top of the step deflects down until it meets the lower wall. Where they meet is the *point of reattachment*. Between the step and reattachment, the fluid circulates in a big eddy. The size of that recirculation bubble grows in proportion to Reynolds number. That is consistent with the presumption that in the inviscid limit, or as $Re \rightarrow \infty$, the flow would consist of a stream above a stagnant zone. The recirculation bubble is driven by viscous diffusion of momentum; high Re means low diffusion, and hence a weaker, longer eddy. At the risk of getting ahead of the story, the *turbulent* backstep is somewhat different. The length of the separation bubble is independent of Re. The difference is that momentum is diffused by turbulent mixing rather than molecular viscosity. Turbulent mixing is relatively independent of Reynolds number.

Consider the two-dimensional, laminar flow over a backward-facing step, as defined in Figure 4.4. At the inflow, a parabolic velocity profile was prescribed with averaged velocity $\bar{u} = 1$ m/s. The step height is 1 m, and the total channel height is $H = 2$ m. The expansion ratio, which is the total height divided by the inlet height, is 2 for this computation. The kinematic viscosity is adjusted to achieve Reynolds numbers of $\bar{u}H/\nu = 250$, 500, and 650.

A structured grid is used for this test case because it allows a simple way of concentrating the grid points near the walls and in the region where separation occurs. The portion near the step is shown in Figure 4.4. The nonuniform grid density

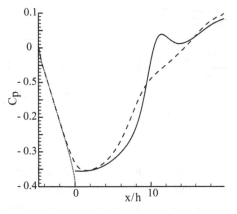

Figure 4.5. Pressure distribution on upper ---- and lower ——— walls. $Re = 650$. The dotted curve is in the entrance channel.

is tailored to resolve regions of steep shear. The full grid contained 15,330 nodes and 15,040 elements.

The growth in length of the separated region as Re increases is readily apparent. At the higher Reynolds numbers, a slender separation bubble is observed on the upper wall. This is not, at first, expected. It occurs because the flow is diverted downward as it rounds the recirculation zone next the the lower wall and reattaches to the surface.

Pressure distributions with $Re = 650$ are shown in Figure 4.5. The pressure on the lower wall has a local maximum at about $x = 11$, which is the point of reattachment. The upper wall separates shortly before this. The dashed C_p curve in Figure 4.5 shows a slight change in slope where that occurs. Generally, the pressure is rising as the flow expands, starting at about 4 step heights downstream. That is where the flow begins to curve down toward reattachment.

Pressure is referenced to the inlet and the pressure coefficient is defined with the bulk velocity:

$$C_p = \frac{p - p_{\text{in}}}{1/2\rho\bar{u}^2}.$$

The theoretical estimate (1.41) is not very accurate; however, it explains why C_p rises above 0 toward the exit. The velocity decreases as the flow expands. Momentum conservation implies a concomitant rise in pressure.

4.1.1.3 Flow around a Circular Cylinder

The most popular illustration of bluff body flow is the two-dimensional circular cylinder. The only analytical solutions are for creeping flow (Chapter 3) and for inviscid, potential flow. The inviscid solution is

$$u = U_\infty \left[1 + \frac{a^2(y^2 - x^2)}{r^4}\right],$$

$$v = -U_\infty \frac{2a^2 xy}{r^4},$$

$$(4.1)$$

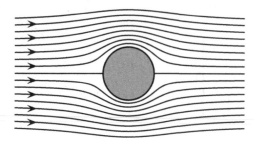

Figure 4.6. Potential flow past a cylinder.

where the flow is from left to right (a standard convention) and a is the radius of the cylinder, centered at the origin. The corresponding streamlines are plotted in Figure 4.6. This is the velocity field that would exist were there no vorticity and no viscosity. This is too idealized to occur in nature.* In front of the cylinder, before the flow acquires vorticity, this solution is a good approximation; behind the cylinder, it is not, as we will see shortly.

The top of the cylinder is at $x = 0$, $y = r = a$. There the velocity is $u = 2U_\infty$, $v = 0$, according to Eq. (4.1). Streamline convergence, seen in Figure 4.6, causes a factor of 2 speedup. At the rear of the cylinder, $x = r = a$, $y = 0$, the velocity is zero. Hence the flow is severely decelerated over the rear of the cylinder, from the top to the rear stagnation point. In that region, the potential flow solution is quite incorrect in the presence of viscosity. No analytical solution exists for the correct flow field. Experiments and the following computations show the intriguing behavior of this flow.

The solution [Eq. (4.1)] gives some guidance in setting up a computational analysis. It provides estimates to use in selecting a suitably large flow domain, when the Reynolds number is not small. The magnitude of the departure from uniform velocity decreases like a^2/r^2. Hence, a domain extending at least 10 radii in all directions is needed for the flow to become uniform to 1%. In Chapter 3 we saw that creeping flow falls off like a/r, so the low Reynolds number domain must be larger in comparison to the radius than the high Reynolds number domain (strictly, the low Re fall off is for flow over a sphere, not a two-dimensional cylinder). At higher Reynolds number, the domain must extend farther in the downstream direction to accommodate the wake.

For definiteness, we will let the velocity entering from the left be $U_\infty = 10$ m/s and the diameter of the cylinder be $d = 0.1$ m. The molecular viscosity is varied to obtain the desired value for the Reynolds number: $\nu = U_\infty d/Re$. Values of Re in the range 1–100 will be considered.

The flow domain is defined in Figure 4.7. The dimensions $30d \times 20d$ for the outer boundary are selected; these are large enough for better than 1% flow uniformity by the criterion mentioned previously. The cylinder is placed $10d$ downstream of the entrance.

---

* Ironically, there is a low Reynolds number apparatus, called a Hele–Shaw cell, which produces potential flow streamlines.

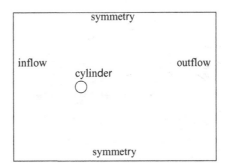

Figure 4.7. Domain for flow over a circular cylinder.

An unstructured grid containing 5,863 nodes and 10,933 elements, presented in Figure 4.8, was created. It contains a layer of rectangular cells adjacent to the surface of the cylinder, as seen in the zoomed panel of Figure 4.8, whereas the rest of the flow domain is meshed with triangles. The layer of regular cells resolves the near-wall viscous region. The unstructured mesh is adequate to capture the flow farther from the wall.

Standard velocity inflow and outflow boundary conditions are applied at left and right ends of the computional domain (Figure 4.7). The cylinder is a no-slip surface, and the top and bottom boundaries are treated as slip, or symmetry, surfaces. Convergence is judged by meeting a residual criterion of 10^{-8} and by the leveling out of forces exerted on the cylinder.

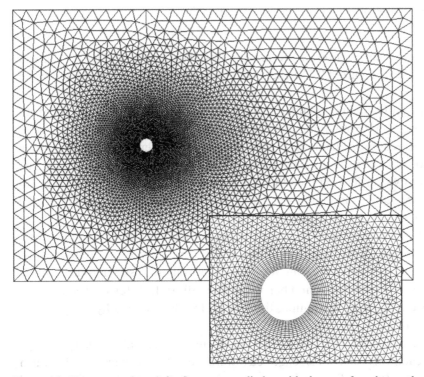

Figure 4.8. Unstructured mesh for flow over a cylinder, with closeup of mesh near the surface.

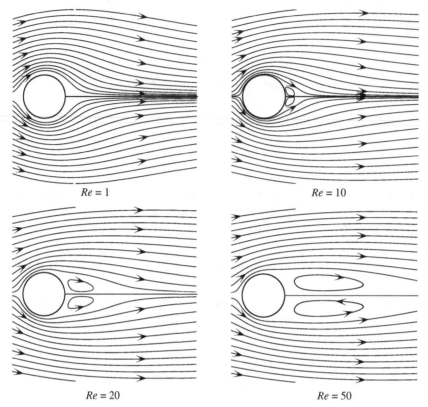

Figure 4.9. Streamlines at various Re in the vicinity of the cylinder.

The downstream portion of the flow at Reynolds numbers of 1, 10, 20, and 50 is presented in Figure 4.9. As the viscosity decreases, the length of the wake increases.

Immediately behind the cylinder, two counterrotating eddies develop. These eddies first appear at a Reynolds number of about 8. At higher Reynolds numbers, their length increases approximately in proportion to Re. This represents an increasing accumulation of vorticity in the lee of the body. Positive vorticity accumulates on one side and negative vorticity on the other. Hence, two recirculating zones, with counterrotation are seen, as in Figure 4.9. These developments are in excellent accord with experimental observations.

Downstream of the eddies, the wake consists of a region of lower velocity. This is evident in the velocity contour plots of Figure 4.10. The low velocity contours grow longer and narrower with increasing Re. The velocity on the upstream side becomes fairly insensitive to Reynolds number when $Re > 10$. In fact, for angles less than about 80° from the front stagnation, the flow is closely described by the potential flow solution (4.1).

We will explain in §4.2 that the wake boundary tends to a parabolic shape, $y \propto \sqrt{x}$ far downstream. This behavior is suggested toward the right side of Figure 4.11, which plots velocity contours at $Re = 50$ in a longer region than Figure 4.10. The superposed

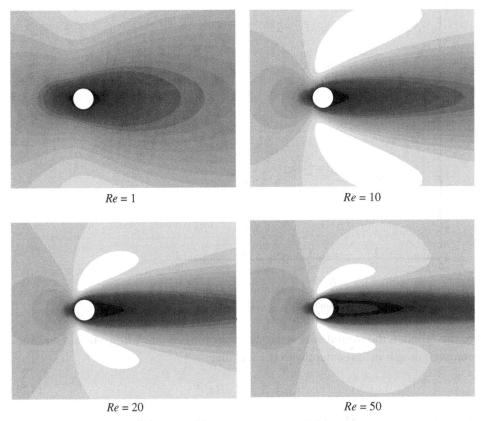

Figure 4.10. Velocity magnitude contours at various *Re* in the vicinity of the cylinder.

velocity profiles help to interpret the contours. The first profile has a small region of reversed flow at its center. Following profiles show the wake spreading and becoming less deep.

By convention, the drag coefficient is defined as the drag force normalized by the dynamic pressure and frontal area:

$$C_D = \frac{D}{1/2 \rho U_\infty^2 A_f}.$$

Figure 4.11. Velocity contours at *Re* = 50 with superposed velocity profiles showing the wake development.

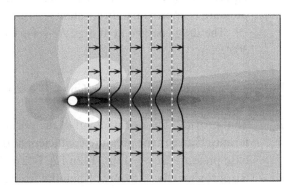

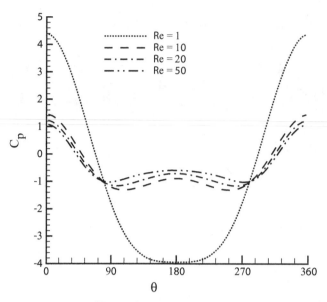

Figure 4.12. Pressure distribution round a circular cylinder.

For the cylinder, the frontal area is the diameter times a unit width: $A_f = d \cdot w$. The computed drag is in agreement with the formula

$$C_D = 1 + 10/Re^{2/3} \tag{4.2}$$

suggested as a fit to data by White (1991). This is a satisfactory fit to experimental data for $5 < Re < 2 \times 10^5$. At $Re = 20$, this formula gives $C_D = 2.35$, whereas our computations give $C_D = 2.17$. In this Reynolds number range, there are no theoretical solutions, and the computed value is within the range of experimental data scatter.

The distribution of surface pressure coefficient around the cylinder is graphed in Figure 4.12. This is normalized by the Bernoulli head, $1/2\rho U_\infty^2$. At larger Re the curves tend toward each other, indicating that inertia is dominating over viscous friction; it tends toward the value $C_p = 1$ implied by Bernoulli's equation at the front stagnation point. Of course, viscosity implicitly is responsible for the wake. At $Re = 50$, separation is slightly before the top of the cylinder, $\theta = 90$, as seen in Figure 4.9. In Figure 4.12, the wake lies in the region $90° \lesssim \theta \lesssim 270°$, where the pressure is approximately constant.

The drag coefficient at high Re is very nearly the integral of $C_p \cos\theta$ from 0 to 360°:

$$C_D = \int_0^{2\pi} C_p(\theta) \cos\theta \, d\theta.$$

That provides the x component of the normalized force. It is approximately the difference between the pressure coefficient at the front stagnation point, $\theta = 0$, and the base pressure, at $\theta = 180°$. Then Eq. (4.2) implies that the base pressure coefficient is approximately $-10/Re^{2/3}$ when the front stagnation point value falls to $C_p = 1$.

Formulas like Eq. (4.2) can sometimes be misleading. In particular, the drag seems to rise steeply at low Reynolds number. This is a consequence of normalizing by the dynamic pressure. For instance, the Stokes drag on a sphere, given by Eq. (3.3) has a drag coefficient equal to $24/Re$. This does not mean that drag tends to infinity as $Re \to 0$; rather, it means that at low Reynolds number this is not an appropriate normalization: pressure is due to viscosity not to inertia. With that warning, Eq. (4.2) describes a decrease of drag coefficient with increasing Re. The formula stops at $Re \approx 2 \times 10^5$ because the boundary layer on the cylinder becomes turbulent at that point. Beyond that, a "drag crisis" occurs (see Chapter 6, page 234).

Asymptotically, Eq. (4.2) gives $D \to 1/2\rho U_\infty^2 A_f$, when $Re \gg 1$. This is exactly the drag caused by flow stagnating on one side of a surface, with the other side, at ambient pressure. That is, the Bernoulli head $P_\infty + 1/2\rho U_\infty^2$ is converted to pressure on one side, whereas the base pressure, on the other side, is P_∞. In practice, the base pressure is slightly subambient and the pressure on the front equals the Bernoulli head only at the stagnation point.

Above a Reynolds number of about 50, an interesting phenomenon occurs: the wake becomes unstable and begins to oscillate in time. To capture that, an unsteady simulation is needed. An experimental observation can be used to estimate the necessary time-step. It is found that when Re is greater than about 200, the *Strouhal number*, defined as

$$St = \frac{d}{U_\infty T}$$

is constant and equal to 0.2. Here T is the period of the wake oscillation. Assuming that 50 time-steps per period are enough to resolve the unsteadiness, the required time-step is estimated as

$$\Delta t = T/50 = d/(50St\, U_\infty) = 10^{-3}\,\mathrm{s}$$

for the present values of $U_\infty = 10\,\mathrm{m/s}$ and $d = 0.1\,\mathrm{m}$.

The value $St \approx 0.2$ is often a good estimate for the period of vortex shedding, even for relatively complex, three-dimensional bluff bodies. However, it should be warned that shedding does not always occur. For instance, a coherent street of vortices is not seen in high Reynolds flow round a sphere. Indeed, three-dimensionality can suppress shedding: a wire wrapped helically around a circular cylinder can suppress vortex shedding. However, in other examples, such as the surface-mounted cube, which is the fifth case of this section, shedding does occur behind three-dimensional bluff bodies.

The total number of time-steps needed for a simulation of shedding can be estimated. The computation must be carried to a stage where periodicity is established. This is called the limit cycle. It is preceded by a transient, during which the solution is converging to the periodic, limit cycle. We invoke the rule of thumb that this will take about 5 to 10 through flow times. The through flow time is the time needed for the flow to enter and exit the domain at the nominal velocity; here, the through flow time is $30d/U_\infty = 0.3\,\mathrm{s}$. With the time-step of $\Delta t = 10^{-3}\,\mathrm{s}$, 5 through

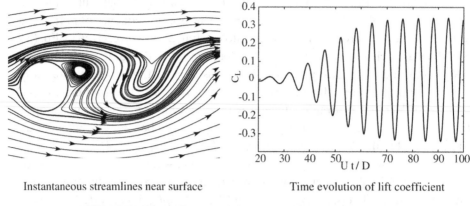

Instantaneous streamlines near surface   Time evolution of lift coefficient

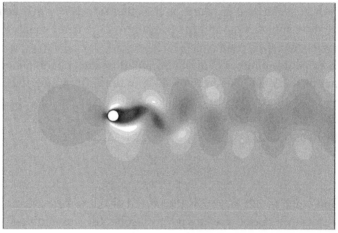

Instantaneous contours of velocity magnitude

Figure 4.13. Vortex shedding behind a cylinder at $Re = 100$.

flow times means 1,500 time-steps. As in the case of steady simulations, the forces on the cylinder are a good indicator of whether the flow has converged.

Accuracy suggests that a second-order time-marching scheme should be invoked. First-order methods can sometimes damp the tendency for vortex shedding to occur.

A time-dependent simulation for $Re = 100$ is presented in Figure 4.13. The lift coefficient oscillates periodically in time, as seen in the figure. The initial transient, and approach to a periodic limit cycle, are seen at the upper right.

Lift is proportional to the circulation round the cylinder. Kelvin's theorem, as discussed in Chapter 1, implies that when a vortex is shed, the cylinder acquires an equal and opposite circulation. Each time a vortex is shed, the circulation round the cylinder increases by an amount equal and opposite to that of the shed vortex. As alternatively signed vortices are shed, the circulation, and concomitant lift, fluctuate between positive and negative values. The lift on a circular cylinder in steady flow is a force per unit width, $F_y = 2\pi\rho\Gamma U_\infty$, directed perpendicular to the approach flow direction (see page 47). Γ is the circulation and ρ is the fluid density.

The drag coefficient, which is not shown, oscillates with a frequency that is twice that of the lift coefficient and of the vortex shedding itself. Drag is due to the difference between the pressure on the front and the back of the cylinder. The pressure on the front is the stagnation pressure of the oncoming flow ($P_\infty + 1/2\rho u_\infty^2$). The pressure at the rear of the cylinder, the base pressure, drops every time a vortex is shed, irrespective of its sign. Thus, it oscillates at twice the shedding frequency. The low base pressure is a consequence of the balance between pressure across the vortex and centrifugal acceleration of fluid circling the vortex. This balance causes the vortex core to be at low pressure (see page 32). The latter varies with the circulation squared.

The velocity contours in Figure 4.13 display the Von Karman vortex street extending downstream of the cylinder. Near the surface, vortices are shed alternatively from above the cylinder and from below it. At the instant shown at the upper left of Figure 4.13 a vortex is leaving the upper side. One half-period later, a similar vortex would be seen leaving the lower side. They array themselves in the wake as a staggered vortex street – the Von Karman vortex street. This array of vortices convects downstream with a speed of about $0.8U_\infty$, shortly downstream of the cylinder. The speed is less than U_∞ because the self-induced velocity of the vortices is upstream; recall from Figure 1.15 the concept of induced velocity of a vortex pair. Here, one must think of an array of oppositely signed vortices, but the idea is similar. Vortices from the upper side of the cylinder convect vortices from the lower side back against the flow and vice versa. Von Karman and Rubach, in 1912, analyzed an idealized version of this configuration; that is how it came to be called the Von Karman vortex street. Von Karman did not discover the phenomenon. He cited experimental studies, starting with Bènard in 1908, that described the phenomenon. Von Karman proposed that the drag on a cylinder could be attributed to the rate of negative momentum added to the wake by shed vortices. The defect of momentum flux in the wake can, indeed, be attributed to an equal and opposite force on the body.

In these computations, a time-averaged drag coefficient value of $C_D = 1.39$ was obtained. The time-averaged lift is zero, but the amplitude of the oscillations of the lift coefficient was $\tilde{C}_L = 0.35$. The nondimensional frequency of the oscillations was $St = 0.17$. These are in reasonable agreement with experiments at this Reynolds number of 100.

4.1.1.4 Flow over a Bump

The previous three examples are two-dimensional. To what extent do they bear on three-dimensional flows? Consider flow over the axisymmetric bump, portrayed in Figure 4.14. The plane of the wall is xz. A mound rises above it in the y direction. Although the geometry has axisymmetry, the oncoming velocity reduces the overall symmetry to reflection in the yz plane passing through the center of the bump.

The reader may wish to speculate on the development of the flow pattern with Reynolds number before going further. Will the flow progress monotonically from upstream to down? How will the wake develop as the Reynolds number increases?

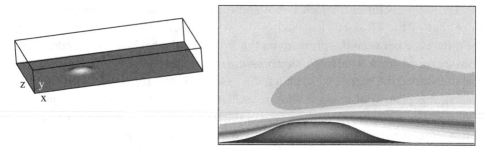

Figure 4.14. A three-dimensional bump on a wall.

Will the streamlines recirculate in eddies? Will the flow pass over or around the bump? Obviously, one expects the flow to separate when the Reynolds number is sufficiently large. However, the vortical layer that leaves the surface is no longer planar. How will it be configured? This axisymmetric geometry retains enough regularity for these questions to be answered. Were the geometry substantially more complex, flow patterns would be rather difficult to anticipate.

Having paused to give the matter some thought, computer visualizations will now be used to provide perspectives on the flow pattern. Before describing the streamlines, a somewhat puzzling perspective will be given. In Figure 4.15, the skin friction vectors on the surface are used to trace flow lines. In experiments, these

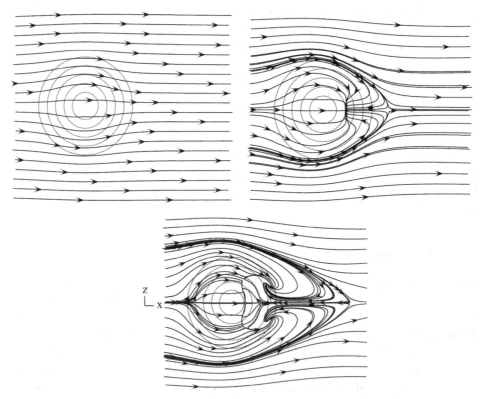

Figure 4.15. Skin friction on the surface. $Re = 1, 100,$ and 500 in the clockwise direction.

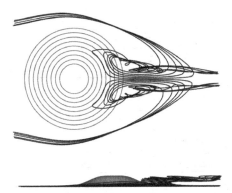 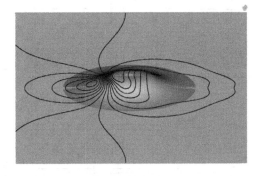

Figure 4.16. Streamlines showing vortices rising from the surface at $Re = 500$. Surface pressure contours are shown at right.

skin friction lines can be visualized by spreading oil, containing fine powder, on the surface and then flowing air over it. The powder forms into streaks, showing patterns like those in Figure 4.15. The topology of skin friction lines will be discussed in Chapter 5, with a few examples from experiments.

At the lowest Reynolds number, $Re = 1$, the flow proceeds monotonically over the hill, in the downstream direction. This is easy to anticipate for the mild bump of Figure 4.14. The particular geometry is an axisymmetric hill. It has been used in experiments that represent turbulent flow in the atmosphere. Here, laminar flow is solved. The shape is specified by

$$f(r) = \frac{h+c}{1+(2r/L)^4} - c \qquad r \le L,$$
$$f(r) = 0 \qquad r > L,$$

where r is the distance from the hill center. The height is h and $c = h/16$. The length L equals $5h$, so that the diameter of the hill is 10 times its height. The top of the hill is a flat plateau, and the maximum slope of its sides is $24°$. This is mild enough to avoid eddying at low Reynolds number (see §3.1.5). The Reynolds number is defined as $Re_h = U_\infty h/nu$.

At $Re = 100$, the pattern becomes more interesting. The skin friction lines reverse direction on the lee of the bump. They are drawn into a short line segment beyond the crest of the hill. This seems to be a sink. These surface stress lines are equivalent to streamlines just above the surface. The sink is where the flow is leaving the surface. Farther downstream, just beyond the end of the bump, the friction lines return to the downstream direction.

When Re is increased further, to 500, the ends of the line segment at which the flow direction reverses become spirals. These are symptoms of three-dimensional, separated flow. The form of that separated flow is somewhat difficult to infer from these intriguing surface stress patterns. Once gross features of the flow are understood, the underlying pattern of shear stress can be deciphered.

A selected set of particle trajectories is shown at the left of Figure 4.16. They are viewed from above, looking down on the hill at the upper left of the figure; at

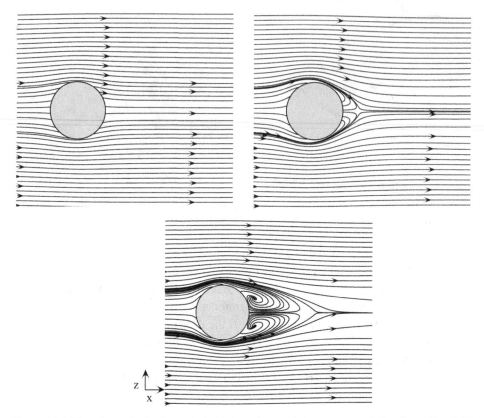

Figure 4.17. Flow lines of velocity projected into a plane, $y = $ constant, parallel to the wall, at 25% of the hill height. $Re = 1, 100, 500$.

the bottom left, they are viewed from the side. The trajectories are paths of fluid elements per Eq. (2.23). They show how a pair of vortices in the lee of the hill causes particles to spiral upward from the surface. The trajectories originate upstream, circumvent the hill, then reverse direction as they are drawn into the wake. They proceed back, toward the hill, and then flow upward along helical paths, finally convecting downstream at a higher elevation. The spirals seen in Figure 4.15 at $Re = 500$ are the footprint of these vortices. The convergences of the surface stress lines are where the particle trajectories flow up, away from the wall.

The surface pressure contours at the right of Figure 4.16 show a high level on the fore side and are bunched near the front of the plateau. That is where the flow accelerates in consequence of streamline convergence. The aft part of the hill is at nearly constant pressure.

One final perspective on the flow. Any planar section, parallel to the lower wall, will intersect the hill in a circle if the section is below its top, and it will be free of geometry if it is above the top. It might be speculated that sections below the crest of the hill will look similar to two-dimensional flow round a circular cylinder, seen in Figure 4.9. That is a reasonable starting point, but it needs modification. Streamlines in a constant y plane are shown in Figure 4.17. Behind a cylinder in two-dimensional flow (Figure 4.9) the recirculation is a closed region that is not entered by streamlines from upstream. Behind the hill, streamlines in a y plane enter the wake from upstream. These are the streamlines of the (u, w) velocity vector, with v set to zero.

The three-dimensional flow will not stay in the plane. The flow that enters the wake spirals up, out of the plane. Flow that appears to enter or exit from the body is actually moving up or down its slope. Despite these misleading aspects, views of the flow projected onto constant y planes provide an analogy between two- and three-dimensional fields. Some similarity in how the wake develops can be seen.

These computations were on an unstructured tetrahedral mesh, with a layer of prisms near the lower surface. It contained 777,736 elements and 304,727 nodes. The size of the outer box defining the computational domain (Figure 4.14) is somewhat arbitrary. The inflow condition is plug flow $u = 1$ in nondimensional units. The length of the box affects the solution because a boundary layer develops on the wall upstream of the hill. A different length will alter the flow profile approaching the hill. In the computations presented here, the outer box has dimensions $-3L < x < 6.5L$, $0 < y < 1.3L$, and $-3L < z < 3L$. Recall that the height of the hill is one-fifth of its length $h = L/5$. The box extends well above the hill to avoid confining the flow. The slip condition was used on the top boundary.

4.1.1.5 An Obstruction on a Wall

If the three-dimensional object is more abrupt than the previous example, other features emerge. The example considered here is a cube sitting on a surface. As the flow encounters the obstruction, it diverts around it, wrapping boundary layer vorticity about the object, in a necklace vortex. The necklace vortex was encountered schematically in Figure 1.19, page 43. Stretching and rotation of vorticity concentrates it into a distinct vortex, looped around the cube.

Figure 4.18 shows a small necklace at the lowest Reynolds number of 50. It increasingly envelops the cube as Re increases. Particle trajectories in Figure 4.19 form a helical braid as they spiral down the vortex. These trajectories maintain a distance from the body. Trajectories starting a bit higher from the bottom wall are sucked into the wake and spiral up in two wake vortices, similarly to Figure 4.16. This apparent pair of vortices, one either side of the symmetry plane, can be viewed as the two legs of an arch; as portrayed in Figure 1.24. By connecting them so, the law that vortex lines cannot end inside the fluid is obeyed.

The side view of velocity contours in the symmetry plane in Figure 4.20 shows the vortical region ahead of the cube and the wake region behind it. Trajectories that impinge on the front of the cube move down and around the front vortical region to emerge in one of the two vortex legs in the wake. They are the trajectories that spiral upward in Figure 4.19.

Returning to Figure 4.18, and $Re = 1,000$, we see a more complex pattern than expected. In fact, this is an instantaneous view of an unsteady flow. Vortices are being shed from the cube. They do not form a long vortex street, as in the case of

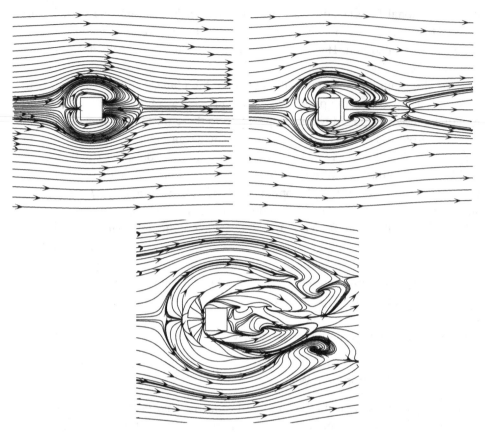

Figure 4.18. Surface stress lines in flow over a cube on a wall. $Re = 50, 200, 1,000$ clockwise from upper left.

Figure 4.13; they are seen in the lee of the body for a few cube diameters downstream. After that the flow returns to a nearly steady state. Two-dimensional bluff bodies produce a longer, clearer vortex shedding pattern than do three-dimensional bodies.

The periodic shedding from the surface mounted cubes is of arch vortices. The legs of the arch are parallel to the vertical sides of the cube. They are shed alternatively from the left and right sides, causing periodic unsteadiness. At the right side of the lowest panel of Figure 4.18, a pair of spirals is merging with the trajectories that loop round the sides. This seems to show the arch vortex merging with the necklace

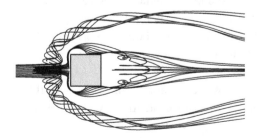

Figure 4.19. A necklace vortex, and evidence of an arch vortex, in flow past a surface mounted cube, seen from above, looking down.

Figure 4.20. Contours of velocity magnitude in the xy symmetry plane.

vortex. This final example emphasizes how much more convoluted flow patterns can be than the geometry in which they take place.

4.1.2 Secondary Circulation

We may abstract one notion from three-dimensional flows: that is the concept of *secondary flow*. Secondary flows are quite common in any but the simplest geometry. The term *secondary* presupposes a primary direction. The primary flow might be directed through a duct from inlet to exit, along its axis. A pressure drop drives this flow. The cross-sectional plane is perpendicular to the main flow. Secondary circulation is circulating motion in that plane. If the duct is straight there will be no secondary circulation.[*] It arises in curved ducts. Figure 4.21 is a schematic of a U-shaped duct.

The classic analysis of secondary flow was conducted by W. R. Dean in a series of articles in the late 1920s. He considered a circular pipe with a centerline having a constant radius of curvature – in other words, a torus; furthermore, a very thin torus. The ratio of the radius of circular cross section to the radius of curvature, a/R, was assumed to be small. These simplifying assumptions make analysis possible; computer simulation demands none of them. Nevertheless, in broad terms, the analysis describes the basic phenomenology that is seen in modern computer simulations.

The scaling invoked by Dean gives some feel for the characterization of secondary flow. Refer to Figure 4.21 for the basic notation. Let the geometry be a pipe of constant cross-sectional dimension, a, and constant curvature, R. The axial flow, down the center of the curved pipe, is in the θ direction. The centrifugal acceleration of the main flow is balanced by an inward, radial pressure gradient, which directs the flow along its curved path.

$$\frac{\partial p}{\partial r} = \frac{u_\theta^2(y, r)}{R + r}. \tag{4.3}$$

---

[*] That is true of laminar flow; even in a straight duct, turbulence can produce secondary circulation if the cross section is not round.

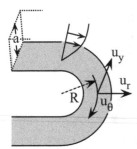

Figure 4.21. Defining sketch for flow in a curved duct.

The primary flow, u_θ, is a function of y and r – in the coordinates of Figure 4.21 – because of the no-slip boundary conditions. For instance, in the small a/R limit studied by Dean, the primary flow is approximately Poiseuille flow in a round pipe. Here, we consider a square section pipe. The primary velocity satisfies no-slip on $y = 0$ and a and $r = R + 1/2a$ and $R - 1/2a$. The centrifugal pressure, Eq. (4.3), has a gradient in the y direction. That will drive a u_y component of velocity. By definition, a secondary flow must exist because the primary velocity consists only of u_θ.

The order of magnitude of the secondary circulation can be estimated to determine how important it is. The estimate follows by assuming a balance between the induced pressure gradient and the inertial acceleration.

Both components of the pressure gradient induced in the cross-sectional plane have a similar magnitude, which is implied by Eq. (4.3). We denote it by

$$\nabla p \sim \frac{\bar{u}_\theta^2}{R},$$

where $\bar{u}_\theta$ is the cross-sectionally averaged velocity. The magnitude of secondary velocity will be denoted u_{sec}. The convective derivative, $u \cdot \nabla u$, is of magnitude u_{sec}^2/a, because its variations occur in the cross plane and hence, over the scale a. The balance between pressure gradient and acceleration is estimated as

$$\frac{u_{\text{sec}}^2}{a} \sim \frac{\bar{u}_\theta^2}{R}.$$

Thus, the secondary flow is of magnitude

$$u_{\text{sec}} \sim \sqrt{\frac{a}{R}}\bar{u}_\theta. \qquad (4.4)$$

This assumes, implicitly, that a is small compared to R. It suggests that the secondary velocity becomes comparable to the primary velocity as a becomes comparable to R. It will be weaker in a thin pipe.

An important parameter arises from comparing inertia and viscous terms in the cross plane. The viscous stress gradient, $\nu \nabla^2 u$, is of magnitude $\nu u_{\text{sec}}/a^2$. The ratio of inertial to viscous terms scales as

$$(u_{\text{sec}}^2/a)/(\nu u_{\text{sec}}/a^2) = \frac{u_{\text{sec}}a}{\nu} \sim \sqrt{\frac{a}{R}}\frac{\bar{u}_\theta a}{\nu}.$$

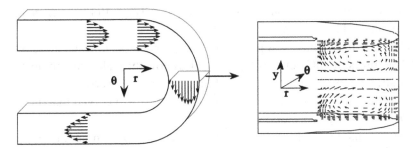

Figure 4.22. Primary and secondary flow in a U-shaped duct. Secondary flow is shown in a square cross section at the middle of the curved portion.

This defines the Dean number. By convention, in circular pipe flow

$$Dn \equiv \sqrt{\frac{a}{R}} \frac{2\bar{u}_\theta a}{\nu}. \tag{4.5}$$

Again, $\bar{u}_\theta$ is the averaged velocity and a is the pipe radius. The maximum velocity in Poiseuille flow through a circular pipe is twice the average. Hence, the second factor in the Dean number is the Reynolds number based on centerline velocity. The Dean number serves as a Reynolds number for the secondary flow; however, it is worth defining only when a/R is small.

The computation of flow in a U-shaped duct (Figure 4.22) illustrates the nature of secondary circulation. At left is the primary flow in plan view. At right is an edge-on view; the U-shaped walls are directed perpendicularly to the page and velocity vectors are plotted in the square cross section at the apex of the U. The secondary flow consists of a pair of streamwise vortices, evidenced by the circulatory velocity vectors.

Secondary flow is generated by a combined effect of pressure gradient and viscosity. In a constant radius of curvature bend, a balance between inward pressure force and outward centrifugal acceleration causes the primary radial pressure gradient. In the main stream, $dp/dr \sim \bar{u}_\theta^2/R$. This pressure gradient will be felt on the plane walls. But as the wall is approached $u_\theta \to 0$, due to the no-slip boundary condition. The centrifugal acceleration is no longer large enough to balance the inward radial pressure gradient. The unbalanced pressure drives flow in the radial direction, u_r, inward, toward the center of curvature. This occurs in a layer next to the the plane walls. In that layer the pressure force is balanced by viscous friction.

Thereby an inward radial velocity develops in the slower-moving fluid near the upper and lower walls of the curved section. As this radial secondary flow encounters the inner wall, it turns upward from the lower wall, and downward from the upper wall. These two streams collide at the center plane and then turn outward, flowing into the center of the duct. This outward stream crosses the duct and meets the outer wall of the channel. That turns it upward, to replenish the flow along the upper endwall, and downward, to replenish the flow along the lower endwall. In concert, these legs form a pattern of secondary circulation consisting of two cells in the cross plane, one occupying the space between the center plane and the upper wall and the other lying between the center plane and the lower wall. The pattern of velocity vectors is seen in Figure 4.22.

The secondary circulation has a few practical consequences. For one, it increases resistance. A given pressure difference will produce less flow through a curved duct than through a straight one of the same length and section; conversely, a larger pressure drop is required to generate a particular volume flux in a curved than in a straight pipe. One might reason as follows.

Let Q be the volume flow rate and Δp be the pressure drop from inlet to exit. Then $\Delta p\, Q$ is the rate at which work is done by the pressure force. Between inflow and exit, energy must be dissipated by viscous friction at this same rate, under steady-state conditions. Denote the cross-section-averaged rate of dissipation per unit volume by $\bar{\varepsilon}$. To balance energy input with dissipation at a prescribed flow rate, Q, the requisite pressure drop is

$$\Delta p = \frac{\bar{\varepsilon} V}{Q},$$

where V is the volume of the pipe. The total dissipation is the sum of that due to the primary flow, plus an addition from the secondary circulation:

$$\Delta p = \frac{[\bar{\varepsilon}_p + \bar{\varepsilon}_{\sec}] V}{Q}.$$

Hence, with fixed Q, the pressure drop is increased by secondary flow. Stated otherwise, the frictional resistance $\Delta p/Q$ is larger in a curved than in a straight pipe, in consequence of the secondary circulation.

In Dean's analysis, the secondary circulation is a small perturbation to Poiseuille flow in a circular pipe. From Eq. (1.26), $\Delta p Q = 8\mu \bar{u}_\theta^2 V/a^2$. Thus, in a straight pipe

$$\bar{\varepsilon} = \frac{8\mu \bar{u}_\theta^2}{a^2} \equiv f Q^2,$$

where f is the proportionality between dissipation and Q^2; the definition $Q = \bar{u}_\theta \pi a^2$ converts between averaged velocity and volumetric flow rate. For a small Dean number, Eq. (4.5), the ratio of friction coefficient in curved and straight pipes is predicted to be

$$\frac{f_c}{f_s} = 1 + 0.03 \left(\frac{Dn}{96}\right)^4.$$

Thus, the effect is small if $Dn < 96$. At large Dn, laboratory data are fit by

$$\frac{f_c}{f_s} = 0.097 Dn^{1/2} + 0.556.$$

Frictional loss increases in proportion to $Dn^{1/2}$ at a large Dean number. Secondary circulation then causes a substantial increase in the pressure drop required to produce a particular volume flow rate through a curved pipe.

Whenever a main stream is turned by a pressure gradient, the flow in viscous wall layers will lie at some angle to the primary flow. At least, that is true if the Reynolds number is not low. The reason is the same as that behind the scaling of Dean flow: in the primary stream inertia balances pressure gradient. If the streamlines are curved, a component of pressure gradient perpendicular to the main stream balances centripedal acceleration. In the viscous region near to walls, that component of pressure gradient drives a secondary circulation.

Another example occurs in a turbine passage. The geometry is that of Figure 2.18, which should be regarded as a y-plane section of a row of blades, held between two plane endwalls. The pressure is low on the upper, convex surface of the blade and high on the lower, concave surface. A pressure gradient points from the upper surface of one blade across the passage to the lower surface of the next. This is the pressure gradient that turns the main stream in its curved path.

The endwalls also experience this pressure gradient, but the flow speed there is lower due to viscosity; pressure and centrifugal forces are imbalanced. Gas flows from the high-pressure surface of one blade toward the low-pressure surface of it neighbor, along the end walls. On meeting the other blade, it turns and flows along the blade span, away from each of the end walls. These counterflowing streams meet and turn out, away from the surface, across the passage. They complete the two circuits of the secondary flow – just as in the case of the U-duct (Figure 4.22). In the context of turbines, these are called passage vortices.

4.2 Wakes, Jets, and Mixing Layers

What type of analysis can provide concepts for describing flow at intermediate Reynolds number? In the previous, numerical examples reference was made to wakes and separated shear layers. Figure 4.1 identifies the latter as mixing layers. Studied in isolation, mixing layers, wakes, and jets are categorized as boundary-free shear layers or just "free shear layers." In the laboratory, an isolated mixing layer is created by sending two streams tangentially along a plate. They come into contact after the trailing edge to produce the mixing layer. Methods to produce the generic free shear layers are implied by the schematic Figure 4.23. Wakes, jets, and mixing layers are building blocks for understanding many features of flow in engineering geometries. Simple analyses of how these shear layers are generated and evolve provide the framework we seek for describing complex flows.

Velocity profiles of the elementary free-shear layers are illustrated by Figure 4.23. Jets are characterized by their momentum excess, wakes by their momentum deficit, and mixing layers by the velocity change across the layer. The wake is a deficit caused by the drag on an upstream body; the jet is an excess of momentum injected from a nozzle. They can be characterized by the magnitude of the velocity excess or deficit as a function of downstream distance, x. They can also be characterized by a thickness, $\delta(x)$, as illustrated in Figure 4.23. The thickness can be defined as the value of y at which the velocity excess or deficit is one-half of its maximum. The wake of a drag-producing body was illustrated in Figure 4.11. That figure shows how the wake thickness increases and its velocity deficit decreases with downstream distance. One might speculate that thickness and deficit are inversely related. Let us consider how this might occur.

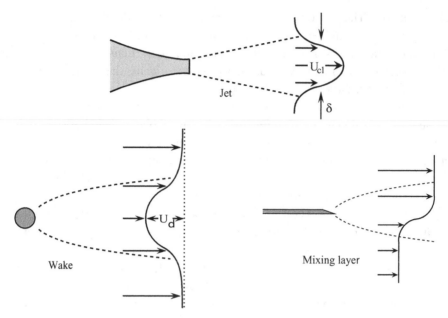

Figure 4.23. Free shear layers schematics.

4.2.1 Scaling Shear Layer Evolution

In their pure form, wakes, jets, and mixing layers are regarded as thin shear layers, with approximately parallel flow: that is, the flow is nearly independent of the x direction, with velocity predominately in the same, u, direction. They are shear layers in the sense that the shear, $\partial u / \partial y$, is nonzero only within the layer or the vorticity, which is approximately $-\partial u / \partial y$, exists only within a thin zone. Thus, we think of the spreading of these layers as being accomplished by diffusion of vorticity in the y direction, normal to the layer. The diffusion coefficient is the molecular viscosity, ν; this is the kinematic viscosity, which has dimensions of length^2/time.

However, in the momentum balance, the transverse velocity, v, cannot be ignored. That is because of the thinness of the layer. Large gradients exist in the transverse, y direction, so a small v will still lead to significant transport.

A simplification called the thin layer approximation can be invoked. The primary approximation is to ignore x derivatives of any given quantity in comparison to y derivatives of that same quantity. Thus the steady Navier–Stokes and continuity equations are reduced to

$$\rho\left(u\frac{\partial u}{\partial x} + v\frac{\partial u}{\partial y}\right) = -\frac{\partial p}{\partial x} + \mu\frac{\partial^2 u}{\partial y^2},$$

$$0 = \frac{\partial p}{\partial y}, \tag{4.6}$$

$$\frac{\partial u}{\partial x} + \frac{\partial v}{\partial y} = 0.$$

In the first equation, $\mu \partial^2 u/\partial x^2$ has been dropped from the exact x-momentum equation. That is the only simplification. It expresses the dominance of diffusion across the layer over diffusion along the layer.

The second of Eqs. (4.6) is essential to the thin shear layer approximation. It states that the pressure does not vary transversely to the flow. The pressure inside the shear layer is the same as that of the ambient fluid, outside the shear layer. This idea is so frequently invoked that it warrants underlining. Pressure varies little in the direction perpendicular to the flow because there is negligible inertia or friction to support pressure differences. If the pressure in the free stream is $P_\infty(x)$, that pressure is felt, as a function of x, inside the shear layer too: $p(x, y) = P_\infty(x)$ for all y.

Let us consider the downstream evolution of shear layer thickness. We are not concerned with solving Eqs. (4.6); our purpose is to infer scaling relations – formulas that relate thickness to downstream distance and velocity scales to length scales. Scaling relations are determined up to a constant of proportionality. The symbols $\propto$ and $\sim$ replace equality. The first indicates proportionality; the second indicates the form of dependence. In both cases, the symbol indicates that numerical constants of proportionality have been dropped.

In the simplest case, the ambient flow is uniform, with zero pressure gradient. Then the first of Eqs. (4.6) describes diffusion of x velocity, with the kinematic viscosity, $\nu = \mu/\rho$, playing the role of diffusion coefficient. Let the uniform ambient velocity be U_∞. For the jet, this can be zero, so first consider a wake or mixing layer – for the mixing layer, U_∞ can be defined as the average of the ambient velocities above and below the layer. The layer grows by diffusion. Were the layer evolving in time, then at time t its thickness would be $\delta \propto \sqrt{\nu t}$ on dimensional grounds. Kinematic viscosity has the dimensions of ℓ^2/t; hence, δ is the thickness to which the vorticity layer has diffused at time t.

When the layer evolves downstream, the effective diffusion time is x/U_∞. Again, solely on dimensional grounds, the shear layer thickness, $\delta(x)$, should grow as

$$\delta(x) \propto \sqrt{x\nu/U_\infty}. \tag{4.7}$$

The square-root dependence on distance is symptomatic of diffusional spreading.

A jet with no ambient velocity scales differently. Consider a flow that exits a nozzle into an ambient that is at rest. Equation (4.7) is applicable, but not with the velocity U_∞, which is zero. Other considerations are required to obtain a suitable velocity. What is needed is a conserved quantity to replace U_∞.

If there is no pressure gradient, then no forces act on the jet, once it leaves the nozzle. Its momentum flux is constant, independent of x. That can be used to develop the scaling. Momentum flux has the form

$$J \sim \rho U_{cl}^2 A, \tag{4.8}$$

where U_{cl} is the centerline velocity and A is a representative cross-sectional area. For a round jet, the thickness, δ, is a radius. The representative area is proportional that of a circle of radius δ: that is, $A \propto \pi \delta^2$. The centerline velocity is not a constant; it varies with downstream distance. But conservation of momentum flux (4.8) implies that velocity and thickness are connected by

$$U_{cl}(x)\delta(x) \sim \sqrt{J/\rho}, \tag{4.9}$$

where the right side is constant. Centerline velocity is inversely proportional to thickness; as the jet spreads, its centerline velocity falls.

Now, U_∞ in Eq. (4.7) is replaced by the centerline velocity to give the diffusional growth

$$\delta(x) \propto \sqrt{x\nu/U_{cl}}.$$

Thickness can be eliminated by using Eq. (4.9), giving the jet centerline velocity as

$$U_{cl} \propto \frac{J}{\mu x}. \tag{4.10}$$

From Eq. (4.9), the thickness corresponding to this centerline velocity is found to be

$$\delta(x) \propto x\nu\sqrt{\frac{\rho}{J}}. \tag{4.11}$$

Thus we find that the round jet spreads in proportion to x, whereas wakes and mixing layers spread like $\sqrt{x}$. The coefficient of proportionality is one over the momentum flux Reynolds number,

$$Re_J = \frac{\sqrt{J/\rho}}{\nu}.$$

Equation (4.11) is restated: $\delta \propto x/Re_J$.

The same argument applied to a two-dimensional jet gives

$$\delta(x) \propto \frac{\mu^2}{J\rho}^{1/3} x^{2/3},$$
$$U_{cl} \propto \frac{J^2}{\mu\rho}^{1/3} x^{-1/3}. \tag{4.12}$$

The derivation will be left as an exercise.

Momentum conservation also can be applied to the wake. Given formula (4.7) for δ, momentum considerations imply a certain downstream development of the wake deficit. The deficit is due to drag force on the body that produced the wake: Figure 4.11 is an illustration. The wake is carried downstream at approximately the free-stream speed. The defect velocity is defined as $U_d = U_\infty - U_{cl}$, as in Figure 4.23. The drag force is represented as

$$F = 1/2\rho U_\infty^2 C_D a \cdot w, \tag{4.13}$$

where C_D is the drag coefficient of the body and $a \cdot w$ is its frontal area. Balancing the deficit of the momentum flux by the drag on the body provides a scaling for the velocity defect.

A proper accounting invokes a control volume, as in §1.6.1. However, there is a subtlety. A natural control volume would be bounded by planes upstream and downstream of the body, extending to $\pm\infty$ in y. But the momentum flux across both planes is then infinite because of the nonzero free-stream velocity. There is a simple solution: the first line of Eq. (4.6) is rewritten as

$$\frac{\partial[u(u - U_\infty)]}{\partial x} + \frac{\partial[v(u - U_\infty)]}{\partial y} = \nu\partial_y^2 u,$$

invoking zero pressure gradient and continuity. The first term suggests that the deficit in momentum flux per unit width of a two-dimensional wake should be defined as

$$J \equiv \int_{-\infty}^{\infty} \rho u(U_\infty - u)dy.$$

The term in parentheses is the velocity deficit. If the deficit is small, this integral scales as

$$J \sim \rho U_\infty U_d \delta.$$

Equating this momentum defect to the drag (4.13) gives

$$\rho U_\infty U_d \delta \sim 1/2\rho U_\infty^2 C_D a.$$

Hence, the wake scaling is

$$\frac{U_d}{U_\infty} \propto \frac{C_D a}{\delta}.$$

Substituting expression (4.7) for δ gives

$$\frac{U_d}{U_\infty} \propto C_D\sqrt{\frac{aRe_a}{x}}, \qquad (4.14)$$

where $Re_a = U_\infty a/\nu$. The centerline velocity falls as the square root of distance. Similar reasoning applied to an axisymmetric wake gives

$$\frac{U_d}{U_\infty} \propto C_D\frac{aRe_a}{x}. \qquad (4.15)$$

The scalings (4.10) and (4.14) describe the behavior far downstream of the nozzle or body that produced the jet or wake. The details of how the wake or jet was produced are immaterial. Close to the source the shear flow does depend on the particular way in which it was created. It evolves downstream to a generic state that depends only on the the net momentum excess (J) or deficit (C_D) imparted to the flow.

4.2.2 Entrainment

A corollary to the scaling (4.11) of jet growth is that flow is *entrained* into a jet. This means that there is a secondary flow within the ambient fluid, which is directed into the jet. What is the cause of entrainment? If the momentum flux, proportional to $\rho U_{cl}^2 A$, is conserved, then the mass flux, proportional to $\rho U_{cl} A$ cannot be conserved. It must increase in proportion to $1/U_{cl}$. For a round jet, this is proportional to downstream distance, x.

Of course, mass conservation is not violated. The additional mass flux is a consequence of surrounding fluid being drawn into the main body of the jet. In a computational domain, this takes the form of a component of velocity entering normal to the boundary, as illustrated in Figure 4.24. The secondary, entrainment flow is weak, on the order of the main velocity, U_{cl}, times the spreading rate, $d\delta/dx$, of the jet. From the formula (4.11), is can be seen that the latter is small if the Reynolds number is appreciable. The scaling is

$$v \sim U_{cl}/Re_J,$$

assuming Re_J to be large. Indeed, all the scalings developed in this section invoke an assumption that the shear layers are thin, which, as a formality, is met when Re_J is large.

The scalings are used to design the flow domain and boundary conditions for CFD analysis. The boundary conditions must allow for entrainment; the domain must be large enough to contain the spreading shear layer and not inhibit its growth. The estimates of spreading give a guideline on how long and how wide the domain must be.

4.2.3 Free and Impinging Round Jets

A laminar, incompressible axisymmetric jet emerges from a circular orifice of diameter d. The inflow velocity, U_{in}, is uniform across the orifice. The Reynolds number is defined by $Re = U_{in}d/v$. Computations were carried out in the range $Re = 10 - 100$.

After a short distance from the exit, the jet should spread in accord with the theory of §4.2.1. In other words, the jet should become self-similar, in the sense that the centerline velocity, U_{cl}, should evolve as $J/(\mu x)$, and the thickness $\delta(x)$ should be proportional to $x\mu(\rho J)^{-1/2}$. J is the momentum flux; for a round, plug flow jet,

$$J = \frac{\rho U_{in}^2 \pi d^2}{4};$$

see §1.6.1. A nondimensional plot can be made of $\tilde{U} \equiv U_{cl}/U_{in}$ as a function of $\tilde{x} \equiv x\mu U_{in}/J \propto xv/U_{in}d^2$. A nondimensional statement of the scaling argument is

$$\tilde{U} \propto \tilde{x}^{-1}. \tag{4.16}$$

One might conclude that the specific choice for the values of J and v in this simulation is irrelevant. Although that is true, special care is needed when defining the flow domain for a given value of Reynolds number. Taking the initial jet diameter as the unit of length, the jet thickness, δ/d, increases as $x/d\, Re$. Thus, at a fixed x/d the jet will be thicker at lower Re. Hence the domain should be wider at lower Reynolds number to contain the jet. At higher Reynolds number the jet will be thinner, and a finer grid will be needed to resolve the flow.

A box of size $20d \times 20d$ was chosen for the present Reynolds number range. At the exit, the jet thickness will be on the order of $20/Re$. With Re in the range

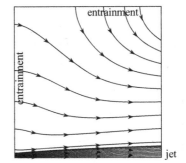

Figure 4.24. Axisymmetric jet: streamlines at $Re = 20$.

10–100, the computational box is wide enough to contain the jet. A simple 109×137 structured grid was used for this computation. It was refined in the region of the jet.

Streamlines are presented in Figure 4.24. These show the upper half of the jet along the lower edge of the figure, and the weak entrainment, driven by the jet, in the upper portion. The part of the left boundary that is above the jet and the top boundary are defined as a constant pressure surfaces. A uniform velocity is prescribed on the portion of the left boundary where the jet enters the domain. The need to allow the jet to entrain fluid through the constant pressure boundaries is obvious. Other examples of jet entrainment will be presented below.

Figure 4.25 consists of velocity profiles at various downstream positions at $Re = 20$ and 100. The development from the plug flow, prescribed at the inlet, into rounded, jet profiles occurs rapidly – more so at the lower Reynolds number. The jet spreads and the maximum velocity decreases as the flow proceeds downstream. The Reynolds number dependence is consistent with the scaling 4.11: a given location at $Re = 20$ scales to a location five times farther downstream at $Re = 100$.

The effectiveness of the scaled form (4.16) is seen more simply in Figure 4.26. That figure plots the jet centerline velocity versus scaled downstream distance. It can be seen that after an initial distance, near the orifice, the jet centerline velocity for all three cases scales perfectly and agrees excellently with the analysis, shown by the dotted line. The distance to reach the theoretical scaling is about $x/d = 0.05\,Re$.

The free jet, exiting from an orifice, is the simplest example of this type of flow. More complexity is added when jets impinge on a surface, are confined by lateral

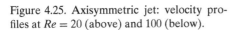

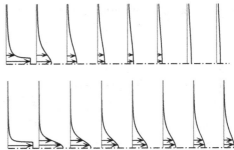

Figure 4.25. Axisymmetric jet: velocity profiles at $Re = 20$ (above) and 100 (below).

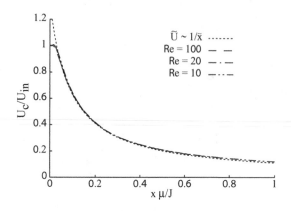

Figure 4.26. Jet centerline velocity, $Re = 10$, 20, and 100.

walls, blow tangent to the surface, and so on. Terminology such as *impinging jets*, *confined jets*, and *wall jets* refer to these factors. The scaling analysis of jet growth is no longer applicable when the jet interacts with a wall. However, it still gives an idea of jet development before impingement and leads qualitatively to expectations about jet evolution in other circumstances.

A computational domain for an impinging jet is illustrated in Figure 4.27. Two possibilities are indicated for the introduction of the jet. The upper boundary can be partly inflow, where the jet enters, with the remainder being constant pressure, where flow is entrained. Alternatively, a nozzle can be included, as indicated by the dashed rectangle. The jet exits at the bottom of the nozzle, which is just a straight pipe in this case. The flow is axisymmetric. The remaining free boundaries of the domain are set to constant pressure, outflow type conditions. This allows fluid to be entrained into the domain or to exit from the domain across these boundaries.

The case with the nozzle is presented in Figure 4.28. The left edge of the domain is the centerline of the jet. Rotating the streamlines about that edge would produce the three-dimensional, axisymmetric flow field. The ratio of height of the jet exit above the impingement surface to jet diameter, h/d, is 2 in this computation. Figure 4.28 shows the initial entrainment, near the nozzle. This induced flow joins the main stream as it flows toward the wall.

After impingement, the flow turns and follows the wall. That portion of the flow is a radially expanding wall jet. The presence of viscous drag on the surface reduces the momentum flux; hence, previous inferences about the growth of a free jet, which were based on momentum conservation, no longer apply.

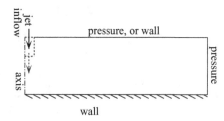

Figure 4.27. Flow domains for impinging jet computations.

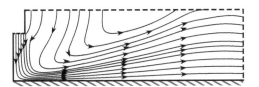

Figure 4.28. Impinging, unconfined jets. $Re = 100, h/d = 2$

Figure 4.29 shows the distribution of friction and pressure coefficients on the impingment wall. The pressure is a maximum at the stagnation point and falls rapidly in the radial direction. The pressure coefficient,

$$C_p \equiv \frac{P - P_{\text{ambient}}}{1/2\rho u_{\text{in}}^2},$$

is less than unity because of the viscous loss in the jet. When h/d is increased to 10, C_p is only 0.3 at the stagnation point (Figure 4.30). There is greater viscous spreading prior to impingement.

The flow makes a rapid transition from an impinging jet, stagnating on a wall, to a radially expanding wall jet. The latter, being driven by the volumetric flow, has with very little associated pressure gradient. The pressure coefficient falls nearly to zero within two jet diameters in the case of Figure 4.29.

The skin friction vanishes at the stagnation point because the flow is perpendicular to the wall. It rises as the flow accelerates radially, away from the stagnation point, as seen in Figure 4.29. The friction coefficient is normalized as

$$C_f \equiv \frac{\tau_w}{1/2\rho u_{\text{in}}^2}.$$

Consider the jet along the wall after it has impinged and turned direction. Because this is an axisymmetric flow the mass flux in the wall jet is proportional to

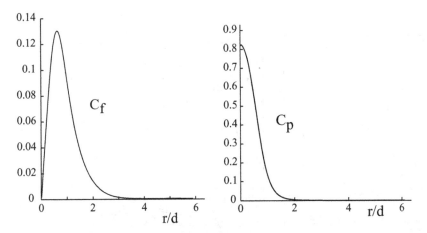

Figure 4.29. Pressure and skin friction coefficients for impinging jet with $h/d = 2$. The pressure on the free boundary is also shown.

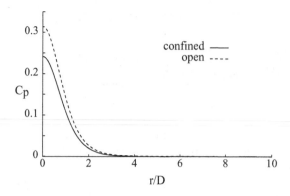

Figure 4.30. Pressure coefficient for impinging jet with $h/d = 10$. The confined and unconfined cases are compared.

$u_m \delta r$, where $\delta(r)$ is the thickness of the jet shear layer, perpendicularly to the wall. $u_m(r)$ is a representative velocity, say, the maximum as a function of y. Proceeding radially outward, the sectional area of the flow domain increases in proportion to $2\pi r$. Mass conservation suggests that as the jet flows radially outward, u_m will fall because r is increasing, although the precise behavior is not obvious because of entrainment and growth of the layer thickness.

Indeed, it can be shown that the maximum velocity of an ideal (self-similar), axisymmetric wall jet falls as $u_m \sim r^{-3/2}$. The thickness increases as $\delta \sim r^{5/4}$. The radial momentum flux in a layer of thickness δ is proportional to $2\pi r (u^2 \delta)$. In this case, it falls like $r^{-3/4}$: momentum is not conserved; it is depleted by frictional drag on the surface. The balance between friction and momentum extraction,

$$2\pi r \tau_w \sim \frac{\partial(2\pi r u^2 \delta)}{\partial r},$$

implies that the wall friction falls like $\tau_w \sim r^{-11/4}$. This is consistent with the rapid decrease of C_f with increasing r, seen in Figure 4.29. Even though the particular

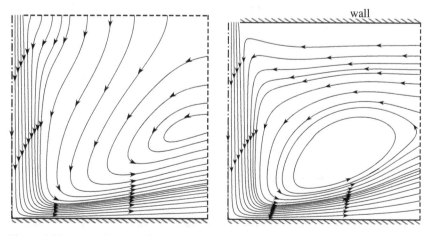

Figure 4.31. Impinging jet, with and without a confining, upper wall. Confinement redirects the entrained flow. $h/d = 10$, $Re = 100$.

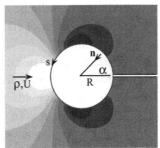

Figure 4.32. Flow past a cylinder with a splitter plate. Pressure contours.

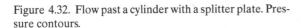

formulas (4.10) and (4.11) for the velocity and width of the jet do not apply, qualitatively, the jet continues to spread and to entrain fluid as it flows along the surface.

The precise form of the flow field depends on the height of the jet above the wall compared to its diameter or h/d. Figure 4.31 is for $h/d = 10$. In this case, there is some inflow at the right boundary where the wall jet exits the domain, which illustrates that the constant pressure boundary condition permits inflow. Indeed, that constant pressure condition also was used on the upper entrainment boundary in all computations. In steady flow, the exit condition on velocity is zero normal derivative (§2.2.4); this condition allows either inflow or outflow.

Entrainment flow can be redirected. For instance, an ejector nozzle consists of a collar around a jet, which channels the entrainment flow into the direction of the main stream. This can be for the purposes of mixing or thrust enhancement. In the present context, a simple illustration is to add a confining wall to the top of the domain. That was done in the right panel of Figure 4.31. The pattern of secondary flow changes. The entrained flow follows the upper, impermeable wall and creates a recirculating eddy next to the jet. The flow is not altered dramatically, but this example illustrates how the secondary flow is influenced by the geometry.

The pressure coefficient is plotted in Figure 4.30. In the confined geometry, the stagnation pressure is lower than in the unconfined case. Both fall well below the value of unity, which would occur in the absence of viscous losses.

EXERCISES

4.1 *Scaling.* Derive Eq. (4.12) and Eq. (4.15).

4.2 *Wake.* The velocity profile behind a bluff body in water is approximated by straight lines. The centerline velocity is 0.6 m/s. It increases linearly above and below the centerline to reach 1 m/s at ±1 cm and remains constant there after. What is the drag on the two-dimensional body.

4.3 *Surface force versus Reynolds number.* Numerical computations of the flow over a circular cylinder with a splitter plate extending from its rear have been carried out for a range of Reynolds numbers. The plate on the back side of the cylinder prevents vortex shedding, keeping the flow steady. The figure shows the geometry and contours of velocity magnitude.

Surface data on the upper half of the cylinder are provided for three Reynolds numbers: $Re = U_\infty D/\nu = 10, 50, 250$. The data in the following table include the curvilinear coordinate s [m], defined in Figure 4.32, the static pressure p[Pa], and the x-component of surface shear stress $\tau_{w,x}$ [Pa]. The origin for s is at the rear of the cylinder.

Using the tabulated data, compute the total drag force. Compute the individual contributions of pressure and of friction. How does the ratio of these components change with Reynolds number?

In the figure, $\boldsymbol{n} = (n_x, n_y)$ is the unit inward normal vector. The radius of the cylinder is $R = 0.5$ m, whereas $\rho_\infty = 1\,\text{kg/m}^3$ and $U_\infty = 1\,\text{m/s}$.

What is the lift versus Reynolds number?

	$Re = 10$		$Re = 50$		$Re = 250$	
s (m)	p (Pa)	τ_x (Pa)	p (Pa)	τ_x (Pa)	p (Pa)	τ_x (Pa)
.0795	−.5650	−.0020	−.4151	−.0010	−.3557	−.0001
.2386	−.6224	−.0045	−.4316	−.0086	−.3575	−.0017
.3976	−.7155	.0388	−.4769	−.0100	−.3655	−.0044
.5567	−.7909	.1661	−.5488	.0170	−.3887	−.0062
.7158	−.7926	.3604	−.6303	.0945	−.4588	.0097
.8748	−.6498	.5279	−.6323	.1968	−.5483	.0617
1.0339	−.3359	.5608	−.4496	.2477	−.4689	.1045
1.1929	.0969	.4149	−.0872	.1982	−.1465	.0911
1.3520	.5044	.1841	.3049	.0908	.2495	.0425
1.5608	.7577	.0119	.5579	.0060	.5110	.0030

4.4 *CFD at intermediate Re.* Compute the flow past a cylinder with splitter plate, corresponding to exercise 4.3 for a similar range of Reynolds numbers. Plot streamlines. How does the separation length vary with *Re*?

Because the flow is symmetric, set the domain size to $[-5D, 10D] \times [0, 5D]$. Use an unstructured grid and avoid exceeding about 40,000 triangles. One hundred grid nodes along one-half of the cylinder should provide adequate resolution. Apply a higher-order scheme for the convection operator (such as QUICK). What is the appropriate boundary condition for the top boundary?

5 High Reynolds Number Flow and Boundary Layers

Boundary layers are an elemental concept of high Reynolds number flow. They are a framework for discussing viscous fluid dynamics by separating the flow into distinct regions. That is the essential nature of boundary layer theory. The seminal ideas were described by Prandtl in 1904. He recognized that viscous flow along surfaces could be divided into two regions: a vortical layer next to the wall and a potential flow farther from the surface. Modern theories have expanded that to multilayered structure; but the basic notion always is of a thin, vortical layer next to the surface and an inviscid outer flow.

The boundary layer concept brought clarity to the puzzle of the high Reynolds number limit. High Reynolds number can be interpreted as low viscosity. Is inviscid flow the correct limit? Without viscosity, fluid flows freely over a surface, slipping relative to the wall. Hence, the tangential velocity is discontinuous between the stationary wall and the flowing fluid. The shear is infinite. Adding the smallest amount of viscosity would cause an infinite stress. Inviscid flow cannot be the correct high Re limit of viscous flow.

Any amount of viscosity will diffuse the velocity discontinuity: the fluid velocity will be brought smoothly to zero at the stationary wall. Even at the highest Reynolds numbers, viscous stresses cannot be ignored. How, then, is high Reynolds number flow to be constructed?

This puzzle is solved by recognizing that viscous influence is confined to a very thin layer next to the wall. The layer of viscous influence becomes increasingly thin as Re becomes increasingly large.

The mathematical term for the formation of thin layers is *singular perturbation*. High Re fluid flow is the archetypal singular perturbation problem.

To illustrate in a simple way, consider the equation

$$\frac{1}{Re}\frac{\partial^2 u}{\partial y^2} - u = -1$$

as $Re \to \infty$. The important feature is that the coefficient of the highest derivative tends to zero in this limit. Setting the first term identically to zero, it might seem that $u = 1$ is the limiting solution. But if $u = 0$ at $y = 0$ is the boundary condition, the full solution is

$$u = 1 - e^{-y\sqrt{Re}}$$

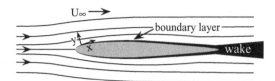

Figure 5.1. Schematic of boundary layer developing on a slender body.

for $y > 0$. At any given y, the inviscid solution $u = 1$ is the high Reynolds number limit, indeed. But for any finite Re, no matter how large, the solution departs from this if y is small enough. For example, the value of $u = 1/2$ is attained at $y = \log 2/\sqrt{Re}$. This y can be very small; nevertheless, half of the decrease from $u = 1$ to $u = 0$ occurs between this point and the wall at $y = 0$. Another way to say this is that a boundary layer exists, with thickness proportional to $1/\sqrt{Re}$.

The paradigm for boundary layer theory is as follows: the velocity is brought to zero across a thin layer. Viscous stresses remain paramount in that layer, even if the viscosity tends to zero. To accommodate this high Reynolds number limit, the thickness of the boundary layer decreases with increasing Reynolds number.

To bring these ideas to bear on fluid dynamics, we must consider a layer that extends spatially along the surface of a body. Away from the body, the high Reynolds number limit is inviscid flow, but the flow is fully viscous within the thin layer.

Figure 5.1 is a schematic of this structure. It represents the inner region by the black area, and the outer region by the streamlines. The boundary layer begins, with nearly zero thickness, at the leading edge and thickens downstream. It leaves the surface to form a vortical wake, trailing the body. It was seen in §1.8.3 that vorticity is produced at the wall. It diffuses out to produce the vortical layer. The new element of boundary layer theory is the assumption that the vortical layer is very thin.

There is only one flow field; the boundary layer does not exist as an independent entity. It is an identifiable portion of the flow field, but it does not have a clear-cut border, separating it from the rest of the domain. Even though it is not a sharply defined region, one can, and we will, define a thickness of the boundary layer. In that manner, the thin zone of high vorticity can be identified and its properties analyzed. If computational analysis is to succeed, the grid must adequately resolve this zone.

5.1 Formal Definition of the Boundary Layer

The boundary layer is a thin layer of vorticity that lies along a surface. Previously, in Chapter 4 (page 157), we have seen that a thin layer of vorticity diffuses a distance $\delta \propto \sqrt{\nu\ell/U_\infty}$ across the flow after traveling a distance ℓ downstream. The ratio of layer thickness to downstream distance,

$$\frac{\delta}{\ell} = \sqrt{\frac{\nu}{U_\infty\ell}} = Re_\ell^{-1/2}, \tag{5.1}$$

becomes small at high Reynolds number. This is the sense in which the boundary layer is thin.

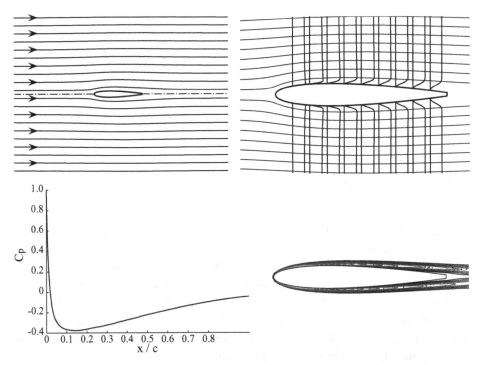

Figure 5.2. Boundary layer in flow over a streamlined surface. Streamlines and velocity profiles; pressure coefficient along surface; contours of vorticity.

To understand the concept of a boundary layer, the flow is divided, notionally, into two regions. The outer region consists of irrotational, potential flow (§5.6), indicated by the streamlines in Figure 5.1. At first, ignore the inner region, indicated by the black zone in the figure. The stream far from the wall moves freely over the surface. It is a uniform approach flow that is obstructed by the body and diverts around it. The streamlined velocity tends to a nonzero value as the surface is approached. The limiting velocity is denoted $U_\infty(x)$, where x is the coordinate locally tangent to the wall, y will be the normal direction. $U_\infty(x)$ is the velocity atop the boundary layer.

The no-slip condition is not met by the streamlined flow; no-slip is enforced by viscous friction. The velocity profile must be extended from U_∞ to 0. The inner zone is a viscous boundary layer that is inserted beneath the irrotational flow.

Within some distance, δ, above the surface, the flow is retarded by viscosity, ultimately being brought to rest at the surface. This is seen in the velocity profiles superposed at the upper right of Figure 5.2. A blow-up of the velocity in the vicinity of the wall would look like that shown in Figure 5.3. Actually, this figure is the Blasius velocity profile, which is the solution for the boundary layer equation on a flat plate. The vertical coordinate in the figure, η, is proportional to y/δ.

If the tangential velocity component is of the form $U(y/\delta)$, then the vorticity is $-\partial U/\partial y$, to a very good approximation. The velocity shear is of order

$$\frac{\partial U}{\partial y} \sim \frac{U_\infty}{\delta} \sim Re^{1/2}\frac{U_\infty}{\ell}.$$

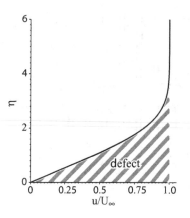

Figure 5.3. The Blasius boundary layer velocity profile is shown by the solid curve.

This is large because $Re \gg 1$. The boundary layer is a region of high velocity gradient because its vorticity is confined in a layer of thickness δ. For instance, consider flow over an airplane wing lying along the x axis. Within the thin boundary layer, the velocity gradient in the y direction will be $Re^{1/2}$ times larger than that in the x direction. As we are considering laminar flow, relevant values of Re might be as high as 10^5; hence, the gradient in the wall normal direction could be two orders of magnitude higher than in the wall tangent direction. A real airplane wing might have $Re \sim 10^8$. Then the boundary layer will be turbulent but still extremely thin.

The notion of inner and outer zones arises naturally in computations. Figure 5.2 shows a prototypical case. The outer flow is irrotational, the inner is vortical. A uniform flow approaching a slender body will remain uniform, except in the vicinity of the surface. The condition of impenetrability diverts the flow within a region that scales on the thickness of the body. Thereby the outer, potential flow is established. The upper-left pane of Figure 5.2 is a perspective of the potential flow. It is nearly uniform, with a small deviation round the airfoil.

Now zoom in to view the inner zone. Next to the surface, viscous stress brings the velocity to zero. At the top right of Figure 5.2, velocity profiles have been superposed on the streamlines. The region of shear near the surface is the boundary layer. The boundary layer region stands out more obviously in the plot of vorticity, at the lower right.

Streamlines divert around the body. In doing so, they spread apart near the stagnation point and converge toward the wall, farther along the rounded surface – see Figure 5.20 for a clear example. Where they spread, the velocity is lower than the incident stream. The velocity is brought to zero at the front stagnation point. Where the streamlines converge, the velocity accelerates; by Bernoulli's law (1.38), the pressure falls. Pressure is highest at the stagnation point, where kinetic energy has been converted into pressure, and it falls as the flow speeds up along the surface. The lower left of Figure 5.2 is a graph of the pressure coefficient versus distance along the airfoil. The pressure coefficient is defined by Eq. (1.39) on page 25. It starts from unity at the stagnation point and falls rapidly. This is the region of favorable pressure gradient. After $x/c = 0.15$ C_p climbs slowly; this is the region of adverse gradient.

In §1.8.3, Eq. (1.63) draws a connection between gradients of surface pressure and production of vorticity. The accelerating pressure gradient at the leading edge injects negative vorticity into the flow – negative because Eq. (1.63) requires that vorticity increases toward zero in the irrotational, free stream. The vortical layer is thin at the leading edge, and the pressure gradient is very large there. Vorticity is being pumped into the flow at a high rate. Recall the observation, made in §1.8.3, that the leading edge is the source of vorticity on a flat plate. With a rounded leading edge, the source is spread out, but still very large, at the leading edge, because $|\nabla p|$ is large.

The vortical region shown in Figure 5.2 thickens with downstream distance. Vorticity is diffusing from the surface, into the fluid. Boundary layer theory predicts that its thickness will be a factor of $Re^{-1/2}$ smaller than the airfoil chord. The vortical zone does not have a sharp border. However, its thickness can be estimated as about $5Re^{-1/2}$ – this is based on the 99% thickness of the Blasius profile; see exercise 5.4. Figure 5.2 was computed for a chord Reynolds number of 10^4. At the trailing edge, the vortical layer thickness should be about 5% of the chord.

However, that depends somewhat on the distribution of surface pressure. As the thickness of the surface decreases toward the rear of the airfoil, streamlines diverge and the flow slows down. The adverse pressure gradient causes the wall to become a source of positive vorticity. This counteracts the negative vorticity injected toward the front of the airfoil. If the positive source is sufficiently strong, it can bring the vorticity to zero near the surface. This is the criterion for separation, as will be seen in §5.1.2. It is not unusual for the trailing edge of an airfoil to be on the verge of separation, especially if the flow is laminar.

The reader might have noticed a previous remark that the vortical layer is thin at the leading edge, and wondered why it does not start from zero thickness. It would on an infinitely thin, flat plate. However, the rounded edge is a stagnation point. Vorticity diffuses from the surface, into the impinging flow, to create a finite thickness layer. A balance is reached in which the layer has a finite thickness, even at the leading edge. The stagnation point boundary layer is called the Hiemenz layer (Batchelor, 1967; White, 1991). If the potential flow velocity approaches zero as $u = -\alpha x$, where x distance from the stagnation point, then the thickness is about $2\sqrt{\nu/\alpha}$ from the classical, Hiemenz solution. That is, the solution for the idealized case of impingement onto an infinite, plane wall. If the leading edge is approximated by an ellipse of principle axes a and b, in the flow and transverse directions, then $\alpha = U_\infty(a+b)/b^2$; this can be derived from the potential flow solution given in Milne-Thomson (1968). To the extent that a is approximately the chord of the airfoil, c, and that b is approximately its thickness, t, the Hiemenz layer thickness is

$$2t\sqrt{\nu/U_\infty c} = 2t/Re^{1/2}.$$

In the present case, of a chord Reynolds number equal to 10^4, this is about 2% of the airfoil thickness. The boundary layer starts from this thickness on the nose and grows to a height of about 5% of the airfoil chord by the trailing edge.

The classical theory of the boundary layer invokes a systematic approximation to the Navier–Stokes equations that is easier to analyze than the full set of equations. In its more advanced form, boundary layer theory invokes an asymptotic framework, called the method of matched asymptotic expansions (VanDyke, 1975). For present purposes it is sufficient to summarize some results of boundary theory that can guide computational analysis based on the full Navier–Stokes equations.

5.1.1 Growth of the Boundary Layer

To be precise, a concrete definition of the boundary layer thickness is needed. Two that are common are its *displacement* thickness and its *momentum* thickness. Actually, these should be called the mass deficit and momentum deficit thicknesses: they measure the amount by which the mass, or momentum, carried in the boundary layer lies below that of a uniform flow.

The displacement thickness, δ_* is the area denoted by striping in Figure 5.3. This is

$$\delta_* = \int 1 - \frac{u(y)}{U_\infty} dy. \tag{5.2a}$$

For the particular curve in Figure 5.3, the integral turns out to be

$$\delta_* = 1.72\sqrt{vx/U_\infty}. \tag{5.3}$$

The thickness grows as the square root of x. At a fixed position, c, which could be the midchord of an airfoil, the thickness normalized by c is proportional to one over the square root of Reynolds number

$$\frac{\delta_*(c/2)}{c} = 1.72 \times \sqrt{1/2} \left(\frac{U_\infty c}{v} \right)^{-1/2}.$$

A nice way to look at the definition of displacement thickness is to recognize that it is the first moment of the vorticity:

$$\delta_* = -\frac{1}{U_\infty} \int y\omega dy. \tag{5.2b}$$

The integrand in Eq. (5.2b) is nonzero only in the boundary layer because the vorticity is confined to that region. Recall that $\omega = -\partial u/\partial y$ in the boundary layer. After noting that $\omega = \partial(U_\infty - u)/\partial y$ because U_∞ is a function of x, integration by parts recovers Eq. (5.2a).

The crosshatching in Figure 5.3 indicates the amount by which the velocity falls beneath U_∞. Hence, the displacement thickness measures the deficit in volume flow within the boundary layer. If the displacement thickness grows with x, it can be inferred that fluid flows out of the top of the boundary layer: the increase of δ_* with downstream distance means that the volume flow within the boundary decreases. Mass conservation then requires a flow out of the boundary layer equal to the decrease in flow carried inside the boundary layer. This is called the displacement effect. Physically, it is a statement that the oncoming flow rides up over the boundary layer. In a case like Figure 5.1, the streamlines would not follow the surface; rather, they would follow the surface plus boundary layer. The effective surface is the actual surface, shifted up to $y = y_{\text{surface}}(x) + \delta_*(x)$.

The second measure of boundary layer thickness is the momentum deficit, Θ. This is defined as

$$\int [1 - u(y)/U_\infty]u(y)/U_\infty dy. \tag{5.3a}$$

It can be written in terms of vorticity as

$$\Theta = -\frac{1}{U_\infty} \int \left[2\frac{u(y)}{U_\infty} - 1\right] y\omega dy. \tag{5.3b}$$

Again, phrased in terms of vorticity, the integrand of Eq. (5.3b) vanishes externally to the boundary layer. Of course, this assumes that the outer part of the flow is irrotational. For the Blasius profile (Figure 5.3), this integral has the value

$$\Theta = 0.664\sqrt{\nu x/U_\infty}. \tag{5.4}$$

We have been intentionally vague about the limits of integration in Eqs. (5.2b) and (5.3b). They are usually denoted as 0 to ∞, with the understanding that $y \to \infty$ refers to a value above the boundary layer but just so. It does not refer to the top of the domain. The range of integration can be thought of as the black zone in Figure 5.1. When written in terms of vorticity, the integrals can also be understood to range over the region of rotational flow ($\omega \neq 0$) next to the surface.

$\rho U_\infty^2 \Theta$ measures the amount by which the momentum flux lies below that of a uniform stream. It is the momentum flux deficit. The momentum deficit grows in consequence of viscous drag at the wall and increases or decreases in consequence of acceleration by pressure gradients. In the absence of the latter, the momentum budget is simply

$$\rho U_\infty^2 \frac{d\Theta}{dx} = \tau_w, \tag{5.5}$$

where τ_w is the drag on the wall. A retarding, frictional force removes momentum from the flow and increases the deficit thickness, Θ. For instance, the net drag, per unit width, on a flat plate of length L is the integral of the wall stress

$$\int_0^L \tau_w dx = \rho U_\infty^2 [\Theta(L) - \Theta(0)].$$

This expresses a balance between the net frictional drag on the plate and the reduction in the momentum carried by the flow. In the case of Eq. (5.4), the drag is $0.664\rho\sqrt{U_\infty^3 \nu L}$, by the above formula.

In the boundary layer $\tau_w = \mu \partial_y u$ at $y = 0$. For the Blasius boundary layer, Eq. (5.4), $d\Theta/dx = 0.332\sqrt{\nu/xU_\infty}$; hence, the nondimensional wall stress is

$$C_f \equiv \frac{\tau_w}{1/2\rho U_\infty^2} = \frac{0.664}{Re_x^{1/2}}, \tag{5.6}$$

where $Re_x = U_\infty x/\nu$.

In the presence of a pressure gradient, it is found that the deficit of momentum flux grows according to

$$\frac{d(\rho U_\infty^2 \Theta)}{dx} = \tau_w + \delta_* \frac{dP_w}{dx}. \tag{5.7}$$

A retarding pressure gradient has $dP_w/dx > 0$. It increases the deficit.

Equation (5.7) invokes the assumption that the pressure gradient is not a function of y. The pressure is felt across the whole boundary layer because it is very thin. In particular, the free-stream pressure is felt at the wall; that is, $P_w = P_\infty$. The latter is assumed to be given by Bernoulli's equation

$$P_\infty = P_0 - 1/2\rho U_\infty^2.$$

Again, referring to Figure 5.1, P_0 is the stagnation pressure at the leading edge of the airfoil. Bernoulli's equation allows Eq. (5.7) also to be written as

$$\frac{d\Theta}{dx} = 1/2 C_f - \frac{\delta_* + 2\Theta}{U_\infty} \frac{dU_\infty}{dx}. \tag{5.8}$$

The pressure gradient

$$\frac{dP_w}{dx} = -U_\infty \frac{dU_\infty}{dx} \tag{5.9}$$

has been used.

To illustrate these formulas, we will follow the time-honored practice of evaluating them with a simple shape assumption. Adopt the velocity profile

$$u = U_\infty \quad y > h,$$

$$u = U_\infty \sin(\pi y/2h) \quad y \leq h.$$

Then the integrals (5.2b or a) and (5.3b or a) are readily evaluated to find

$$\delta_* = h - \frac{2h}{\pi} = 0.363h; \quad \Theta = \frac{2h}{\pi} - \frac{h}{2} = 0.137h.$$

The ratio $\delta_*/\Theta = 2.65$ is surprisingly close to the value 2.59 that is obtained by integrating the exact Blasius profile, shown in Figure 5.2. h is analogous to the 99% boundary layer thickness, which is defined as the value of y at which $u = 0.99U_\infty$. For the Blasius profile

$$\Theta = 0.134\delta_{99}.$$

This is remarkably close to the value $\Theta = 0.137h$ cited above.

This example can be taken further by invoking (5.7) for the case of zero pressure gradient. The wall stress is $\mu du/dy|_0 = \mu\pi U_\infty/2h$. Then

$$\frac{d\Theta}{dx} = \frac{\nu\pi}{2hU_\infty} = \frac{\nu}{\Theta U_\infty}\left(1 - \frac{\pi}{4}\right).$$

Integrating this for Θ gives

$$\Theta = \sqrt{\frac{\nu x}{U_\infty}}\sqrt{2 - \frac{\pi}{2}} = 0.655\sqrt{\frac{\nu x}{U_\infty}}. \tag{5.10}$$

The exact Blasius profile gives 0.664 for the constant. The closeness of the numerical values is coincidental.

This exercise illustrates that momentum thickness increases in consequence of skin friction. It also illustrates that momentum thickness is smaller than displacement thickness. That is an obvious consequence of the definitions (5.2a) and (5.3a) because the velocity inside the boundary layer decreases from its value in the free stream to zero at the wall: hence $u/U_\infty < 1$.

5.1.2 Pressure Gradients and Separation

Equation (5.7) provides a qualitative understanding of how the boundary layer grows. In an accelerating stream the pressure falls. Then $dP_w/dx < 0$ implies that the boundary layer will grow less rapidly than under a constant stream. Conversely, in an adverse pressure gradient, $dP_w/dx > 0$, the decelerating stream will lead to an increased growth of the boundary layer.

Those inferences are not quite as obvious as has just been implied. The pressure gradient will alter the surface stress. In an accelerating flow C_f increases; however, not by enough to counter the reduction in boundary layer growth rate caused by the direct effect of pressure. Similarly, an adverse pressure gradient will cause C_f to fall, but the growth still is increased by adverse pressure gradients. The correctness of these assertions can be verified by a rough analysis.

Assume that wall shear (du/dy) is proportional to U_∞/Θ, so that the wall *stress* is proportional to that times viscosity – say, $\tau_w = \alpha \mu U_\infty/\Theta$. α is a constant of proportionality. Then Eq. (5.7) is estimated as

$$\frac{d\Theta}{dx} = \frac{\alpha \nu}{\Theta U_\infty} - \frac{(\delta_* + 2\Theta)}{U_\infty} \frac{dU_\infty}{dx}. \tag{5.11}$$

In the zero pressure gradient case, $dU_\infty/dx = 0$, integrating this recovers $\Theta \propto \sqrt{x\nu/U_\infty}$, as illustrated by Eq. (5.10). If U_∞ obeys a power law, $U_\infty \propto (x + x_0)^\beta$ it follows from Eq. (5.11) that $\Theta \propto (x + x_0)^{(1-\beta)/2}$. Hence, in accelerating flow, for which $\beta > 0$, Θ will grow more slowly than in decelerating flow, where $\beta < 0$. In the literature on boundary layer theory, this case where U_∞ follows a power law is associated with the name Faulkner–Skan.

Some profiles of velocity versus distance from the wall, representative of zero, adverse, and favorable pressure gradients (ZPG, APG, FPG) are plotted in Figure 5.4. The favorable gradient increases du/dy at the wall and hence raises the skin friction. Conversely, a decelerating pressure gradient thickens the boundary layer and decreases the wall shear. The shear increases initially above the wall but falls toward zero at the top of the profile. It follows that the APG profile must have an *inflection point*, where du/dy reaches a maximum. The leftmost profile in Figure 5.4 is initially concave downward, at small η. Above some point it becomes becomes convex upward. That point is the point of inflection. An inflection in the velocity profile is a destabilizing factor, according to theory (see Chapter 6). Thus, the APG profile has a tendency to develop unsteadiness and to transition to a turbulent state.

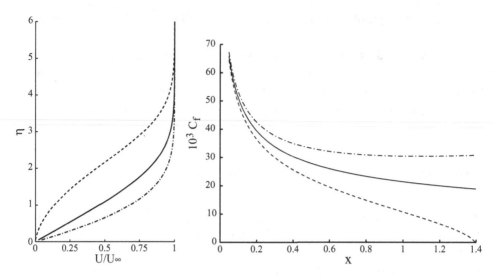

Figure 5.4. The effect of pressure gradient on the boundary layer velocity profile and friction coefficient. (———) Zero; (– – – –) adverse; (—·—) favorable.

The right side of Figure 5.4 illustrates the downstream evolution of friction coefficient under zero, adverse, and favorable pressure gradients. C_f in favorable pressure gradient is above that in zero pressure gradient; in adverse pressure gradient, it is below. The APG case is continued to the point at which C_f vanishes. The reduction of C_f in a decelerating pressure gradient is accompanied by growth in boundary layer thickness. The extent of the vortical layer increases until it breaks away from the surface at the point of separation, where C_f goes to 0.

That the point of separation is where $C_f = 0$ can be understood by reference to Figure 4.1. It is clear from that schematic that $du/dy > 0$ upstream of separation. Downstream of separation, the flow is reversed, $du/dy < 0$. Near to the surface, the flow is in the positive x direction before separation and the negative direction after separation. Where the positive and negative directed streams meet, they turn and flow away from the wall. Hence, the flow separates from the surface where $du/dy = 0$ at $y = 0$.

If the flow separates, finer grid resolution is needed to capture the vortical layer. If the mesh is too coarse, the layer will be spread by numerical diffusion. The detached shear layer would then be artificially thickened by the numerical error.

A separation can be from a smooth wall or at a salient edge. For instance, the flow off a step separates at the edge of the step, Figure 1.9 being an example. Then the location is known and the mesh can be refined in that vicinity. But when the separation is from a smooth, curved surface, the location is not known in advance. For instance, the separation from a circular cylinder in laminar flow is at about 80° from the front stagnation point; it does not reach the top (90°) point. In Figure 4.9, for $Re = 50$ the flow clearly leaves the surface, but the precise location of separation is not evident. Its location is where the tangential skin friction reverses direction, which is about 80° from the stagnation point on the left side of the cylinder.

The separating layer forces the oncoming flow away from the surface; it rides up, over the detaching layer. It was previously mentioned in connection with Figure 5.1 that the external stream is displaced by the attached boundary layer. This is referred to as an inviscid–viscous interaction. When the flow is attached it is a weak interaction, with the outer flow streamlines being displaced slightly away from the body. When the flow separates the interaction is strong; the outer flow is severely disturbed by the separating shear layer. That is a challenging case to compute because the flow does not follow the surface. Accurate gridding requires some insight into the flow – possibly the grid can be adapted to the flow as the computation proceeds.

Some element of an analysis of separation is constructed by considering the velocity very near the surface. To a first approximation, at a small height y above the wall, u equals $y\partial_y u(0) = y\tau_w/\mu$. The mass flux (per unit width) at a height $h(x)$ is

$$\dot{m} = \int_0^h \rho u \, dy \approx 1/2 h^2 \tau_w / \nu. \tag{5.12}$$

Then the thickness, h, of a constant mass flux streamline is proportional to $\sqrt{\dot{m}/\tau_w}$. As τ_w decreases in an adverse pressure gradient, the streamline moves away from the wall. Eventually, as $\tau_w \to 0$ the streamline erupts, and leaves the near-wall vicinity.

A second approximation uses the fact that at a wall, the momentum balance is $\mu \partial^2 u/\partial y^2 = dp_w/dx$. Then integrating the two-term Taylor series expansion

$$u \approx \frac{\partial u(0)}{\partial y} y + 1/2 \frac{\partial^2 u(0)}{\partial y^2} y^2$$

gives

$$\dot{m} = \int_0^h \rho u \, dy = 1/2 h^2 \left(\frac{\tau_w}{\nu} + \frac{h}{3\nu} \frac{dp_w}{dx} \right). \tag{5.13}$$

From this, the slope at separation can be evaluated. The separation streamline in Figure 4.1 is the line of zero mass flux. From Eq. (5.13) this is the line

$$\tau_w + \frac{h}{3} \frac{dp_w}{dx} = 0.$$

If the separation point is called x_s, then near separation the wall stress passes linearly through zero: $\tau_w = (x - x_s) d\tau_w/dx$. The slope of the streamline is $h/(x - x_s)$:

$$\frac{dh}{dx} = -3 \frac{d\tau_w/dx}{dp_w/dx}. \tag{5.14}$$

On approaching separation from the left, τ_w will be decreasing, so the numerator of this expression is negative. Thus for h to increase dp/dx must be positive: separation cannot take place in a favorable pressure gradient. The angle of separation depends on the local flow behavior. In Eq. (5.14), the gradients of pressure and skin friction set the separation angle.

Separation might involve complete detachment of the flow, as occurs from blunt afterbodies or from airfoils at high angle of attack. This is termed massive separation.

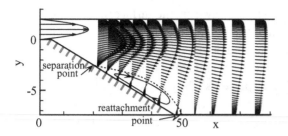

Figure 5.5. Separation from the wall of an asymmetric plane diffuser. Note the large aspect ratio of the figure; the y direction was expanded by a factor of 3.5 for display.

But the shear layer may return to the surface, after a small excursion, as in Figure 5.5. The zone of recirculating flow is called a separation bubble. A slender bubble can be called a marginal separation.

The y axis in Figure 5.5 has been magnified for clarity of display; the ramp angle is about $10°$. This computation was done on a simple grid, with H-topology (§2.1.3). At the inlet, a fully developed channel flow profile was prescribed. The Reynolds number based on inlet height was 2×10^4. That is high enough for the flow to be fully turbulent. A closure model (Chapter 6) was solved to account for turbulent mixing of mean momentum. A laminar flow would not be stable at this ramp angle and Reynolds number.

In some applications, a marginal separation might not be deleterious. For instance, aerodynamic properties of an airfoil with a small separation bubble may be satisfactory. A massive separation causes a large loss of lift and certainly is not satisfactory.

5.1.3 Gridding

Were the computational domain filled with a uniform mesh, the number of cells would be excessive. Suppose 10 cells are needed to resolve the boundary layer. In two dimensions their size would be $\delta/10 \times \delta/10$. If the body scale is c and the domain extends to $10c$ in each direction, the number of cells will be of order $100c/\delta$. This scales as $100\sqrt{Re}$ giving

$$N \sim 10^4 Re$$

for the size of a two-dimensional mesh. At a modest Reynolds number of 10^3, the number of cells is of order 10^7. That is two or three orders of magnitude larger than necessary.

A more satisfactory strategy is to invoke the two region structure of boundary layer theory. The outer, potential flow requires a resolution comparable to that needed to resolve the geometry. If a cell dimension of $c/30$ is needed to capture the surface and the domain is $10c$ in length, about $300 \times 300 \sim 10^5$ cells will capture the potential flow.

If the mesh is oriented tangentially to the walls, the vortical layer next to the surface can be resolved if the cell width is $\delta/10$ in the direction normal to the surface and $c/30$ in the tangential direction. Ten cells are needed across the layer. If the layer

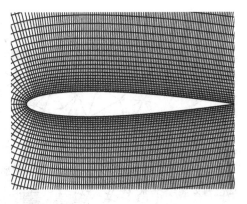

Figure 5.6. Grid for flow round the airfoil.

extends across the whole domain, 300 cells are needed in the other direction, giving a total of 3,000 boundary layer cells. This is small compared to the number in the potential flow region. The whole computation can be done with $N \sim 10^5$ – much less than naively filling the domain with a uniform mesh.

There are two problems with this strategy. First, the cell size jumps discontinuously from $\delta/10$ to $c/30$ at the top of the boundary layer. The size ratio is $c/3\delta \sim 1/3\sqrt{Re}$. This is typically larger than 10. Such a jump in cell size produces a very poor quality mesh (see §2.1.5). The second difficulty is the potentially high aspect ratio of the boundary layer cells. The aspect ratio is also $1/3\sqrt{Re}$. At high, laminar Reynolds numbers, around 10^5, the aspect ratio can exceed 100. A good algorithm can handle this. However, high aspect ratio cells can cause stiffness. The computation may converge quite slowly and multigrid acceleration may become inefficient. This is especially likely when the high aspect ratio cells are distant from the surface, for instance, in the wake.

The problems can be alleviated by smoothly expanding the cell size. Suppose that the wall normal dimension is to expand from $\delta/10$ to $c/3$ in 10 steps. At each step let the size expand by a factor of r. At the nth step, the size is

$$\frac{\delta}{10}r^{n-1}.$$

After $n = 10$ expansions

$$\frac{\delta}{10}r^9 = \frac{c}{3}.$$

Thus $r = (10c/3\delta)^{1/9} \sim Re^{1/18}$. At $Re = 10^4$ the expansion factor is about 1.7. The cell size now grows smoothly. High aspect ratio cells still occur near walls, but an expanding mesh will reduce the aspect ratio away from the surface. The flow round the airfoil in Figure 5.3 was computed on the expanding C-grid of Figure 5.6.

For unstructured meshes, the region near the wall in which the cell size expands is usually filled with prisms. The outer, potential flow is commonly meshed with tetrahedrons. A bump is meshed in this way in Figure 5.7. The prismatic, boundary layer mesh is more regular than the outer region, tetrahedral mesh. Its cells follow

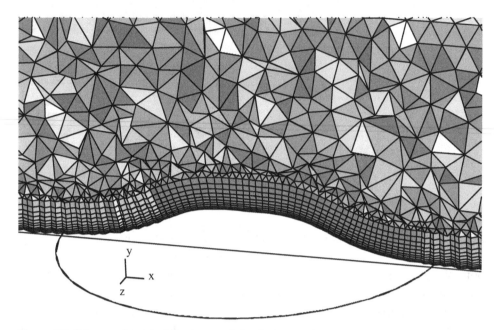

Figure 5.7. Prisms and tetrahedrons in a mesh for flow over a bump.

the surface, being of fairly high aspect ratio next to the wall. Their size expands, making a smooth transition to the lower aspect ratio mesh in the outer region.

The base of the prism layer is defined by a surface triangulation. The mesh of Figure 5.7 was grown from the surface portrayed in Figure 5.8. The volume fills with cells by propagation from the underlying surface mesh. The region to be filled with prisms is defined by inflating the surface. Various inflated surface techniques have been developed for generating boundary layer grids. All are designed to produce near surface cells that follow the geometry and are long and thin to capture the steep flow gradients normal to the surface.

The form of a prismatic wall layer grid is shown more clearly by Figure 5.9. The surface is a sphere. The base of the lowest prism is a triangle in the discrete,

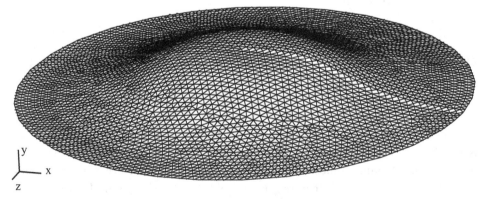

Figure 5.8. Triangulated surface over which prisms grow.

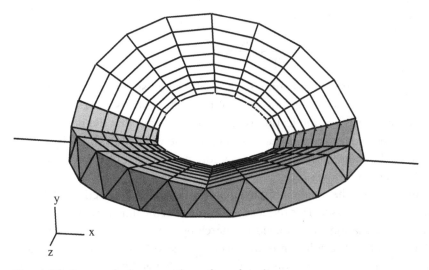

Figure 5.9. Layers of prisms near the surface of a sphere.

triangulated representation of the sphere. Only the near-wall portion of the volume mesh is shown. The prisms transition to tetrahedrons in a larger region than that of the figure. The orientation of the prisms is evident in this case. Actually, this grid was designed to capture a spherically expanding domain, rather than a boundary layer; but it nicely exemplifies the manner of prismatic near-wall meshing.

5.2 Approximate Equations

The formal, mathematical boundary layer theory invokes systematic approximation to the Navier–Stokes equations. The approximations stem from the thinness of the boundary layer and the nearly tangential direction of the flow. In consequence of the tangentiality, the pressure is approximately independent of the y direction. That is a bit like the parallel flow approximation of §1.5. In consequence of the thinness, gradients of a given quantity are much larger in the y than in the x direction. In particular, viscous diffusion in the streamwise direction can be ignored relative to that in the wall normal direction.

The y-momentum equation becomes simply

$$\frac{\partial p}{\partial y} = 0. \tag{5.15a}$$

The pressure is equal to $p_\infty(x)$ at all y, through out the layer. Conversely, the free-stream pressure equals the wall pressure.

The x-momentum equation is changed only by neglecting streamwise diffusion:

$$\frac{\partial u}{\partial t} + u\frac{\partial u}{\partial x} + v\frac{\partial u}{\partial y} = -\frac{1}{\rho}\frac{\partial p}{\partial x} + v\frac{\partial^2 u}{\partial y^2}. \tag{5.15b}$$

The v velocity is found from continuity,

$$\frac{\partial u}{\partial x} + \frac{\partial v}{\partial y} = 0. \tag{5.15c}$$

The Blasius profile, portrayed in Figure 5.2, is a self-similar solution to Eqs. (5.15b) and (5.15c) with $dp/dx = 0$ (Batchelor, 1967; White, 1991; exercise 5.4).

The most informative simplification is Eq. (5.15a). It says that the pressure felt through out the boundary layer is that imposed by the outer flow; it is given, not solved inside the boundary layer. Pressure is a forcing function. It is obtained from the external, potential flow via Eq. (5.9).

The remainder of the x-momentum equation describes convection with the stream and diffusion across it. This reduced description of the fluid mechanics allows the u-momentum equation to be solved by marching downstream. Starting with a given profile $u(y)$ at x_0, the solution is advanced successively to $x_0 + \Delta x$, $x_0 + 2\Delta x$, and so forth. At each step momentum diffuses and is changed by the applied pressure gradient. Equations that can be solved by the marching method are called *parabolic*. In the simplest case, the boundary layer thickness grows as $\delta \sim \sqrt{x}$. The curve $y(x) = \delta$, describing the height of the boundary layer, is the parabola $x \propto y^2$.

The full Navier–Stokes equations are elliptic. They cannot be solved by marching. What happens downstream feeds back upstream via the pressure (in subsonic, flow). Indeed, in boundary layer theory, ellipticity is present but only in the *outer* region. There the equations must be solved simultaneously at all points throughout the domain. However, there is a simplification: it is that viscosity can be neglected in the outer region. Without viscosity, vorticity does not diffuse into that region. In most applications of boundary layer theory, the outer region is irrotational – which is a great simplification.

In a full CFD simulation these simplified equations are not used; the complete Navier–Stokes equations are solved without approximation. Boundary layer theory describes aspects of the solution. Not only that, the zones identified in the asymptotic theory guide grid generation. Substantial velocity changes occur across a layer of thickness δ. Fine grids are needed to resolve them. With structured meshing, long, thin cells are needed inside the boundary layer. With unstructured meshing, it is common to use layers of prism shaped cells near the wall – as was described in §5.1.3.

Equation (5.7) is a ready consequence of the boundary layer equation (5.15b). Exercise 5.1 asks the reader to reformulate the convection term. That reformulation, combined with Bernoulli's equation for the free-stream pressure allows Eq. (5.6) to be written as

$$\frac{\partial u(u - U_\infty)}{\partial x} + \frac{\partial v(u - U_\infty)}{\partial y} + (u - U_\infty)\frac{\partial U_\infty}{\partial x} = v\frac{\partial^2 u}{\partial y^2}.$$

Now integrate each term with respect to y from the wall, $y = 0$, to the free stream, $y \to \infty$. The first is $-\partial_x(U_\infty^2 \Theta)$; the second vanishes; the third is $-\delta_* U_\infty \partial_x U_\infty/dx$; and the right side contributes $-\tau_w$. The reader may confirm that these results agregate to give Eq. (5.6).

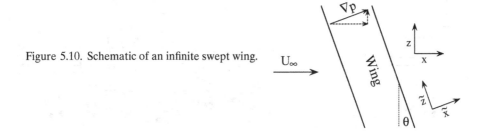

Figure 5.10. Schematic of an infinite swept wing.

5.3 Three-Dimensional Boundary Layers

The geometry in Figure 5.3 is a one-dimensional curve. The upper and lower surfaces of the airfoil are defined by a curves of the form $y(x)$. The flow field is two dimensional: $u(x, y)$, $v(x, y)$. If the surface were two dimensional, $y(x, z)$, then the flow field would be three dimensional. The boundary layer on a geometry like Figure 5.8 is three dimensional. We have already seen in Chapter 4 that quite complex flow patterns can be produced by flow over shapes as simple as that of Figure 5.8.

Let us step back, and consider the simplest of three-dimensional boundary layers. The wing in Figure 5.3 can be imagined to be the cross section of a cylinder that extends indefinitely out of and into the page. It lies on the z axis. Now rotate this cylinder by an angle of θ about the y axis. Leave the incident flow unchanged. In plan, we have Figure 5.10. This is called the infinite swept wing. It provides an understanding of a primary feature of three-dimensional boundary layers: the cross-flow velocity. Within the boundary layer, the flow direction in the xz plane differs from that above the boundary layer. The difference between the flow direction at any height and that of the external stream is the cross flow.

The origin of cross flow can be understood in two ways. These correspond to the two coordinate systems, x and $\tilde{x}$, of Figure 5.10. A pressure gradient is generated as the incident flow encounters the wing. The component of velocity normal to the wing stagnates on the surface and is diverted above and below it. By Bernoulli's equation, a pressure gradient is produced in the ambient stream. Within the thin shear layer approximation, that pressure is felt through the boundary layer. This is a pressure of the form $p(\tilde{x})$.

The component of incident velocity that is parallel to the direction of independence, $\tilde{z}$, is unaltered in the free stream. Hence, no pressure gradient along the wing is caused; the pressure gradient is perpendicular to the axis of the wing. Denote it by $dp_\infty/d\tilde{x}$. Then its components in the xz coordinate system are

$$\frac{\partial p}{\partial x} = \frac{dp_\infty}{d\tilde{x}} \cos\theta, \quad \frac{\partial p}{\partial z} = \frac{dp_\infty}{d\tilde{x}} \sin\theta.$$

The x- and z-momentum equations in the thin layer approximation are

$$\frac{Du}{Dt} = -\frac{1}{\rho}\frac{\partial p}{\partial x} + v\frac{\partial^2 u}{\partial y^2},$$

$$\frac{Dw}{Dt} = -\frac{1}{\rho}\frac{\partial p}{\partial z} + v\frac{\partial^2 w}{\partial y^2}.$$

(5.16)

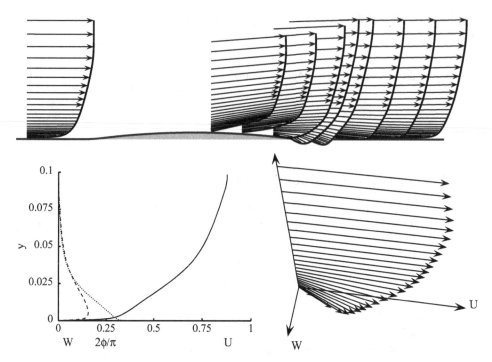

Figure 5.11. Three-dimensional boundary layer in flow over a swept bump. Main (———) and cross flow (– – – –). The flow angle is shown by a dotted line. The spiraling direction of the flow is portrayed at the right.

Each component feels a different pressure gradient. They are also subjected to different boundary conditions. At the surface, $u = w = 0$. If the body is slender, then the free-stream condition is $u \to U_\infty$ and $w \to 0$. Thus, the ratio w/u will vary with height above the surface. Define the flow angle by $\phi(y) = \tan^{-1}(w/u)$. It is zero in the free stream, but nonzero near the surface.

An example of primary velocity, cross-flow velocity, and flow angle is provided in Figure 5.11. These are computed profiles for a turbulent boundary layer flowing over an infinite swept bump. The bump can be imagined to extend into and out of the page at an angle of 45° to the plane of the page. A cross section through the geometry and a few flow profiles are displayed at the top of the figure.

One of the skewed profiles is replotted at the bottom of the figure next to profiles of u, w, and the flow angle. The velocity vector rotates across the boundary layer, away from the ambient direction. The flow angle – the dotted line in the graph at the lower left of Figure 5.11 – increases from 0° at $y = 0.1$ to about 30° at the surface. Hence, the surface shear stress is not colinear with the free-stream velocity.

The cause of the flow direction skewing with height is more readily seen by working in the $\tilde{x}$ system of coordinates. The boundary layer equations become

$$\frac{D\tilde{u}}{Dt} = -\frac{1}{\rho}\frac{\partial p}{\partial \tilde{x}} + \nu\frac{\partial^2 \tilde{u}}{\partial y^2},$$

$$\frac{D\tilde{w}}{Dt} = \nu\frac{\partial^2 \tilde{w}}{\partial y^2}. \tag{5.17}$$

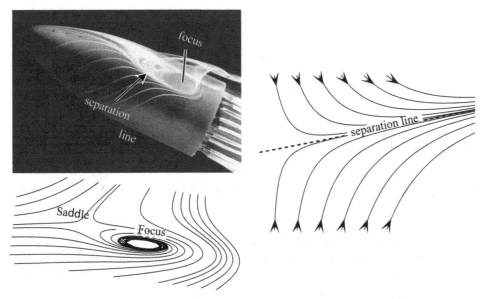

Figure 5.12. An example of three-dimensional separation, with a schematic of its singular points. Picture from Délery et al. (2001). At right, surface streamline convergence onto a separation line.

In the approach flow, $\widetilde{U}_\infty = U_\infty \cos\theta$ and $\widetilde{W}_\infty = -U_\infty \sin\theta$. The second of Eqs. (5.17) describes diffusion of the free-stream velocity into the boundary layer. This equation contains no pressure gradient; it is the same as the equation of heat diffusion between two fixed temperatures. $\tilde{w}$ diffuses from $\widetilde{W}_\infty$ at the top of the boundary layer to zero at the surface.

Cross flow is one feature that distinguishes three from two-dimensional boundary layers. They are also distinguished by characteristics of separation.

5.4 Separation in Three Dimensions and Critical Points

The observation that separation occurs at a point of vanishing skin friction is only true in two-dimensional steady flow. In unsteady flow, the skin friction can vanish at some instant, even though the vortical layer does not separate from the surface. The oscillatory boundary layer, discussed in §1.5.2, is a simple case in point: in every cycle the skin friction will pass through zero, but there is no separation. However, that is the less fascinating of the two caveats.

The more interesting exception occurs in three dimensions. Instead of requiring a point of zero skin friction, flow can erupt from the surface where surface streamlines converge. As they converge, their trajectories become tangent to a *curve* of separation. The nose cone at angle of attack, shown in Figure 5.12, provides a concrete illustration of three-dimensional separation. The boundary layer leaves the surface in a sheet rising from the separation line. In this case the sheet is wrapping into a cylindrical vortex as it leaves the surface. In the typical three-dimensional separation, the boundary layer leaves the surface along a curve. The surface skin

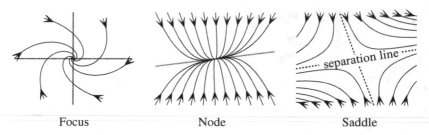

Focus   Node   Saddle

Figure 5.13. Types of singular points for three-dimensional separation.

friction vector is tangent to that curve, but τ_w does not vanish, except at isolated points.

Recall the argument leading to Eq. (5.12). The mass flux between the wall and a nearby height, h, defined a streamline. For the flow to remain attached, the line has to remain near the wall. If the flow is three dimensional, the mass flux must be defined for a stream tube. Let the tube's cross section be a rectangle of height h and width w. The width is appended to the result (5.12) to give

$$\dot{m} = 1/2h^2 w \tau_w / \nu. \tag{5.18}$$

The height of the constant mass flux stream tube is $h \propto 1/\sqrt{w\tau_w}$. Now there are two ways that the streamline can erupt from the wall: either $\tau_w \to 0$ or $w \to 0$. The latter is the new separation criterion that occurs in three dimensions. As $w \to 0$ surface streamlines converge onto a separation line. This is illustrated at the right of Figure 5.12. The flow toward the separation line is accompanied by flow out of the plane of the page.

Eruption in consequence of $w \to 0$ accounts for most of the separation line. However, $\tau_w \to 0$ can also occur. The isolated points where τ_w does vanish are called *critical points*. An analysis shows there to be three generic types of critical point: *saddle, focus,* and *node*. They are shown in Figure 5.13. In general, the singular points will be distorted versions of these forms. The point of zero stress is at the origin. The curves with arrows indicate the flow pattern in the vicinity of each singular point. These patterns define the type: they spiral into, or out of, a focus; emanate from a node; and sweep past a saddle, approaching it and then being repelled from it.

The three types are derived mathematically from the linear approximation $\tau = (x - x_s) \cdot \nabla \tau|_s$ to the surface stress near its zero at x_s. The classification is based on the eigenvalues of the matrix of derivatives, $\nabla \tau$: its eigenvalues are either complex, real, and same signed or real and oppositely signed for the three categories of Figure 5.13 (Panton, 1997).

The saddle point and focus are patterns that can be identified in visualizations of surface skin friction lines in CFD analysis. For instance, the separation line illustrated at the right of Figure 5.12 can be understood as the right half of the saddle singularity, illustrated in Figure 5.13. It is a principle of topology that the left half of the saddle point streamlines also must exist if the body is closed. The saddle point might be just in front of the circle surrounding the focus in the picture at the left of Figure 5.12. A

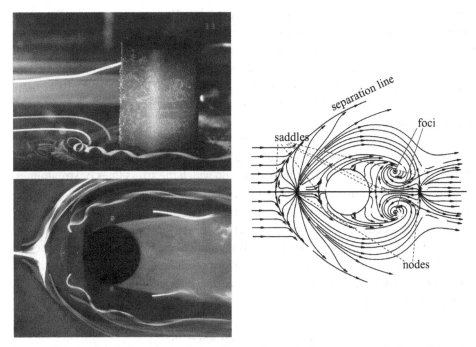

Figure 5.14. Singular points, separation, and reattachment in flow round a cylinder–wall junction; Délery et al. (2001).

sketch of the likely flow pattern is provided below the picture, with the saddle point labeled.

Nodal singularities (Figure 5.13) are associated with attachment, not separation. For instance, where the flow impinges, at the front of the blunt object in Figure 5.12, the streamlines would form a node, centered on the point of stagnation.

Saddle, focus, and node are the language for discussing three-dimensional separation and reattachment. A wonderful example of this is shown in Figure 5.14, from Délery et al. (2001). This a wind tunnel visualization of the flow round a cylinder attached to a wall. A horseshoe-shaped vortex wraps around the front of the cylinder, trailing behind it as a pair of counter-rotating vortices. This is seen in the lower-left photograph. The upper-left photograph shows how the flow spirals around the vortex, leaving the surface at its front, returning toward the surface at its rear. Hence, we expect to see separation and reattachment lines, as, indeed, the drawing of the surface streamline pattern shows. This experimental example is topologically equivalent to the computational example in Figure 4.19. The latter may help the reader to make sense of the flow patterns.

In the streamline sketch at the right of Figure 5.14, the approach flow separates from the wall on a curve emanating from a saddle point. That is the front of the horseshoe vortex. It reattaches to the surface on a curve emanating from the front nodal point. In fact, a second saddle and separation line can be seen, just next to the cylinder. The flow from that separation line spirals into two foci at the rear of the cylinder. These can just be seen in the lower-left photograph. They represent vertical vortices that probably form an arch in the lee of the cylinder. These surface streamlines and singular points give a basis for understanding the complex three-dimensional flow that produces them. It requires some extrapolation and some intuition for the vortex distributions and for the separating shear layers that would produce particular patterns on the surface. The photographic components of Figures 5.14 and 5.12 help in this regard. They give an idea of the flow features that produce the surface stress lines at the right.

Singular points can be used to characterize a surface algebraically. Nodes and foci are assigned an index of $+1$, saddles an index of -1. The sum of the indices of all singular points is a property of the surface on which the stress lines or streamlines are drawn. This gives rise to topological rules. On a plane surface, the indices of the singular points must sum to zero. Actually, that is true only if the stress lines become straight at the edges of the plane; in other words, if the flow at ∞ is a constant velocity. If a rule exists, then it must be that the indices sum to zero, because a set of straight lines, with no singularities, can be drawn on the plane surface. Hence, if the sum is a property of the surface, it must be zero. This can be stated as a constraint on all possible patterns that can occur:

$$N_{\text{nodes}} + N_{\text{foci}} - N_{\text{saddle}} = 0.$$

For instance, the flow past a vortex will contain a nodal point and a saddle point, like the line drawing at the lower left of Figure 5.12. The sum is 1 node -1 saddle $= 0$.

On a plane containing closed objects, the critical point indices sum to the negative of the number of objects. For instance, in Figure 5.14 they sum to -1 because there is one object on the plane. That figure contains five saddles, two foci, and two nodes. Alternatively, the circle could be contracted to a point so that it becomes a third node. Then the indices would sum to zero, corresponding to the plane with no objects on it.

On a closed, uninterrupted surface, the indices sum to $+2$. For instance, the lines of longitude on a sphere have two nodes, one at each pole. This topological rule applies to a body like the nose cone in Figure 5.12. However, it would be rather difficult to identify all the singular points in that case. Some of them are out of view, on the other side of the body. The idea of characterizing the three-dimensional separated flow in terms of its critical points and separation lines goes beyond the topological rules. Critical points can be identified by type, whether or not all of them can be located. Then one can cite saddle points of separation or nodal points of attachment to describe flow patterns.

5.5 High Reynolds Number Flow over a Spheroid

The flow over a prolate spheroid, at angle of incidence, provides an example of three-dimensional, separated flow. The flow field is shown in Figure 5.15. It is incident from below the spheroid. As the flow passes over the body, it separates from the surface, and vortex sheets leaving the separation lines roll into detached vortices. The particle

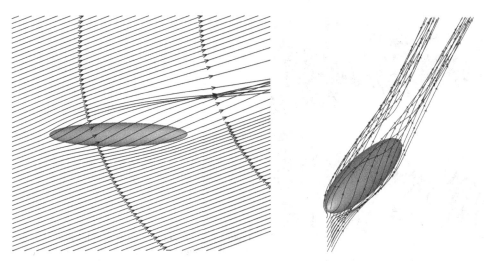

Figure 5.15. Flow round a spheroid at angle of incidence. Flow lines and vortices rolling up at three-dimensional separation line.

trajectories at the right of Figure 5.15 show the vortices separating, symmetrically, from the two sides of of the body.

Further understanding of the flow field is obtained by plotting the surface stress field. Trajectories along surface stress lines are displayed in Figure 5.16. They diverge

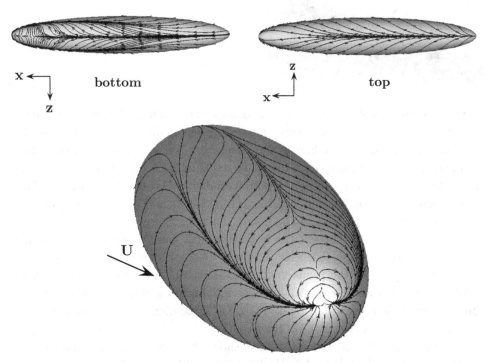

Figure 5.16. Surface stress lines converge onto a separation line. The surface viewed from above and from below. The lowest figure is a view from downstream, looking back toward the body.

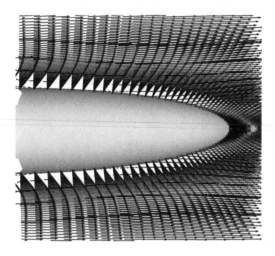

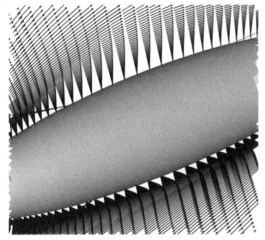

Figure 5.17. Boundary layer profiles near the surface of a spheroid. At top is a horizontal slice; at bottom a vertical slice through the spheroid.

from the symmetry plane and converge onto a separation line. The symmetry line is a line of attachment: the stress lines point outward from the symmetry plane. In the view of the bottom of the spheroid, the attachment line is the only feature visible. In the topside view of the spheroid, the flow out from the symmetry plane is seen to coalesce onto lines of separation on both lateral sides. Vorticity leaves the surface along these separation lines and rolls up into the detached vortices. The end view shows the pattern of attachment and separation quite clearly.

Zooming in on the surface, the boundary layer can be extracted by plotting velocity vectors in planes that slice the body. The upper pane of Figure 5.17 is a slice perpendicular to the symmetry plane. Boundary layer profiles are seen to evolve from a full form at the left to inflectional profiles characteristic of adverse pressure gradient flows at the right. These are three-dimensional velocity vectors that are turning out of the plane of the page.

A slice parallel to the symmetry plane shows the difference between the upper and lower sides of the spheroid. In the lower pane of Figure 5.17, the profiles below the body are full, representative of attached flow, whereas those on top are shallow and suggestive of a boundary layer approaching separation.

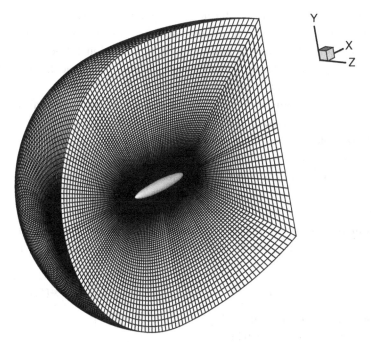

Figure 5.18. O-mesh for spheroid.

These computations are of laminar flow at a Reynolds number of 6,000, based on the axial length of the spheroid and incident velocity U_∞. The angle of attack was chosen to be 20°, with the intention to illustrate the three-dimensional boundary layers and flow separation at the back portion of the spheroid.

The 6:1 spheroid was placed in the center of the domain with a spheroidal outer boundary, truncated on the downstream end (see Figure 5.18). For this Reynolds number, the laminar flow remains steady and symmetric in the spanwise direction, thus the computation was performed on one half of the domain and reflected across the symmetry plane for display. The grid displayed in Figure 5.18 shows the entire computational domain. The planar face is the symmetry boundary.

The grid is structured, with O-topology in two directions. Figure 5.18 is a perspective view of the O-O-topology. This figure shows both the global structure and the high resolution in the vicinity of the spheroidal surface. The latter enables the computation to resolve fully the boundary layers. The grid contains 1,036,800 computational cells (1,045,029 nodes), with two-thirds of them located in the boundary layer. That division of resources is symptomatic of the expense of resolving surface shear layers in high Reynolds number, separating flow.

5.6 Potential Flow

The irrotational motion of fluids was studied extensively in classical literature, much of it written in the 19th century. The authors include some of the great physicists of that era: Kelvin, Rayleigh, Helmholtz, and Laplace among others. Potential theory is developed encyclopedically in Milne-Thomson (1968), thoroughly in Batchelor

(1967), and comprehensively in many other references. Although its role has receded in the era of computational fluid dynamics, it remains indispensable to aerodynamic theory. In the present chapter, potential flow applies to the outer region in boundary layer theory. Thus, its role is as an element of high Reynolds number fluid mechanics. It describes the portion of the flow that is effectively inviscid.

The term *potential flow* remains integral to the language of fluid dynamics. Irrotational flow theory has already been met in this text (in §1.8.4 and §4.1.1.3). It was cited on page 48 to explain the relation between circulation and lift. On page 137 irrotational flow round a circular cylinder was contrasted to real, viscous bluff body flow. In that case, a discrepancy between the two was seen primarily on the aft part of the cylinder.

Potential flow is irrotational flow; that is where the name originates. The terminology follows from the observation that irrotational means zero vorticity or $\nabla \wedge \boldsymbol{u} = 0$. If its curl vanishes, the velocity can be equated to the gradient of a potential:

$$\boldsymbol{u} = \nabla \phi. \tag{5.19}$$

The identity $\nabla \wedge (\nabla \phi) = 0$ ensures irrotationality.

If the fluid is incompressible, $\nabla \cdot \boldsymbol{u} = 0$ becomes

$$\nabla^2 \phi = 0. \tag{5.20}$$

This is Laplace's equation of classical physics. Solving it, subject to boundary conditions, provides the velocity field via Eq. (5.19). Note that there is no role for the Navier–Stokes equations; they are preempted by the assertion of irrotationality. If the Navier–Stokes equations are regarded as the law of vorticity evolution, they have been eliminated by asserting that the vorticity is zero.

Powerful methods exist for solving Laplace's equation (5.20). They can be found in Milne-Thomson (1968) and in many texts on classical mathematical physics. The approach most conducive to the present objectives is inverse solution via sources and sinks. That is not the most elegant mathematically, but it is a very physical approach. Hence, it serves as a conceptual tool with which to reason about potential flow fields.

5.6.1 Point Sources

The flow approaching a blunt-nosed body diverts over and under the body. In two dimensions, a stagnation point and stagnation streamline divide the flow into that passing over and that passing under. This can be described by a simple analytical and conceptual model: a point source in a uniform approach flow.

The point source injects fluid into the domain. It flows radially outward from the source, with a uniform angular distribution. Consider a sphere around a source in three dimensions or a circle around a source in two dimensions. Let M be the volumetric, or area, flow rate in three or two dimensions. The normal velocity integrated over any surface that encloses the source must equal M. Then the radial velocity at

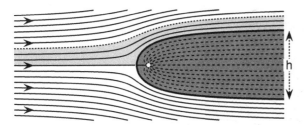

Figure 5.19. Streamlines in flow about a point source. The dark shaded region represents a blunt body; the lightly shaded region represents a curved ramp.

the sphere or circle must satisfy $4\pi r^2 u_r = M$ or $2\pi r u_r = M$ to conserve mass. Hence, the radial velocity produced by point sources is

$$u_r = \frac{M}{4\pi r^2} \quad \text{or} \quad u_r = \frac{M}{2\pi r} \tag{5.21}$$

in three and two dimensions. The radial direction is x/r or, in component form, $(x, y, z)/r$. Hence, the Cartesian velocity vectors are

$$(u, v, w) = \frac{(x, y, z)M}{4\pi r^3} \quad \text{and} \quad (u, v) = \frac{(x, y)M}{2\pi r^2}. \tag{5.22}$$

Now add a uniform velocity $(U_\infty, 0)$ to the two-dimensional source. Because the potential flow equation (5.20) is linear, the velocities simply add:

$$(u, v) = (U_\infty, 0) + \frac{M}{2\pi r^2}(x, y). \tag{5.23}$$

The corresponding streamlines are shown in Figure 5.19. A stagnation point exists where $(u, v) = (0, 0)$ or where

$$(x, y) = \left(\frac{-M}{2\pi U_\infty}, 0 \right).$$

Far downstream, as $r \to \infty$ in formula (5.23), the flow returns to the uniform velocity $(U_\infty, 0)$. The injected mass occupies a region of height h; hence, the area flux $U_\infty h$ equals the rate of injection M.

Thus, it is seen that the flow field is the same as that past a flat slab with a rounded nose (Figure 5.19). The equivalent mass flux is found from the slab thickness to be $M = U_\infty h$.

We have modeled the flow past a blunt body by an equivalent flow past a point source. The flow field in Figure 5.19 is simply the vector sum of a uniform velocity and a radial outflow from the origin. The sum is stated algebraically as Eq. (5.23). Let us consider how this vector sum leads to the flow field.

The radial outflow decreases in magnitude as $1/r$. Very near the origin, the second term of Eq. (5.23) is dominant, and the streamlines are radial. That region lies in the darkened area of Figure 5.19, which is inside the body. Far from the origin the first term of Eq. (5.22) dominates and the flow is uniform with straight streamlines – for instance, in the flow approaching from the left. At intermediate distances, a vector sum gives the flow direction. Along the negative x axis the flow

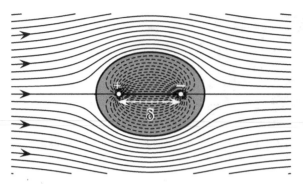

Figure 5.20. Streamlines in flow about a Rankine body.

from the source is in the negative x direction. The incident flow is in the positive
x direction. The stagnation point is where they sum to zero. Upstream of that U_∞
dominates and the flow is toward the right.

Above the stagnation point the source velocity vector is angled back and up. The
upward component is not canceled by the incident velocity, so the fluid has a positive
v velocity. Below the stagnation point, the source vector points down and back.
Adding the horizontal incident vector does not cancel the downward component of
velocity. The reader might select points in the flow and reason how the vector sum
of point source plus incident velocity generates the local flow direction.

The body per se is defined by a streamline. Indeed, any streamline could define
a body – although it should be chosen so that the source lies inside the surface, not
in the flow, where it would be unphysical. That is because the boundary condition
is the no-penetration criterion, that the flow is tangential to the surface. This is the
inviscid condition. The viscous, no-slip condition is met by inserting a boundary
layer between the potential flow and the wall. Thus, we are addressing the outer
flow region of boundary layer theory – see §5.2. Any streamline in the upper half of
Figure 5.19 defines the wall of a ramp beneath a potential flow. The dotted streamline
and lightly shaded region is an example.

Let us continue with the limited, but instructive, method of creating potential
flow from a distribution of sources. Now add a sink of strength: M at $x = \delta$. In place
of Eq. (5.23), we have

$$(u, v) = (U_\infty, 0) + \frac{M}{2\pi} \left[\frac{(x, y)}{r^2} - \frac{(x - \delta, y)}{r_\delta^2} \right], \tag{5.24}$$

where $r_\delta^2 = (x - \delta)^2 + y^2$ is the distance to the sink. This produces a symmetric, closed
body, the Rankine ovoid. Streamlines are plotted in Figure 5.20. The body is closed
because the total source strength is zero. The sink produces a stagnation point at
the downstream end of the body. Again, the velocity at any point is a vector sum
of contributions, now from the source, sink, and uniform flow. Midway between
source and sink, the v velocity vanishes because the source and sink make opposing
contributions in that direction. Their contributions to the u component reinforce;
hence, the u velocity is larger than the incident flow.

The limit $\delta \to 0$ produces a potential flow round a circular cylinder. In that sense, the flow is generated by a uniform stream and a source–sink doublet – that is, a source and sink extremely close to one another. The velocity is obtained from Eq. (5.24) in the small δ limit. It was already quoted on page 137:

$$(u, v) = (U_\infty, 0) + \frac{M\delta}{2\pi r^4} \left(y^2 - x^2, -2xy\right).$$

If a is the radius of the cylinder, the stagnation point should be at $x = -a, y = 0$. This is accomplished by a source of strength $M\delta = 2\pi U_\infty a^2$, giving the velocity components of Eq. (4.1)

$$(u, v) = (U_\infty, 0) + \frac{U_\infty a^2}{r^4} \left(y^2 - x^2, -2xy\right). \tag{5.25}$$

The corresponding streamlines are shown in Figure 4.6. At the top of the cylinder, $(x, y) = (0, a)$, the velocity is $(u, v) = (2, 0)U_\infty$. Hence, the flow speeds up by a factor of 2. It accelerates in consequence of streamline convergence.

A similar construction produces flow round a sphere in three dimensions. Exercise 5.13 asks the reader to show that the flow accelerates by a factor of 3/2 in that case.

According to Bernoulli's theorem [Eq. (1.38)], the pressure falls as the flow accelerates. The pressure is highest at the stagnation point and falls as the flow accelerates to the middle of a cylinder. If we let $x = a \cos\theta$ and $y = a \sin\theta$ on the surface of the circular cylinder, then Eq. (5.25) shows that $(u^2 + v^2) = 4U_\infty^2 \sin^2\theta$. Bernoulli's equation then gives the pressure coefficient

$$\frac{P - P_\infty}{1/2\rho U_\infty^2} = 1 - \frac{u^2 + v^2}{U_\infty^2} = 1 - 4\sin^2\theta. \tag{5.26}$$

The front stagnation point is at $\theta = 0$. The top of the cylinder is $\theta = \pi/2$. The pressure coefficients are 1 and -3, respectively. The top point experiences a suction relative to the ambient pressure.

However, the pressure is symmetric top to bottom and left to right; hence, it exerts no net force on the cylinder. Pressure acts normal to the surface, with the total force

$$\boldsymbol{F} = -\int_0^{2\pi} [P(\theta) - P_\infty](\cos\theta, \sin\theta)a \, d\theta = (0, 0). \tag{5.27}$$

There is neither an x nor a y component of force.

However, at any finite Reynolds number, no matter how high, the drag will not vanish. According to Eq. (5.26), the pressure falls from the stagnation point at $\theta = 0°$, reaching a minimum at $\theta = 90°$. Upon proceeding beyond the top of the cylinder the pressure rises; the flow is strongly decelerated. The boundary layer is subject to a large adverse pressure gradient and separates. This has been illustrated in §4.1.1.3. Unless the Reynolds number is low, a vortical wake develops on the lee side of the cylinder. Then the flow is not irrotational. Solution (5.25) predicts its own demise.

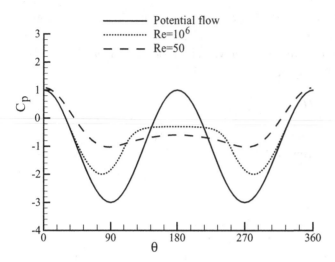

Figure 5.21. Pressure distribution around a cylinder according to potential flow and at finite Reynolds numbers.

In the separated region, the pressure is approximately constant. Its value is termed the *base pressure*. Figure 5.21 shows typical pressure distributions compared to potential flow. They differ most importantly in the region between 90° and 270°, which is the lee side. The base pressure rises with Reynolds number but is always subambient. The drag on a cylinder is a consequence of the force imbalance produced by the high pressure on the front, stagnation, region minus the base pressure. An empirical formula for the resultant drag coefficient is Eq. (4.2).

5.6.2 Images

One could proceed to consider more sources and sinks; indeed, bodies can be constructed from a continuous distribution of sources and sinks. If the net strength sums to zero, the body will be closed. We need not pursue that here: it is a well-established method of solving Laplace's equation. However, there is one more insight to be gleaned.

Consider two sources separated in the y direction. One is at $y = h$ and the other is at $y = -h$. Add their velocities vectorally. On $y = 0$, the v velocities are equal and opposite, summing to zero. The u velocities are equal, summing to twice the value of an isolated source. Showing this mathematically is simply a matter of adding formulas for sources located at $y = \pm h$:

$$(u, v) = \frac{M}{2\pi} \left[\frac{(x, y - h)}{r_h^2} + \frac{(x, y + h)}{r_{-h}^2} \right]. \tag{5.28}$$

On $y = 0$,

$$(u, v) = \frac{M}{2\pi} \left[\frac{(2x, 0)}{x^2 + h^2} \right],$$

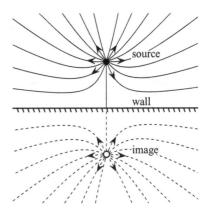

Figure 5.22. A point source and its image in a plane boundary.

v vanishes, and u is doubled. A representative set of streamlines is plotted in Figure 5.22.

No flow crosses $y = 0$; it can be interpreted as an impermeable wall. This is the *method of images*. If the body is represented by a distribution of sources within the region $y > 0$, a plane wall at $y = 0$ can be represented by a distribution of sources placed at the mirror-image points to the real sources.

This idea is not restricted to source representations. The potential flow past a body next to a plane wall can be constructed by reflecting the body in the surface and then ignoring the wall. This is as useful conceptually as it is analytically. Suppose the body in Figure 5.19 were placed above a wall, parallel to it. Would the stagnation point move up, down, or stay at the center of the nose?

To answer this, imagine an image body situated an equal distance below the wall and then ignore the wall. Velocities add vectorally. The image body produces a positive v velocity above itself. Hence, at the position of the real body, the image will contribute a positive v component. At the nose, now $v > 0$: it is no longer a stagnation point. For the velocity to vanish, the real body must contribute a negative v to cancel the image contribution. Hence, the stagnation point moves down to the lower side of the real body.

This simple example is a framework for qualitative reasoning about ground effects in aerodynamics. The upwash in front of a wing will be decreased when it is near the ground because of a contribution from its mirror image in the ground. If the lifting wing is modeled by a vortex in uniform flow, the image is an opposite signed vortex, also in uniform flow (exercise 5.14).

5.6.3 Point Vortices and Circulation

Although drag force owes its existence to violation of potential flow, lift forces require no such violation. Lift was met on page 47.

The potential vortex was defined by Eq. (1.54). The point vortex is like the point source, except its velocity field is in the angular direction rather than in the radial direction and its strength is the circulation, not the volumetric flow rate.

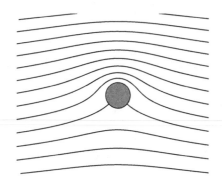

Figure 5.23. Streamlines for flow over a cylinder with circulation.

The point vortex has circular streamlines, so it provides a potential flow outside the cylinder defined by any one of them: that is

$$u_\theta = \frac{\Gamma}{2\pi r}, \quad r > a.$$

Adding this to Eq. (5.25) provides flow around a cylinder with circulation. An example of streamlines is shown in Figure 5.23. The mathematical formula for the velocity field is

$$(u, v) = (U_\infty, 0) + \frac{U_\infty a^2}{r^4} \left(y^2 - x^2, -2xy\right) + \frac{\Gamma}{2\pi r^2} (-y, x). \tag{5.29}$$

Evaluating this on the surface, obtaining the pressure from Bernoulli's equation, and integrating as in Eq. (5.27) proves that the lift force (per unit width) is $-\rho\Gamma U_\infty$, consistently with Eq. (1.64). Note, that Eq. (5.29) has a stagnation point at angle $-\alpha$ from the centerline, such that

$$\Gamma = 4\pi a U_\infty \sin\alpha, \tag{5.30}$$

as is found by setting $x = a\cos\alpha$, $y = -a\sin\alpha$ and equating the velocity to zero in Eq. (5.29).

This may be translated into an understanding of lift on airfoils. Imagine that the cylinder in Figure 5.23 is stretched along a diameter passing through the rear stagnation point. The cylinder is elongated in a direction lying at angle α to the incident flow. Then the trailing edge is sharpened to produce the airfoil shape of that in Figure 5.24. If the streamlines are similarly transformed, the flow field at the top of that figure is obtained.

Suppose there were no circulation, but the direction of stretching was at the same angle α. Then the streamlines prior to transformation are those of Figure 4.6. Because the axis of stretching does not pass through the stagnation point, the resultant flow pattern is the lower part of Figure 5.24. The stagnation point ends up on the upper surface, above the direction of elongation. The flow would encircle the trailing edge, experiencing a strong acceleration as it rounds the sharp corner, followed by a very rapid deceleration to the stagnation point. That would provoke a separation. Vorticity would be shed from the surface, leaving a flow that leaves the trailing edge

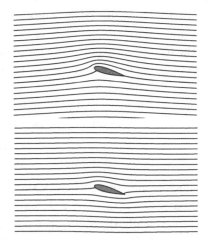

Figure 5.24. A wing with and without circulation.

tangentially. The vortex will have a circulation exactly equal and opposite to the circulation that is left on the airfoil. Thus, the requirement for the flow to leave the trailing edge tangentially sets the circulation, just as in Eq. (5.30). The lift force may be expressed

$$F_y = 1/2\rho U_\infty^2 A C_L,$$

where C_L is the coefficient of lift and A is the wing area span times width. For a slender, symmetric wing, it turns out that $C_L = 2\pi \sin \alpha$.

5.6.4 Stream Function

Potential flow is so named because the velocity derives from a potential function. Were the contours of constant potential plotted, they would lie *perpendicular* to the direction of the velocity. The velocity parallels the gradient of the potential.

In two dimensions, a function whose contours are *parallel* to the flow can be defined. It is named the *stream function*. Indeed, the contours of Figures 5.19 and 5.20 are level surfaces of a stream function. Arrows were added to the contours to imply a flow direction.

The stream function is denoted ψ (x). If its level surfaces are perpendicular to those of ϕ, then $\nabla \psi \cdot \nabla \phi = 0$ or, by Eq. (5.19),

$$u\frac{\partial \psi}{\partial x} + v\frac{\partial \psi}{\partial y} = 0. \tag{5.31}$$

This states that the gradient of ψ is perpendicular to the velocity vector. Equation (5.31) is met by defining the relation between ψ and u as

$$\frac{\partial \psi}{\partial x} = -v; \quad \frac{\partial \psi}{\partial y} = u. \tag{5.32}$$

This way of defining the stream function automatically satisfies the continuity equation (5.15c).

The streamlines can be thought of in two ways: in Chapter 2 they were introduced as the trajectories (2.23); here they are identified as curves defined by ψ = constant. The former concept is more intuitive, and more general. The latter is applicable only in two dimensions, but it plays a major role in classical fluid dynamics.

The equivalence of these two interpretations follows from Eq. (5.31). It is shown by proving that particle trajectories are curves of constant ψ. Differentiating $\psi(X, Y)$ = constant with respect to time gives

$$\frac{dX}{dt}\frac{\partial\psi}{\partial x} + \frac{dY}{dt}\frac{\partial\psi}{\partial y} = 0,$$

which is the same as Eq. (5.31). X, Y is the position of a particle moving with the fluid; $dX/dt = u$, $dY/dt = v$. Thus ψ is a constant of the motion of fluid particles. Consequently, the trajectories traced out by integrating equations (2.23) with a given velocity field are curves of constant ψ.

Consider the point source, Eq. (5.22). An integral of

$$u = \frac{\partial\psi}{\partial y} = \frac{Mx}{2\pi(x^2 + y^2)}.$$

is

$$\psi = \frac{M}{2\pi}\tan^{-1}(y/x). \tag{5.33}$$

This is consistent with the formula for v in Eq. (5.22). The streamlines are lines $x/y =$ constant, which are radii pointing out from the origin.

The stream function of a uniform flow in the x direction is $\psi = U_\infty y$. Combining the source and a uniform flow gives the stream function plotted in Figure 5.19:

$$\psi = U_\infty y + \frac{U_\infty h}{2\pi}\tan^{-1}(y/x). \tag{5.34}$$

Figure 5.19 is simply a contour plot of this function. The reader can consult Milne-Thomson (1968) for many more analytical examples of stream functions.

Although a vector stream function can be defined in three dimensions, it is primarily of value in two dimensions. Only in two dimensions does it provide the flow lines. However, its utility is not restricted to potential flow. Definition (5.32) enforces incompressibility $\partial_x u + \partial_y v = 0$, but not irrotationality. If the velocity is incompressible and rotational, that is, if $\partial_x v - \partial_y u = \omega \neq 0$, then ψ satisfies Poisson's equation

$$\nabla^2\psi = -\omega. \tag{5.35}$$

This is derived by inserting (5.32) into the vorticity formula. Given a vorticity distribution and a geometry, streamlines can be mapped by solving Eq. (5.35) and plotting contours of ψ. This is a quick way to map particle trajectories for flow visualization.

The boundary condition to Eq. (5.35) is that surfaces must be streamlines. Because streamlines are level curves of ψ, the boundary condition is that ψ = constant on walls. This is the nonpenetration condition, whereby the flow is tangential to the wall.

Figure 5.25. Flow accelerates where streamlines converge, as at an opening corner, and it
decelerates where they diverge.

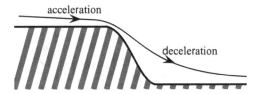

Another understanding of ψ follows from definition (5.32): the mass flux between two streamlines is constant in two dimensions. The reasoning is as follows.
The velocity is parallel to the streamlines by definition. Hence, the gradient of ψ is
perpendicular to the velocity. The magnitude of this gradient equals the magnitude
of the velocity:

$$|\nabla \psi| = \sqrt{\left(\frac{\partial \psi}{\partial x}\right)^2 + \left(\frac{\partial \psi}{\partial y}\right)^2} = |\boldsymbol{u}|.$$

If $\delta\psi$ is the change of ψ between two streamlines and d is their separation, then
$|\nabla \psi| = \delta\psi/d$ giving

$$\delta\psi = |\boldsymbol{u}|d. \tag{5.36}$$

The right side is the volume flux. Equation (5.36) states that difference of ψ between
any two streamlines equals the flux carried in the channel that they define.

A corollary is that the velocity magnitude varies inversely with distance between
lines of constant ψ. For instance, at the upper corner of the step in Figure 5.25, the displayed streamline draws near to the wall. The wall itself is another streamline. Hence,
the distance between these two streamlines decreases, so the velocity increases. A
high velocity occurs at the upper, rounded corner. Then, as the streamline passes the
upper corner and approaches the lower wall, it first moves away from the surface. The
spacing increases and the velocity falls. A low velocity occurs at the base of the step.

Generally, a corner that falls away from the approaching flow will produce
streamline convergence and flow acceleration – as at the top of the step in Figure 5.25.
By Bernoulli's equation, a low pressure occurs at such a corner. In the limit of a sharp
corner, potential flow theory produces an infinite acceleration and an infinite suction
pressure.

Where the wall turns upward into the flow, the streamlines diverge and the
velocity decreases on approaching the wall. This could be illustrated by reversing
the arrows in Figure 5.25, regarding it as flow from right to left. At a sharp upward
corner, the velocity of potential flow theory falls to zero, and the pressure rises to the
stagnation pressure. The rising pressure will cause the boundary layer on the wall to
separate before the corner is reached.

5.6.5 Added Mass

An intriguing example of unsteady potential flow analysis is provided by Lighthill's
(1975) theory of fish swimming. A fish swims by wriggling its tail. Doing so, it imparts

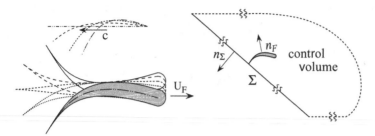

Figure 5.26. Swimming fish.

momentum to the water that it leaves behind. The backward momentum of the water propels the fish forward by conservation of net momentum – or the principle of action and reaction. At first sight, this is rather puzzling. As the tail wags back and forth, it seems that the shed momentum should go back and forth, oscillating the position of the fish's body, but not producing persistent movement. That is not correct.

The tail motion must be thought of differently. It helps to start by thinking of a long, slender fish, like an eel, or maybe a swimming snake. These animals propagate sinusoidal waves along their body, from head to tail. The tail does, indeed, wag back and forth, but it is the wave moving back, along the body that provides the water with backward momentum and the fish with forward thrust.

The eel's movement is called the *anguilliform* mode of swimming. It is typified by the undulatory body motion. These are relatively slow swimmers. The faster fish make a more abrupt movement of the aft part of their body and their tail. Although it may seem that they are wagging the tail back and forth, they, too, can be seen to send a wave along their body, as illustrated in Figure 5.26. The intersection of the curved body with a horizontal line propagates backward with velocity c. The theory shows that the swimming velocity is less than this speed. That enables the fish to add momentum to the water as it flows over the fish's body and so to propel itself forward by reaction [Lighthill (1975) contains an extensive description of various modes of swimming].

The basic fluid mechanical analysis of swimming by reaction is inviscid and irrotational. It invokes the idea of added mass. When a body moves through a fluid, its momentum is supplemented by that of the fluid that it displaces. This is accounted for by adding an effective mass to the body's mass.

A body moving with velocity V creates an additional kinetic energy $1/2\mathcal{M}_A V^2$ in the water. The subscript A denotes added mass. This is a result of potential flow theory (Milne-Thomson, 1968). Added mass is rigorous in that context. If the body accelerates, in addition to its own inertia, it will be resisted by the inertial force $\mathcal{M}_A dV/dt$ of the fluid.

Returning to modeling a swimming fish: a cross section of the fish, in the plane perpendicular to its backbone, can be approximated by an ellipse in translation. The potential flow round an ellipse is shown in Figure 5.27. Two frames of reference are portrayed: at left the frame is moving with the ellipse; at right the frame is stationary and the ellipse is moving. The ellipse moves down with velocity $-V$. Relative to its

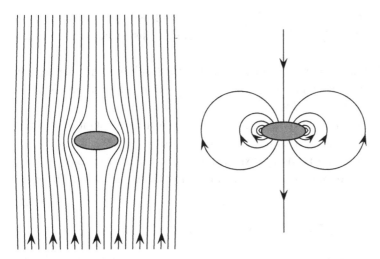

Figure 5.27. Potential flow around a translating ellipse in the frame of the surface and in a stationary frame.

frame, the fluid moves up at velocity V. It flows around the body, as in the left side of Figure 5.27. Shifting to the fixed frame, the flow pattern becomes that at right in Figure 5.27. This is simply a matter of adding the velocity, $-V$, to the figure at left. Perhaps the view at right looks less familiar than the perspective at left. It is the view that is relevant here. Fluid is pushed ahead of the downward-translating ellipse and is drawn in behind it. The water has a net downward momentum that is imparted to it by the section of the fish.

A rather elegant analysis of added mass can be found in Milne-Thomson (1968). Essentially, the energy of the fluid defines the added mass:

$$1/2 \mathcal{M}_A V^2 \equiv \int_V 1/2 \rho |\boldsymbol{u}|^2 d\mathcal{V}. \tag{5.37}$$

The integral on the right is the kinetic energy of the fluid; on the left it is represented by the added mass and translational velocity of the body. For potential flow, as at the right of Figure 5.27, the divergence theorem allows the integral to be rewritten as

$$\mathcal{M}_A = \frac{\rho}{V^2} \int_S \phi \, \hat{n} \cdot \nabla \phi \, d\mathcal{S}.$$

For a circular cylinder, the last term of Eq. (5.25) corresponds to $\phi = U_\infty a^2 x / r^2$ and the added mass is found to be

$$\mathcal{M}_A = \pi a^2 \rho.$$

In the case of an ellipse moving perpendicular to one of its principal axes, of radius a, the added mass again is $\pi a^2 \rho$. This is true for all eccentricities; it does not depend on the radius, b, of the second principal axis. The area of the ellipse is πab; hence, the added mass is *not* equal to the volume displaced by the body. Indeed, added mass is a fluid dynamical property, not a fluid static property.* It is largely a coincidence that for a circular section the added mass does equal the displaced volume times density.

Now, a control volume analysis can be applied to the fish.[†] The time-dependent control volume is shown in Figure 5.26. The dashed portion is at a large distance. Velocity and pressure are negligible on this part of the boundary. A plane passing perpendicular to the tail further bounds the control volume; this tail plane is denoted by Σ. The inertia of the water is assumed to be generated only by the lateral movement of the cross section. This is an approximation, valid if the section does not vary too rapidly along the body. Let s be distance along the fish's backbone. The inertia is

$$\mathcal{I} = \int_0^\ell \mathcal{M}_A(s)V(s)\hat{n}_F(s)ds.$$

The notation is that used previously: $\mathcal{M}_A(s)$ is the added mass of the elliptic cross section at s and $V(s)$ is the local velocity of the ellipse relative to the fluid. $\hat{n}_F(s)$ is the vector normal to the fish's body, locally at s, as in Figure 5.26. Each elliptic section contributes to the total inertia of the water, according to its added mass and its translational velocity.

The momentum balance (1.44) gives

$$\dot{\mathcal{I}} = -\int_\Sigma p(\boldsymbol{x}_\sigma)\hat{n}_\Sigma d\sigma + \int_\Sigma \rho\boldsymbol{u}u_\Sigma(\sigma)d\sigma + \boldsymbol{F}, \tag{5.38}$$

where $\boldsymbol{F}$ is the drag force exerted by the fish on the water. After invoking Bernoulli's equation for the pressure, it evolves that

$$\dot{\mathcal{I}} = 1/2\mathcal{M}_A(\ell)V(\ell)^2\hat{n}_\Sigma + \mathcal{M}_A(\ell)V(\ell)\hat{n}_F(\ell)u_T + \boldsymbol{F}. \tag{5.39}$$

Note that the right side involves only properties at the tail ($x = \ell$) of the fish, in this simplified analysis. It can be argued that this is why the cross sections of very efficient fish are small in the early part their aft portion, expanding rapidly to a large tail. This reduces energy losses due to the wagging motion while maintaining a high added mass at the tail.

The time average of Eq. (5.39) determines the mean swimming speed. Let c be the velocity at which the wave propagates back along the spine. The water is moving at velocity $-U_F$ relative to the fish. Thus, the velocity of the wave is $c - U_F$ relative to the water, and the velocity with which the elliptic cross section pushes into the water is $V = (c - U_F)$ times the slope of the fish's spine. If the tail oscillates sinusoidally in time with slope $\alpha \sin(\omega t)$, the simple formula

$$T = 1/4\alpha^2\mathcal{M}_A\left[c^2 - U_F^2\right] \tag{5.40}$$

---

* The displaced volume, times the density difference between the body and the fluid, times gravity, is the hydrostatic buoyancy force.

[†] Readers with no interest to see the fish example to its finish can skip to the next section.

emerges. It is seen that positive thrust is produced only if the wave velocity is greater than the swimming speed; alternatively, because thrust balances because drag during steady swimming, the ratio of swimming speed to wave speed is

$$\frac{U_F}{c} = \sqrt{1 - \frac{4D}{c^2\alpha^2\mathcal{M}_A}}.$$

High speed argues in favor of large added mass and large angular deflections of the tail.

EXERCISES

5.1 *Momentum balance.* The overall momentum balance (5.5) is derived by integrating Eq. (5.15b) from $y = 0$ to $y \to \infty$. As a preliminary step, explain why the the convection term can be written as

$$\frac{\partial u(u - U_\infty)}{\partial x} + \frac{\partial v(u - U_\infty)}{\partial y} + u\frac{\partial U_\infty}{\partial x}.$$

Show how Eq. (5.8) follows.

Assume that the boundary layer profile can be approximated by

$$U = U_s(x)\left[\frac{3}{2}\frac{y}{\delta} - \frac{1}{2}\frac{y^3}{\delta^3}\right]; \ y \le \delta$$

$$U = U_s(x); \ y > \delta.$$

Evaluate δ^*, Θ, and C_f. Using these values and Eq. (5.8) find the evolution of Θ versus x for two cases: (i) $U_s = U_\infty$ is constant, and (ii) $U_s = -K/x$.

5.2 *Drag on a plate.* The second case of exercise 5.1 corresponds to flow into a sink at $x = 0$. Use that solution here and ignore end effects. Derive a formula for the force per unit width required to hold the plate depicted at the right stationary in front of the inlet. The mass flux into the inlet is $\dot{M}$, entering through the angular sector $\pm\pi/8$. How does the force scale with $\dot{M}$?

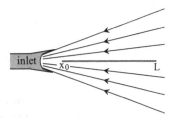

5.3 *Separation.* Equation 5.14 can alternatively be described as requiring $d\tau_w/dp_w$ to be negative if the boundary layer separates. Explain why this implies that two-dimensional separation cannot occur in a favorable pressure gradient.

5.4 *Blasius boundary layer equation.* Assume that the velocity has the functional form

$$u = U_\infty f'(\eta); \quad v = \sqrt{\frac{U_\infty v}{2x}}[\eta f'(\eta) - f(\eta)],$$

$$\text{with} \quad \eta \equiv y\sqrt{\frac{U_\infty}{2xv}}.$$

Show that the x-momentum equation (5.15b) simplifies to

$$f''' + ff'' = 0$$

if $dp/dx = 0$. This is Blasius's boundary layer equation. Why does $dp/dx = 0$ imply that U_∞ is constant?

Despite its simplicity, this equation has no exact solution. A computed solution is plotted in Figure 5.2. Write a computer program to solve the Blasius equation. Evaluate the momentum and 99% boundary layer thicknesses. Show that the wall stress satisfies $\tau_w = \rho U_\infty^2 d\Theta/dx$.

5.5 *Boundary layers and wall jets.* Numerical computations of the boundary layer on a flat plate have been carried out at Reynolds number $Re = U_\infty L/\nu = 4{,}000$. Uniform inflow velocity $U_\infty = 1\,\text{m/s}$ was used and a symmetry plane was placed upstream and downstream of the plate. The length of the plate is $L = 4\,\text{m}$.

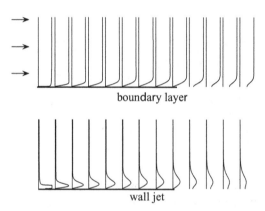

boundary layer

wall jet

Figure 1. Exercise 5.5.

A wall jet over the same flat plate and at the same Reynolds number $Re = U_{\text{jet}}L/\nu = 4{,}000$ has also been computed. In both cases, the fluid density is $\rho = 1\,\text{kg/m}^3$. The corresponding velocity profiles are presented in Figure 1. The jet inflow is placed at the beginning of the plate with $U_{\text{jet}} = 10$ m/s. The thickness of the jet is $H_{\text{jet}} = 0.05$ m. The symmetry plane downstream of the plate is retained. Explain why the scaling [(4.10) and (4.11)] is not applicable here.

The x component of surface shear stress $\tau_{w,x}(\text{Pa})$ for both cases is plotted in Figure 2. Using the data presented in Figure 2, compute the drag force on the flat plate. In the vicinity of the leading edge, assume that the surface shear stress is proportional $1/\sqrt{x}$ in the boundary layer, whereas for the wall jet assume $\tau_w \propto 1/x^{4/5}$.

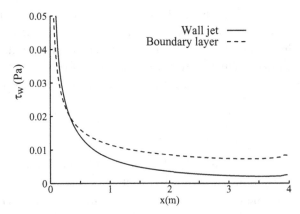

Figure 2. Exercise 5.5.

5.6 *Momentum thickness and drag.* For the boundary layer in exercise 5.5, the velocity profile at the trailing edge of the plate is given in the table. What drag force is implied by these data?

y (m)	U (m/s)	y (m)	U (m/s)	y (m)	U (m/s)
0.10	0.04696	1.60	0.69670	3.50	0.99071
0.20	0.09391	1.80	0.76106	4.00	0.99777
0.30	0.14081	2.00	0.81670	4.50	0.99958
0.40	0.18761	2.20	0.86330	5.00	0.99994
0.50	0.23423	2.40	0.90107	5.50	0.99999
0.60	0.28058	2.60	0.93060	6.00	1.00000
0.70	0.32653	2.80	0.95288	6.50	1.00000
0.80	0.37196	3.00	0.96905		
0.90	0.41672				
1.00	0.46063				
1.10	0.50354				
1.20	0.54525				
1.30	0.58559				
1.40	0.62439				
1.50	0.66147				

5.7 *Compute wall jets.* Compute the wall jet of exercise 5.5. Vary the jet thickness from $H_{jet} = 0.1$ m to $H_{jet} = 0.5$ m. Use structured grids and avoid exceeding 15,000 cells. To allow for the entrainment, use constant pressure boundaries. How does the jet development change as the jet thickness is increased? How should the domain size (both the length and the height) vary with increased thickness of the jet? Make plots of skin friction and compute the drag on the plate.

5.8 *Ekman spiral.* Consider a large tank of water that is being rotated by a motor. In the frame of reference of the tank, the equations governing a fluid flow are

$$-2\Omega v = \nu \partial_z^2 u,$$
$$2\Omega u = \nu \partial_z^2 v,$$

where Ω is the angular velocity and z is the direction normal to the bottom of the tank. These are the equations away from the vertical walls of the tank.

Initially, the water is in solid body rotation, $u = v = 0$. The tank is suddenly slowed down and the bottom velocity becomes, $v = -V_w$, $u = 0$, on $z = 0$. Show that a three-dimensional boundary layer develops with velocity of the form

$$v = -V_w \cos(z/\delta)e^{-z/\delta},$$
$$u = A \sin(z/\delta)e^{-z/\delta}.$$

Find formulas for A and δ. This is called the Ekman layer. Note that a deceleration in the v direction generates a u velocity. That is a consequence of the rotating frame of reference. The velocity direction spirals with height, much like Figure 5.11. Plot the variation of the horizontal velocity vector (u, v) with height.

As an approximation let $V_w = -\Delta\Omega r$, where r is radius from the center of rotation. The tank radius R is assumed to be much greater than δ. What torque does the water exert on the bottom of the tank?

5.9 *Three-dimensional boundary layer.* A three-dimensional boundary layer is subjected to a pressure gradient. Equations (5.16) are solved to obtain the velocity profiles seen in the figure below.

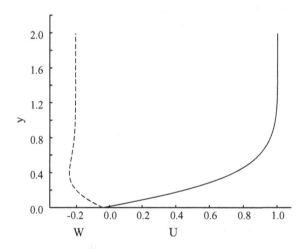

Is the x pressure gradient favorable or adverse?

5.10 *Saddles, nodes, and foci.* Identify saddles, nodes, and foci in all three cases of Figure 4.15.

5.11 *Velocity potential and stream function.* Show that $\phi = x^2 - y^2$ is a solution to Laplace's equation (5.19). What is the corresponding stream function? Plot the streamlines as contours of constant stream function.

5.12 *Potential flow.* Explain how a combination of a point source and a line sink, as at the left of the following figure, represents the flow around an airfoil shaped surface, as at the right of the below figure.

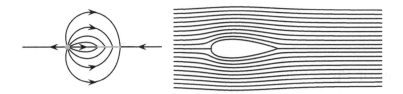

Following the method of Eq. (5.24), generate potential flow round a closed body by summing sources and sinks of your choosing. Explore several source locations and strengths. Plot the streamlines showing the shape of the closed body.

5.13 *Potential flow round a sphere.* Add a three-dimensional doublet to a uniform flow to construct the potential flow solution for a sphere. Show that the maximum surface speed is $U_s = 3/2 U_\infty$. Plot the pressure distribution over the surface.

5.14 *Potential with point vortices.* Recall the velocity field of a point vortex, Eq. (1.54). Plot the streamlines of a vortex in an oncoming, uniform flow. Choose a sign of vorticity that represents positive lift on a wing. Add an image vortex to represent the ground effect; do you expect an increase or a decrease of the lift?

6 Turbulent Flow

It is likely that most questions the reader might pose about turbulent flow have no sat-isfactory answer – questions like, What is its cause? How can equations as innocuous as the Navier–Stokes momentum equations produce such complex solutions? How can we describe it? How do we predict its properties? and so on. The phenomenon is common experience: turbulent eddying is seen in smoke billowing above a large fire, in dust clouds rising from an explosion, in the wake of a fast-moving boat; it is heard in the roar of a jet engine, in the wind rushing over an automobile; it is felt when an airplane bobs up and down in it or when a stiff breeze blows in one's face. Turbulence is an essential element of many processes. A text on fluid mechanics is not complete without a chapter on turbulence. That said, we provide, in this chapter, an introduction to computation of turbulent flow. The reader interested in a more thorough treatment of the subject can consult books entirely devoted to turbulence, such as Pope (2000).

The word *turbulence* conjures up the notion of randomness. It has entered ev-eryday vocabulary, divorced from the field of fluid mechanics. It evokes images of roiling, churning, and disorder. These are valid definitions, but in fluid flow it is often less severe than vernacular usage suggests. A 10% level of velocity fluctuation may be considered to be substantial. Turbulence is best defined as the irregular compo-nent of motion that occurs in fluids when the Reynolds number is sufficiently high. The irregularity may be mild or it may be severe.

The term *eddy* arises repeatedly in discussions of turbulent flow. The reader might feel that, before proceeding further, we should define this term. That feeling must be dispelled: an eddy is a vague notion that appears in many guises. The term invokes an image of vortices or of whorls. However, as commonly used, it refers more often to some type of statistical property that has no graphical representation. The energy spectrum can be described as average energy of eddies plotted against their size. Actually, it is the average energy at a given wavelength plotted versus wavelength (Figure 6.17); wavelength is interpreted synonymously with eddy size.

As described in §6.2.1, the "eddy viscosity" is the primary tool for engineering prediction of turbulent flow. Again, the reference to eddies is notional. The eddy vis-cosity represents stirring by turbulent motion in an averaged sense. The designation "eddy" indicates that properties of turbulence are being modeled, but there is no connection to any representation of eddies, per se. We ask the reader's forbearance of the indiscriminate use of the term *eddy*.

6.1 Computational Approaches

There are two approaches to computing turbulent flow, as has been illustrated in Figure 1.25. Either the Navier–Stokes equations are used to *simulate* the full randomness or only the averaged flow is computed, with the transport properties of the randomness being represented by a *model*. The first encompasses various types of eddy simulation. Most practical computations use the second approach. Eddy simulation is the subject of §6.4. We begin with a discussion of statistically averaged turbulence modeling.

6.2 Statistically Averaged Method

Turbulence is governed by the exact Navier–Stokes momentum equations; but it is of little comfort to realize that turbulence is just a particular type of solution to these laws of fluid dynamics. Its full disorderliness can be directly simulated – the art of direct numerical simulation (DNS) – but that is too expensive computationally and too fine a level of detail for almost all practical purposes. When the need is for timely prediction of the flow fields in practical devices, statistically averaged simulation is usually invoked.

There are good reasons why most turbulent flow prediction methods resort to a statistical description. The average over an ensemble of realizations of turbulent flow is a reproducible, smooth flow field. Quantities like the mean and variance of the velocity provide definite values to use in engineering analysis. It would seem that the smoother, averaged flow also is more amenable to computation than is the instantaneous, random flow. One can see this distinctly in the comparison between a random velocity field and its averaged streamlines in Figure 6.1. The upper part of this figure is a DNS of flow in a channel with ribs on the lower wall; the lower part is zoomed in around one the ribs and shows streamlines of the average of the upper figure. One can see the great simplification introduced by averaging. In this case, the DNS domain contains four ribs. This provides a sufficiently long domain to avoid spurious effects of inflow and outflow boundaries. The DNS flow is three dimensional, time dependent, and not periodic from rib to rib. The averaged flow is two dimensional and steady. It repeats from rib to rib, so the domain need include only one of them, as in the lower panel of Figure 6.1. The averaged description is enormously simpler.

Or, at least, that seems true until one broaches the question of *closure*. Unfortunately, there are no exact equations governing the smooth, averaged (or *mean*) flow field. In one way or other, the smooth mean flow depends on the irregular fluctuations. The latter are responsible for mixing and for diffusing the mean velocity. One way to understand this is to think of the billows of smoke emitted by a chimney. Individual billows occur with some randomness in space and time. Were a time-lapse picture taken, these billows would be smoothed into a continuous plume. That is the gist of Figure 6.1. The upper view is a snapshot of the randomness. The lower is analogous to a time-lapse picture. Indeed, the example of Figure 6.1 shows that

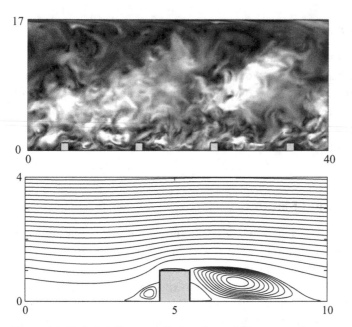

Figure 6.1. Turbulent flow over ribs in a channel. Upper part shows contours of velocity magnitude. Lower part shows streamlines of the averaged flow near one rib.

it can be quite difficult to infer the underlying mean flow from the instantaneous, random field.

If the mean depends on the fluctuations, and only the mean is to be computed, how do effects of the fluctuations enter the analysis? Their role must be represented without their presence. This is called the *turbulence closure problem*. Models must be introduced to represent enhanced mixing caused by fluctuations. This can be approached mathematically, but ultimately a degree of physical reasoning and empiricism is needed. We will start with a mathematical description of averaging.

For historical reasons, the mean velocity is called the Reynolds-averaged velocity and the equations obtained by averaging the Navier–Stokes equations are called the Reynolds-averaged Navier–Stokes equations – abbreviated to *RANS* equations. The historical origin of this terminology is that the same Reynolds (Osborne) who devised the nondimensional parameter that bears his name also introduced the averaging process. Actually, his average was closer to the filtering that is used in large eddy simulation (§6.4), but it is ensemble averaging that is named for him.

The ensemble average of a random function $f(\boldsymbol{x}, t)$ is just the conventional average, expressed as either a sum over discrete samples or as an integral over a continuous signal:

$$\overline{f}(\boldsymbol{x}, t) = \lim_{N \to \infty} \frac{1}{N} \sum_{i=1}^{N} f_i(\boldsymbol{x}, t). \tag{6.1}$$

The average is simply a sum of values divided by their number. f_i could be a set of velocity measurements. N values would be measured, summed, and divided by N.

In an experiment, it may be possible to replace the sum of values may by an average over time. The time average is defined by

$$\overline{f}(x) = \lim_{t \to \infty} \frac{1}{t} \int_0^t f(x; t')dt'. \tag{6.2}$$

Formally, the condition for time and ensemble averaging to be the same is that the process be statistically stationary. If the statistics do not change with time, then the process is statistically stationary; if the statistics do not change with time, they can be computed as a time average.

The qualification that a time average is equivalent to an ensemble average only if statistical stationarity obtains is sometimes misunderstood. In general, the ensemble average (6.1) would require a set of data to be obtained under given conditions. That could mean that the measurements must be made at a fixed time after the start of the experiment and the experiment be repeated N times. In general, this will not be the same as the time average of the data.

If the time at which samples are taken is irrelevant, the process is stationary. Then the experiment can be started, and data can be sampled as the experiment runs, averaging over a long time. Stationarity is violated if the geometry changes with time, as in an oscillating piston, or if the flow contains coherent unsteadiness, as in the case of vortex shedding behind bluff bodies (Chapter 4). It is not unusual to have to distinguish Reynolds averaging from time averaging. The distinction should certainly be kept in mind.

By a simple matter of definition, any random variable can be written as the sum of its average and a fluctuation. Just write the variable as a sum of the two parts:

$$u = U + u', \tag{6.3}$$

where $U \equiv \overline{u}$ is the average velocity and u' is the fluctuation. The fluctuation is defined to be $u' = u - U$. The fluctuation also is defined to be the turbulence. For instance, if the measurements were $u = 5, 4, 6, 9$, then $U = 6$, and the fluctuations are $u' = -1, -2, 0, 3$: by definition, they sum to 0.

The lowest-level turbulence problem can now be identified. It is to predict the averaged flow field U. This is not a problem of physical law but one of description. The momentum and continuity equations apply just as surely to turbulent as to laminar flow. But their solution is smooth and regular in laminar flow and chaotic and irregular in turbulent flow. The statistical description replaces the latter by a smooth, averaged flow. The turbulence problem, as formulated for RANS computation, is to describe the statistics of the velocity field, without access to data on the random flow.

Toward this end, the averaged Navier–Stokes equations can be formed in hopes of finding equations that govern the mean velocity. The momentum and continuity equations of viscous flow, whether turbulent or laminar, are Eqs. (1.14) and (1.15).

If the decomposition (6.3) is substituted into Eq. (1.14) and the result is averaged, the RANS equations

$$\rho\left[\frac{\partial \boldsymbol{U}}{\partial t} + (\boldsymbol{U} \cdot \nabla)\boldsymbol{U}\right] = -\nabla P + \mu\nabla^2\boldsymbol{U} - \underbrace{\nabla \cdot \rho\overline{\boldsymbol{u}'\boldsymbol{u}'}}_{\substack{\text{turbulent} \\ \text{stress}}},$$

(6.4)

$$\nabla \cdot \boldsymbol{U} = 0$$

are obtained. Equations (6.4) for the mean velocity are the same as Eq. (1.14) except for the last, underbracketed term of the momentum equation.

The quantity $\overline{\boldsymbol{u}'\boldsymbol{u}'}$ is the Reynolds stress tensor; in matrix form

$$\overline{\boldsymbol{u}'\boldsymbol{u}'} = \begin{bmatrix} \overline{u'^2} & \overline{u'v'} & \overline{u'w'} \\ \overline{u'v'} & \overline{v'^2} & \overline{v'w'} \\ \overline{u'w'} & \overline{v'w'} & \overline{w'^2} \end{bmatrix}.$$

(6.5)

The Reynolds stress tensor appears to play a similar role to the viscous stress tensor, σ, which was introduced in Chapter 1. Both are second-order tensors, whose divergence appears on the right side of the momentum equation. Recall (page 7) that a second-order tensor has two directions associated with it. Here both directions are those of the velocity fluctuation.

The mean flow equations (6.4) are unclosed because they contain the six unknown components of the Reynolds stress tensor, $\overline{\boldsymbol{u}'\boldsymbol{u}'}$. These derive from the quadratic nonlinearity, $\boldsymbol{u} \cdot \nabla\boldsymbol{u}$, of the Navier–Stokes equations. That seemingly innocuous term is anything but innocuous. Here it is responsible for the lack of closure of the mean flow equations: the number of unknowns is 10 and the number of equations is 4.

The extra term in Eq. (6.4) is the divergence of the Reynolds stress tensor. Although the terminology *stress* is used, this term follows from averaging the convective derivative. So a perspective on the Reynolds stresses is that they represent the averaged effect of turbulent convection. Any understanding of the nature of closure models relies on recognizing that they represent this ensemble averaged effect. Thinking, again, about a time-lapse photograph of a smoke plume, puffs of smokes are convected with an element of randomness. The time-averaged image represents the averaged effect of random convection. In that case, the averaging diffuses smoke concentration. The photographs in Figure 6.2 are a case in point. A short time exposure reveals irregularities in the plume. A time-lapse photograph smooths these to show the averaged state.

In the RANS equations (6.4) random convection, plus averaging, diffuses momentum. If the top image of Figure 6.1 is thought of as a snapshot of the instantaneous velocity field, then the whisps represent the instantaneous chaotic mixing. The Reynolds stress tensor represents the averaged mixing that transforms the snapshot into a time-lapse photograph. Thus it represents a combination of mixing per se and of smoothing by averaging.

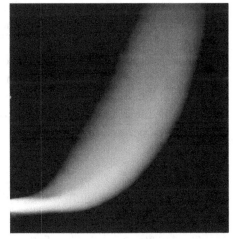

Figure 6.2. Instantaneous and time-averaged visualization of a jet in cross flow. Photographs courtesy of L. K. Su and M. G. Mungal.

6.2.1 Eddy Viscosity Model

The most widely used analytical model of this combined process is the eddy viscosity. It takes the mathematical form of another viscous transport coefficient, ν_T. The subscript T indicates that it models the effect of turbulence. Viscosity smooths a flow field by diffusing momentum; eddy viscosity does the same. However, unlike molecular viscosity (in a uniform temperature fluid), eddy viscosity varies with position in the flow. It can be seen in Figure 6.1 that the scale of the whisps is smaller near the lower boundary than in the bulk of the fluid. Correspondingly, the extent of mixing, and hence the eddy viscosity, decreases as the wall is approached. The model is of a fluid that has a viscosity that is a function of position. The dependence on position is a consequence of variations in size and intensity of the turbulent eddies. Figure 6.3 shows how eddy viscosity varies with distance from a wall.

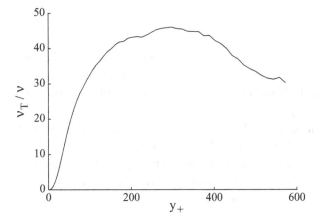

Figure 6.3. Eddy viscosity distribution versus wall distance in plane channel flow.

The eddies are themselves a function of the mean flow, which they predict. In that sense, the viscosity model is nonlinear: the eddy viscosity affects the flow; the flow determines the eddy viscosity.

Transport equations for $\overline{u'u'}$ can be derived, although they too are unclosed. We will not discuss the full equations of the second moment. However, the equation for turbulent kinetic energy requires comment. Several models invoke this equation to predict the mean-squared amplitude of the turbulent fluctuations. The turbulent kinetic energy, per unit mass, is

$$k \equiv 1/2(\overline{u'^2} + \overline{v'^2} + \overline{w'^2}).$$

In equivalent, index notation, $k \equiv 1/2\overline{u_i'u_i'}$.

The exact k transport equation is

$$\frac{\partial k}{\partial t} + U_j \frac{\partial k}{\partial x_j} = \mathcal{P} - \varepsilon - 1/\rho \left[\frac{\partial}{\partial x_j} (\overline{u_j'p'} + 1/2\rho\overline{u_j'u_i'u_i'}) \right] + \frac{\partial}{\partial x_i} \left(\mu \frac{\partial k}{\partial x_j} \right). \quad (6.6)$$

The important symbols on the right are the rate of production, $\mathcal{P}$, and the rate of dissipation, ε. Production does not refer to production of kinetic energy. It refers to the rate at which kinetic energy is transferred from the mean flow to the turbulence. The specific expression for $\mathcal{P}$ is

$$\mathcal{P} = -\overline{u_i'u_j'} \frac{\partial U_i}{\partial x_j}. \quad (6.7)$$

This is the product of turbulent stress times mean rate of strain. Stress times rate of strain equals the rate of working; in this case, it is the rate of working by the mean flow on the turbulence. An equal and opposite term arises in the equation for the energy of the mean flow; it is the rate at which energy is extracted by the turbulence, from the mean flow.

Dissipation refers to viscous dissipation of kinetic energy into heat. The specific expression for ε is

$$\varepsilon = \nu \frac{\overline{\partial u_j'}}{\partial x_i} \frac{\partial u_j'}{\partial x_i}. \quad (6.8)$$

This is a nonnegative quantity; hence, $-\varepsilon$ is a sink of energy in Eq. (6.6). The basic notions are that turbulent energy is produced from mean flow gradients, and that it is dissipated by viscous action. Imbalances between production and dissipation will cause k either to grow or to decay.

The third, bracketed term on the right side of Eq. (6.6) is a transport term. It redistributes k in space. It is usually represented by an eddy viscosity.

How does kinetic energy relate to modeling turbulent transport? Reynolds stress is assumed to be related to rate of strain by analogy to the expression (1.13) for molecular stress. The viscous constitutive form is

$$-\overline{u'u'} = \nu_T(\nabla U + {}^t[\nabla U]) - 2/3\,k\,I.$$

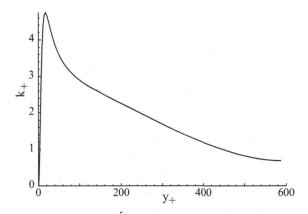

Figure 6.4. Turbulent kinetic energy versus wall distance in plane channel flow.

In equivalent, index notation

$$-\overline{u_i' u_j'} = 2v_t S_{ij} - 2/3k\delta_{ij}\,. \tag{6.9}$$

The mean rate of strain tensor is $S_{ij} = 1/2(\partial_i U_j + \partial_j U_i)$. Simply put, the turbulent stress is modeled by additional viscous transport. With this representation, Eq. (6.7) for the rate of energy production becomes

$$\mathcal{P} = 2v_T |S|^2,$$

where $|S|^2$ is the trace of the matrix squared: $|S|^2 = S_{ij} S_{ji}$.

In the RANS equations (6.4), the underbracketed term becomes

$$\frac{\partial}{\partial x_j}\left[\mu_t \left(\frac{\partial U_j}{\partial x_i} + \frac{\partial U_i}{\partial x_j}\right)\right]$$

plus a term, involving $2/3k$, that can be absorbed into the pressure. The momentum equation is now "closed": the unknown Reynolds stress tensor has been replaced by a function of the mean flow. Hence, the number of unknowns is now equal to the number of equations – which is what is meant by the equations being closed. That is not quite true: the eddy viscosity coefficient is not known.

To implement the eddy viscosity approach, the distribution of v_T in space must be predicted. Let us consider how to accomplish this. The extent and magnitude of mixing is a property of the size and intensity of local eddies. Turbulent energy varies with position in the flow; Figure 6.4 illustrates its variation with distance from the wall in plane channel flow. It starts from zero at the wall, rises steeply to a maximum, and then decreases at further distance from the wall. Generally, the eddying is most vigorous where the mean shear is largest. That is where turbulence is most strongly produced. In the plane channel example of Figure 6.4, the shear is largest at the wall. However, in the immediate vicinity of a wall turbulent energy is damped by viscous friction. At the wall $k = 0$ is the no-slip boundary condition. Above that, the intense shear produces a steep increase of k, to its near-wall maximum.

The spatial distribution of k should be related to that of ν_T. Predicting $k(\mathbf{x})$ is a step toward our goal. The spatial distribution of turbulent energy is determined by Eq. (6.6). Solving it will provide k as a function of position through out the flow domain. There is a qualification: Eq. (6.6) is unclosed. We will not describe the semiempirical methods of closing that equation. Suffice it to say that the eddy viscosity assumption (6.9) closes the transport and production terms, so the only remaining unknown variable is ε.

How a solution for $k(\mathbf{x})$ leads to $\nu_T(\mathbf{x})$ still is not obvious. The following is one line of reasoning that is consistent with the way that eddy viscosity is used.

In §4.2.1, shear layer spreading was related to viscosity by $\delta^2 \sim \nu t$. This could be used to measure the viscosity as $\nu \propto 1/2 d\delta^2/dt$. By analogy, suppose that eddies mix fluid particles by a distance y before they are dissipated by viscosity. Let the eddy lifetime be denoted by T. Then the eddy viscosity can be estimated as

$$\nu_T = 1/2\frac{d\overline{y^2}}{dt} = \overline{v'y},$$

where $v' = dy/dt$. Typically, during their lifetime, eddies mix particles a distance $y \sim vT$, giving

$$\nu_T = \overline{v'^2}T = C_\mu kT, \tag{6.10}$$

where C_μ is a constant of proportionality and k has replaced $\overline{v^2}$ as a measure of the intensity of the velocity fluctuation. Admittedly, this is not a very satisfactory derivation. It is an allusion to the Lagrangian theory of dispersion (Durbin and Pettersson Reif, 2001), which is more satisfactory. In that theory T arises as a correlation timescale. For present purposes, the pertinent observation is that turbulent mixing is characterized both by the energy, k, of the eddies and by their timescale, T.

In two-equation models, a second equation is added to Eq. (6.6), from which the timescale is derived. In the k-ε model an equation is added for ε and the timescale is represented by $T = k/\varepsilon$; in the k-ω model, a new variable is introduced as $\omega = 1/T$. This simply defines ω. A transport equation is then devised for it. First, the transport term in Eq. (6.6) is replaced by eddy viscosity, so that equation is modeled as

$$\frac{Dk}{Dt} = \mathcal{P} - \varepsilon + \frac{\partial}{\partial x_j}\left[(\nu + \nu_T)\frac{\partial k}{\partial x_j}\right]. \tag{6.11}$$

The ε and ω model equations are dimensionally consistent analogies to Eq. (6.11); they take the form

$$\frac{D\varepsilon}{Dt} = \frac{C_{\varepsilon 1}\mathcal{P} - C_{\varepsilon 2}\varepsilon}{T} + \frac{\partial}{\partial x_j}\left[\left(\nu + \frac{\nu_T}{\sigma_\varepsilon}\right)\frac{\partial\varepsilon}{\partial x_j}\right],$$

and $\tag{6.12}$

$$\frac{D\omega}{Dt} = \frac{C_{\omega 1}\mathcal{P}\,\omega}{k} - C_{\omega 2}\omega^2 + \frac{\partial}{\partial x_j}\left[\left(\nu + \frac{\nu_T}{\sigma_\omega}\right)\frac{\partial\omega}{\partial x_j}\right].$$

The right sides are production minus destruction plus diffusion.

For present purposes, the relevant aspects of these formulations are that they contain empirical constants $C_?$ and $\sigma_?$ and that they transport the turbulence scale. The latter enables the model to predict the spatial distribution of turbulence for use in predicting the distribution of eddy viscosity. The eddy viscosity is calculated by Eq. (6.10), either as

$$v_t = C_\mu k^2 / \varepsilon$$

or as   (6.13)

$$v_T = C_\mu k / \omega,$$

depending on whether the k-ε or the k-ω model is invoked.

Other models might predict v_T by a different method, or use a more elaborate approach than Eq. (6.9) to obtain the Reynolds stress. All are in the vein of devising flexible, robust methods to predict the *averaged* mixing caused by turbulent eddies. But it should be appreciated that eddy viscosity provides a limited, pragmatic, representation of turbulent mixing.

Ultimately, these models are buried in the flow solver. They convert a computer code for solving the Navier–Stokes equations into a code for predicting the Reynolds-averaged field. They are quite remarkable in the sense that they enable the mean flow to be predicted without simulating the random, eddying motion. A good deal of effort went into devising equations and setting values of empirical constants. Books by Durbin and Pettersson Reif (2001) and Pope (2000) describe the basis of turbulence modeling.

Some insight is obtained by considering the solution when k and ε or ω are not functions of x. Then there is no diffusion in the k equation; it becomes

$$\frac{dk}{dt} = \mathcal{P} - \varepsilon.$$ (6.14)

If production is greater than dissipation, turbulent energy grows. If production vanishes, there is no choice but for k to decay. Production depends on mean velocity gradients. In their absence, $\mathcal{P} = 0$ and the model reduces to

$$\frac{dk}{dt} = -\varepsilon,$$

with either

$$\frac{d\varepsilon}{dt} = -C_{\varepsilon 2} \frac{\varepsilon^2}{k} \quad \text{or} \quad \frac{d\omega}{dt} = -C_{\omega 2}\omega^2.$$

There is an equivalence between the k-ε and k-ω models in this case. For both the solution is

$$k = k_0 \left(\frac{t}{t_0} + 1 \right)^{-n},$$ (6.15)

with $n = 1/C_{\omega 2}$ or $n = 1/(C_{\varepsilon 2} - 1)$. The standard model constants are $C_{\varepsilon 2} = 1.92$ and $C_{\omega 2} = 5/6$, giving $n = 1.1$ or 1.2. At long times k decays like t^{-n}. The models are designed to do this, and their constants are chosen to make n agree with experiment.

In the presence of mean velocity gradients, $\mathcal{P} = 2\nu_T|S|^2$. For instance, in a linear shear, $u = \alpha y$, the rate of strain is $|S|^2 = 1/2\alpha^2$. Now the solution grows exponentially with time. It is of the form

$$k = k_0 e^{\lambda t}. \tag{6.16}$$

The growth rate is found to be

$$\lambda = |S|(C_{\varepsilon 2} - C_{\varepsilon 1})\sqrt{\frac{2C_\mu}{(C_{\varepsilon 2} - 1)(C_{\varepsilon 1} - 1)}}.$$

The standard model constants are $C_{\varepsilon 2} = 1.92$, $C_{\varepsilon 1} = 1.44$, and $C_\mu = 0.09$. Hence, $\lambda = 0.32|S|$. The constants are $C_{\omega 2} = 5/6$ and $C_{\omega 1} = 5/9$ for the k-ω model. A correspondence $C_{\varepsilon 1} \rightarrow 1 + C_{\omega 1}$ and $C_{\varepsilon 2} \rightarrow 1 + C_{\omega 2}$ can be drawn between the models. Hence k-ω gives $\lambda = 0.18|S|$.

It might seem unphysical for k to grow indefinitely in Eq. (6.16). It is. This is the solution for an artificial situation where the mean velocity has uniform gradient everywhere. For instance, the shear flow $u = \alpha y$ extends to $y = \pm\infty$. Recall that production refers to extraction of energy from the mean flow by the turbulence. In this artificial case, the mean flow has an infinite reservoir of energy. As time progresses, fluid particles disperse to increasing y distances. The velocity difference between their initial and final positions becomes increasingly large. The fluctuating energy can grow without bound.

If the shear were confined, and the velocity did not increase indefinitely, k would not continue to grow. The primary value of the idealized solution (6.16) is to show how the model accounts for net production of k. For energy to be produced $C_{\varepsilon 2} > C_{\varepsilon 1}$ is required. For the model to be properly posed $C_{\varepsilon 2} > 1$ and $C_{\varepsilon 1} > 1$. Decreasing $C_{\varepsilon 1}$ with $C_{\varepsilon 2}$ fixed increases the energy growth rate in homogeneous shear and the spreading rate of free-shear layers.

In short, there is some theory behind the k-ε and k-ω models. It combines with empirical data for their constants to provide a method for predicting the mean flow field.

6.2.2 Computed Example

Consider an application of the statistically averaged RANS method to flow down a backward ramp, as sketched in Figure 6.5. Our theme is the relative insensitivity of turbulent flow to Reynolds number.

The ramp is a circular arc with radius $R = 1.81L$, where L is the length of the ramp. The height of the ramp is $h = 0.3L$ and the height of the channel upstream of the ramp is $H = 1.87L$. A structured computational grid was created, consisting of 161×121 cells. The grid spacing was refined near the lower boundary, with the first cell center at $y^+ < 1$ – this will be defined in Eq. (6.19), and its significance explained thereafter.

In a laminar flow at modest Reynolds number, the streamlines will separate from the curved surface and reattach to the lower farther downstream. As the Reynolds

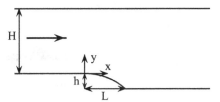

Figure 6.5. Geometry for flow down a backward ramp.

number increases, the length of the separated zone will increase, nearly in proportion to the Reynolds number.

A turbulent boundary layer is more resistant to separation. Once it does separate, the size of the separation zone is relatively insensitive to Reynolds number. These both are consequences of turbulent mixing. In the present RANS simulation, turbulent mixing is modeled by eddy viscosity. The boundary layer is more resistant to separation because eddy viscosity is large compared to molecular viscosity over most of the boundary layer. Eddy viscosity is relatively independent of Reynolds number because it parameterizes mixing by random convection rather than diffusion by molecular action.

The ramp in Figure 6.5 is steep enough to separate even when the boundary layer is turbulent. Representative streamlines are shown in Figure 6.6. A small recirculation region is seen at the downstream end of the ramp.

Flows were computed at two Reynolds numbers: $Re_L = U_\infty L/\nu = 7 \times 10^5$ and $Re_L = 10^5$, differing by a factor of 7. The momentum thickness Reynolds numbers at $x/L = -0.2$ were $Re_\theta = 20,100$ and $Re_\theta = 3,400$, respectively. These values provide a characterization of the boundary layer approaching the ramp. Indeed, the momentum thickness is a more relevant characterization of the flow than Re_L because it controls the tendency to separate.

The computed friction coefficient on the plane wall upstream of the ramp scales consistently with theory for the flat plate boundary layer. An empirical formula for its dependence on boundary layer thickness is $C_f = 0.025 Re_\theta^{-1/4}$, as discussed later

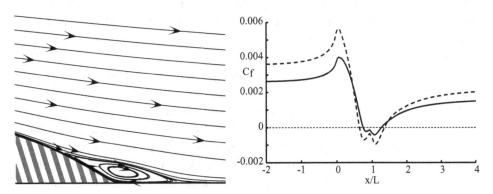

Figure 6.6. Streamlines at the base of the ramp, showing the separation bubble. $Re_\theta = 20,100$ (———); $Re_\theta = 3,400$ (- - - -).

Figure 6.7. Turbulent kinetic energy near the lower wall.

in Eq. (6.27). At the right of Figure 6.6, the friction coefficient is lower for the higher Re_θ.

The recirculation zone is the region of negative C_f in Figure 6.6. Clearly, its size depends only weakly on the Reynolds number. That is in stark contrast to laminar flow computations over the same ramp. The laminar recirculation grows by about a factor of 7 as the Reynolds number increases by that same factor. See Figure 4.4 on page 136 for an illustration.

Contours of turbulent kinetic energy, k, in Figure 6.7 indicate a pocket of high fluctuation bordering the separated zone. Production of turbulence is enhanced in the mixing layer. An adverse pressure gradient, quite generally, will lead to increases of k, whether or not it causes separation.

6.2.3 Limitations of Scalar Eddy Viscosity Models

The principle of material frame indifference states that molecular stresses are uninfluenced by rotation. A liquid contained in a rotating tank is subjected to the same viscous law as in a stationary tank. A corollary is that viscosity laws depend on the rate of strain, not on the rate of rotation.

Such is not the case for turbulence. Turbulent stress in a rotating tank is influenced by the rate of rotation. The Navier–Stokes equations are altered if they are expressed in a rotating frame: Coriolis acceleration must be added. Coriolis acceleration acts on turbulent fluctuations, modifying the Reynolds stress. If Ω is the rate of frame rotation, $\overline{u'_i u'_j}$ is a function of Ω. But the eddy viscosity model, Eq. (6.9) with Eq. (6.13) contains no such dependence.

The same shortcoming exists in flow with curved streamlines. Strong convex curvature can suppress turbulent energy. The effect is not captured by eddy viscosity models. They can give wildly wrong predictions in strongly swirling flow. In the core of an intense vortex, turbulence is suppressed; the flow is quasilaminar. Eddy viscosity predicts that the core remains turbulent.

Figure 6.8 is the example of a cyclone swirler. Flow enters tangentially, as indicated, driving a vortex inside the tube. The contraction stretches and intensifies the vortex. The bottom end of the tube is closed. Fluid exits at the top. The axial velocity contains zones of downward and upward flow. Eddy viscosity predicts downflow near the axis and upflow near the walls – downflow corresponds to a positive axial velocity in the line plot. In reality, the flow is upward near the axis, down in an annulus, and up near the walls. The upflow on the axis occurs because turbulence has been suppressed. The low viscosity in that region provides a path of low resistance for the up flow. The eddy viscosity model erroneously predicts high viscosity on the

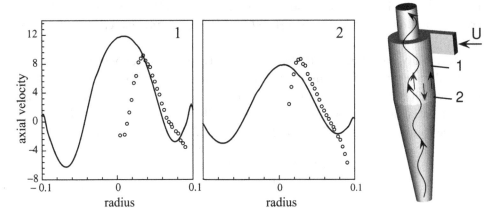

Figure 6.8. The cyclone swirler illustrates how the scalar, eddy viscosity formulation can fail in the presence of strong swirl. The solid line is the eddy viscosity prediction; the symbols are data. Figure courtesy of G. Iaccarino

axis. Fluid is dragged downward by the surrounding flow; there is no path of low resistance.

A physical explanation of the effect of swirl is provided by a stability argument due to Lord Rayleigh. In equilibrium, a radial pressure gradient balances the centrifugal acceleration in a vortex [Eq. (1.47)]:

$$\frac{1}{\rho}\frac{\partial p}{\partial r} = \frac{u_\theta^2}{r} = \frac{L^2}{r^3},$$

where $L = rv_\theta$ is the angular momentum. The question of stability hinges on whether a displaced fluid element experiences a restoring force, driving it back toward its initial location. Let an element initiating at $r - \delta r$ be displaced to r, conserving its angular momentum. When it reaches r it experiences the centrifugal acceleration

$$\frac{L^2(r - \delta r)}{r^3}.$$

Compare this to the pressure gradient:

$$\frac{L^2(r - \delta r)}{r^3} - \frac{1}{\rho}\frac{\partial p}{\partial r} = \frac{L^2(r - \delta r) - L(r)^2}{r^3}$$

$$\approx -\delta r \frac{\partial L^2/\partial_r}{r^3}.$$

If this is negative, the particle will be driven back by the pressure gradient; if it is positive, the particle will be thrown further out by the centrifugal acceleration. Rayleigh's centrifugal stability criterion is that, if

$$\frac{\partial L^2}{\partial r} > 0, \tag{6.17}$$

the curvilinear flow is stable: when angular momentum increases as the radius increases the flow is stable. Conversely, $\partial L^2/\partial r < 0$ is the inviscid criterion for instability.

Instability will intensify turbulence. Instability per se might not be relevant if the flow is already turbulent, but the conditions that make laminar flow unstable are those under which turbulence production is enhanced. In the cyclone swirler, Figure 6.8, the tangential injection ensures that $\partial L^2/\partial r > 0$. Turbulence is suppressed.

The Rankine vortex (1.71) provides an illustrative example. It has

$$v_\theta = \Omega r, \quad r < R$$
$$v_\theta = \frac{\Omega R^2}{r}, \quad r > R.$$

Thus

$$L = \Omega r^2, \quad r < R$$
$$L = \Omega R^2, \quad r > R.$$

The stability condition $\partial L^2/\partial r > 0$ is met for $r < R$. The core of the vortex is a region of stability.

If Ω is large enough, the vortex core will become laminar. This occurs in the tip vortices produced by the wing of a large airplane. Those vortices persist several kilometers behind the aircraft. Eddy viscosity models do not account for suppression in the core; they predict that the vortices diffuse and lose strength too quickly.

Other effects of imposed forces may not be properly represented by a scalar turbulent viscosity. Buoyancy can suppress or enhance turbulence. Stable stratification opposes velocity fluctuations in the direction of gravity, say the y direction. Those components of the Reynolds stress tensor that contain v will be reduced. Although the u and w velocities are not directly affected, their intensities also can be reduced through dependence on v. The need to distinguish how the various components of the Reynolds stress tensor are affected implies that a scalar, eddy viscosity model will not capture the physics. In stratified shear flow, the Richardson number

$$Ri \equiv \frac{-g\partial\rho/\partial y}{\rho(\partial U/\partial y)^2}$$

characterizes the relative strengths of buoyant suppression and shear production. If $Ri > 1/4$ stable buoyancy is overwhelming; turbulence will not be sustained by the shear and the flow reverts to laminar.

A perhaps less compelling shortcoming of eddy viscosity is the erroneous prediction of normal stresses by Eq. (6.9). Consider a parallel shear flow, $U(y)$. Then Eq. (6.9) gives

$$[\overline{u_i'u_j'}] = \begin{bmatrix} 2/3k & -v_T\partial_y U & 0 \\ -v_T\partial_y U & 2/3k & 0 \\ 0 & 0 & 2/3k \end{bmatrix}. \tag{6.18}$$

Thus, erroneously, $\overline{u'^2} = \overline{w'^2} = \overline{v'^2} = 2/3k$. In fact $\overline{u'^2} > \overline{w'^2} > \overline{v'^2}$ is measured in experiments; $\overline{u'^2}$ is on the order of three times larger than $\overline{v'^2}$.

This failing usually is not catastrophic. In shear flows, the mean flow equation (6.4) depends primarily on the shear stress $-\overline{uv}$. The ambition of an eddy viscosity model is to approximate this component fairly. It often does so.

We do not discuss methods for capturing effects that are missed by the eddy viscosity formulation. The curious reader can consult specialist texts on turbulence modeling (Durbin and Petterson Reif, 2001; Pope, 2000). An informed user of turbulence models should appreciate their virtues and demerits.

6.3 Turbulent Shear Flows

Often one attempts to understand a complex flow by identifying certain generic elements. The flow field might include boundary layers, separated shear layers, wakes, and so on. An understanding can be gained by identifying such elements and studying them in isolation. In a practical application, they are not likely to be isolated; they are components of the overall flow. Practical applications invariably involve a multitude of complicating peculiarities. These might be vortices that form at the juncture between a plane wall and an appendage, a pressure gradient transverse to the flow that skews the flow direction, or a large variety of other geometrical and fluid dynamical intricacies. Not all of them can be decomposed into simpler elements, but many can at some level of abstraction.

6.3.1 Boundary Layers

A good deal of our understanding of turbulent flow is based on identifying relevant scales – lengths, times, velocities – that are used to normalize dimensional variables. This is distinct from conventional dimensional analysis because the scales are internal properties of the flow, not externally imposed. By identifying the relevant scales, it is hoped that properties of the mean flow can be inferred. The most successful result of scaling is the logarithmic law. It describes a portion of the velocity profile in turbulent boundary layers. Our discussion begins with wall bounded flow, including the log-law.

6.3.1.1 The Viscous Wall Layer

In wall bounded flow, the skin friction is used as a velocity scale. The surface stress divided by density has the dimensions of velocity squared. We introduce the friction velocity, defined by

$$u_* \equiv \sqrt{\frac{\tau_w}{\rho}}.$$

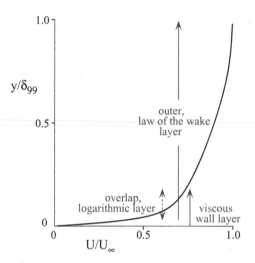

Figure 6.9. Regions of a turbulent boundary layer.

This is regarded as a measure of the order of magnitude of turbulent velocity fluctuations. That is not unreasonable: momentum mixing generates the wall stress. Velocity fluctuations cause the mixing.

Molecular viscosity plays a role near the wall; hence, it provides another parameter. Combining, v, with dimensions ℓ^2/t and u_*, with dimensions ℓ/t provides v/u_* as a length scale and v/u_*^2 as a timescale. These are called *plus units*. That terminology comes simply from the practice of denoting nondimensional variables by a subscript $+$. Plus units normalize a region next to the wall, called the *viscous sublayer*.

Consider the regions marked on Figure 6.9. There is a viscous wall layer next to the surface – also called the *law of the wall* region – and there is an outer region extending to the top of the boundary layer – the *law of the wake* region. Any measure of boundary layer thickness can be used for the outer, large-scale region. The momentum thickness Θ [see Eq. (5.3b)] serves the purpose.

The ratio of the outer to inner region thicknesses is $\Theta/(v/u_*)$. This is a Reynolds number. Its value will be larger than 50 in any fully turbulent boundary layer and typically will be larger than 100. Hence, the outer region is more than an order of magnitude thicker than the inner region. It occupies the bulk of the boundary layer.

However, the steepest shear occurs in the inner region, as can be seen in Figure 6.9. The velocity profile in that figure is at a quite low Reynolds number, with the extent of the viscous region exaggerated for display. In practice, the viscous layer is extremely thin. This presents computational stiffness, as will be discussed later.

The idea of subregions within the boundary layer may seem to have come out of the blue. They should not be taken too literally. Their borders are quite fuzzy. Nevertheless, the concepts of an inner, viscous region, an outer, inviscid region, and an overlap layer common to both provide a powerful framework for understanding of turbulent boundary layers. By inference, they are crucial to knowledgeable computational analysis of wall-bounded turbulent flow.

Nondimensional variables, referred to as *plus units*, are defined by

$$y_+ = yu_*/\nu; \quad u_+ = u/u_*. \tag{6.19}$$

These arise naturally by the following consideration. The no-slip condition $u = 0$ at $y = 0$ suggests that velocity tends linearly to 0 as $y \to 0$. The viscous wall stress is $\tau_w = \mu \partial u/\partial y|_0$ or the velocity gradient is $\partial u/\partial y|_0 = \tau_w/\mu$. Hence, as $y \to 0$,

$$u \to \frac{\tau_w y}{\mu} = u_* \frac{u_* y}{\nu}.$$

In nondimensional form, this reads

$$u_+ \to y_+ \tag{6.20}$$

as $y_+ \to 0$. Moving away from the wall, but still in the viscous wall layer, the velocity is no longer linear. It is assumed to remain a function of the form

$$u_+ = F(y_+). \tag{6.21}$$

This is the law-of-the-wall. It is a statement that this nondimensionalization will collapse profiles of u versus y onto a single curve. In any particular flow, τ_w may vary with position on the wall; it will also vary from one flow to the next. However, in each case, if the velocity and coordinate are nondimensionalized as in Eq. (6.21) the same function $F(y_+)$ will be obtained. This is an assumption of *universality* of the near wall region. It is not always true: for instance, at points where $\tau_w \to 0$ it fails. Nevertheless, it is usually a good assumption.

6.3.1.2 The Log-Layer and Outer Region

The mean boundary layer velocity must become constant in the free stream, $u \to U_\infty$ as $y \to \infty$. Therefore, u_* cannot be the only velocity scale; U_∞ should play a role. Similarly, turbulent mixing must have a pronounced effect on the mean flow; molecular viscosity is no longer the only viscosity scale. Indeed, above the viscous wall layer, molecular viscosity can be ignored in the mean momentum equation.

The mean velocity in the law-of-the-wake region can be represented by

$$u = U_\infty - u_* f(y/\Theta). \tag{6.22}$$

In this formula, $u_* f(y/\Theta)$ equals the departure of U from its free-stream value; it is the amount by which $U(y)$ lies to the left of U_∞ in the left half of Figure 6.10. Asymptotically $U \to U_\infty$ at large y and $f \to 0$.

In a zero pressure gradient boundary layer, the law-of-the-wall occupies the interval $0 < y \lesssim 0.2\delta_{99}$; the law-of-the-wake occupies $40\nu/u_* \lesssim y < \infty$. The specific values $40\nu/u_*$ and $0.2\delta_{99}$ should be treated as representative, not rigorous. These two zones intersect between $40\nu/u_* \lesssim y \lesssim 0.2\delta_{99}$; this is the logarithmic overlap layer (see Figure 6.9).

The representations (6.21) and (6.22) lead to the famous log law. To get rid of U_∞, both of these formulas are differentiated with respect to y. From the law-of-the-wake Eq. (6.22)

$$\frac{\partial u}{\partial y} = -\frac{u_*}{\Theta}f'(y/\Theta) = -\frac{u_*}{y}\left[\frac{y}{\Theta}f'(y/\Theta)\right]$$

and from the law-of-the-wall equation (6.21)

$$\frac{\partial u}{\partial y} = \frac{u_*}{y}[y_+F'(y_+)].$$

Assume that an overlap region exists in which either of them represents the velocity. Then in this region

$$\frac{u_*}{y}\left[-\frac{y}{\Theta}f'(y/\Theta)\right] \simeq \frac{u_*}{y}[y_+F'(y_+)].$$

For these expressions to be equivalent for any y, the bracketed terms must be equal. But they are functions of different variables. They can only be equal if they are constant. Their constant value is written as $1/\kappa$. κ is termed the Von Karman constant. Theodore Von Karman first inferred the log-law from a mixing length argument. His student, Clarke Millikan, give the overlap argument summarized above. Experimentally, κ is found to have the value 0.41 ± 0.02.

The criterion for the wall and wake regions to overlap smoothly has been found to be

$$\frac{\partial u}{\partial y} = \frac{u_*}{\kappa y}. \tag{6.23}$$

This is a statement of the log-law. Integrating it explains this terminology:

$$u = \frac{1}{\kappa}u_* \log y_+ + Bu_*. \tag{6.24}$$

The integration constant, B is found experimentally to be about 5.1 ± 0.4.

Equation (6.24) is motivation to plot velocity profiles on log-linear axes. On a plot of u_+ versus $\log y_+$, B is the intercept and $1/\kappa$ is the slope. When data are plotted in this way, experimental values of the constants are obtained.

A typical plot is illustrated by Figure 6.10. The curves $u_+ = y_+$ and $u_+ = \log(y_+)/\kappa + B$ are shown by dashed lines on the right hand graph. These are the behaviors in the viscous wall layer and in the logarithmic, overlap layer. The extent of the logarithmic layer increases with Reynolds number, consistent with the previous comment that it is the region $40 \lesssim y_+ \lesssim 0.2u_*\delta_{99}/\nu$ in a zero pressure gradient boundary layer. The upper limit of this range is a Reynolds number. When replotted on linear–linear axes, the appearance of the data on the right of Figure 6.10 provides some perspective. That is shown in the left panel of Figure 6.10. The log-law is drawn by the solid curve. In linear–linear coordinates the log region is not particularly distinctive. The steep shear occurs below the log layer; the bulk of the boundary layer lies above it.

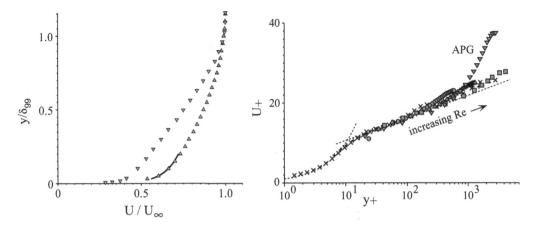

Figure 6.10. Velocity profiles in turbulent boundary layers. The plot at right shows the log-law as a straight line on log-linear axes. The linear–linear plot, at the left, provides another perspective. The log-region is shown by a short, solid line. The symbols are experimental data. ∇ are data for adverse pressure gradient; all others are for zero pressure gradient.

Equations (6.23) and (6.24) play a very important role in turbulent flow computation. They embody a concept of near-wall universality, stated more generally by Eq. (6.21). Irrespective of the geometry or pressure gradients, the velocity near the wall is a function of y_+. The particularities of the flow enter only through their effect on the friction velocity u_*.

This must be qualified. It assumes that the boundary layer is very thin; hence as a flow approaches separation, near-wall universality will not be correct. It assumes that the boundary layer is in a fully developed state; if the Reynolds number is low, that will not be true. A guideline is that the momentum thickness Reynolds number $U_\infty \Theta / \nu$ should be greater than 3,000. If the wall is curved, the radius of curvature must be large compared to Θ. A rule of thumb is that the radius of curvature should be more than 20 times Θ to ignore curvature effects on the turbulence and so on; other caveats could be made. Their gist is that perturbations must not be too large. Forcing by the Reynolds stress gradient must dominate other effects for the log-law to be valid. Because the wall layer is thin, this is often a good assumption.

Above the wall layer, the velocity profile is not universal. A simple example is of a boundary layer subject to a pressure gradient. The effect of an adverse pressure gradient is illustrated in Figure 6.10 by the downward triangles, labeled *APG*. On the log-linear plot, APG causes the data to rise above the log-law. This occurs in the outer, law-of-the-wake region. Even in zero pressure gradient, the outer region velocity rises above the log-law. The difference between the actual velocity and the log-line is called the wake function. That is, the velocity might be represented as

$$u_+ = \frac{1}{\kappa}\log(y_+) + B + \frac{\Pi}{\kappa}w(y/\delta_{99}), \qquad (6.25)$$

where w is the functional dependence of the wake component and Π is a function of pressure gradient that determines the magnitude of the departure from the log-line.

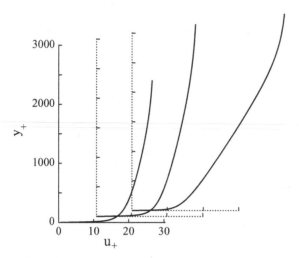

Figure 6.11. A boundary layer is initiated with a zero pressure gradient profile and then computed as it evolves in an adverse pressure gradient.

Π increases with the magnitude of the adverse pressure gradient. An empirical fit to certain data is

$$\Pi = 0.8(\beta + 0.5)^{3/4},$$

where $\beta \equiv (\delta^*/\tau_w)d_x P_\infty$ is a pressure gradient parameter.

The law-of-the-wake formula (6.25) is of historical interest. It provided a concrete velocity profile to work with. However, the function $w(y/\delta_{99})$ depends on the particular flow. Unlike the law-of-the-wall, it has no claims to "universality." Nowadays, the outer boundary layer profile is computed, either by full Navier–Stokes solutions or by simplified, boundary layer computations. The evolution of a boundary layer profile from a zero to an adverse pressure gradient is shown in Figure 6.11. The initially full profile is gradually eroded. The rightmost profile is characteristic of adverse pressure gradients.

6.3.1.3 Growth of Boundary Layer Thickness

It is of interest to contrast the growth of laminar and turbulent layers. The integrated momentum equation (5.8) is equally applicable to laminar or turbulent flow. The steps in its derivation all carry through when transport is by a combination of Reynolds and viscous stresses. The difference enters via the friction coefficient, C_f. Without a full solution for the velocity, the skin friction is unknown. Skin friction represents the rate at which momentum is being removed at the surface by viscous action. Qualitatively, turbulent mixing will increase the rate of momentum transfer to the wall. This is effected by an increase of the shear relative to laminar flow.

Velocity profiles for laminar and turbulent flow are contrasted in Figure 6.12. Turbulent eddies stir the fluid, producing a flatter profile away from the wall. In a thin layer next to the surface, the turbulence is suppressed. This is the viscous sublayer,

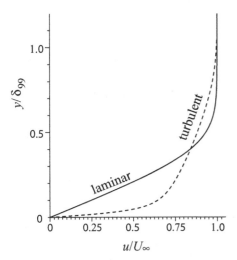

Figure 6.12. Comparison between laminar and turbulent velocity profiles.

where turbulent stirring is reduced to levels comparable to molecular diffusion. The velocity drops rapidly across that layer to satisfy the no-slip condition at the wall. Hence the steep shear.

To a large extent, the ratio of eddy to molecular viscosity, plotted in Figure 6.3 explains why computed turbulent and laminar profiles are so different. Eddy viscosity rises rapidly above the wall to many times the molecular viscosity. High viscosity diffuses momentum to create a flatter profile in the main part of the boundary layer. Steep shear can persist only in the low viscosity region near the wall.

The laminar, Blasius skin friction formula (5.6) can be stated

$$C_f = 0.664^2 \, R_\theta^{-1} = 0.44 \, R_\theta^{-1}, \tag{6.26}$$

where $R_\theta = U_\infty \Theta / \nu$ is the momentum thickness Reynolds number. Experimental measurements in turbulent flow are fit by

$$C_f = 0.025 \, R_\theta^{-0.25} \tag{6.27}$$

in the regime $R_\theta > 1{,}000$. The turbulent friction coefficient is at least an order of magnitude greater than the laminar value at a given momentum thickness Reynolds number. Another data correlation that involves both the momentum thickness and displacement thickness is

$$C_f = 0.246 R_\theta^{-0.268} 10^{-0.678 \delta_*/\Theta}. \tag{6.28}$$

This formula is valid in the presence of pressure gradients, whereas Eq. (6.27) is valid only for zero pressure gradient.

The momentum integral (5.8) provides a contrast between laminar and turbulent growth of boundary layer thickness. With $C_f = 0.025 R_\theta^{-0.25}$ and no pressure gradient, (5.8) becomes

$$\frac{d\Theta}{dx} = 0.0125 \left(\frac{\nu}{U_\infty \theta} \right)^{1/4} \tag{6.29}$$

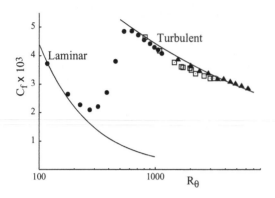

Figure 6.13. Friction coefficient in a transitional boundary layer.

and has solution

$$\frac{\theta}{x} = 0.036 \left(\frac{U_\infty x}{\nu} \right)^{-1/5}. \tag{6.30}$$

The thickness grows a bit less than linearly, as x to the power 4/5, with the slope decreasing as the x Reynolds number to the power $-1/5$. This can be compared to the laminar boundary layer, for which

$$\frac{\theta}{x} = 0.664 \left(\frac{U_\infty x}{\nu} \right)^{-1/2}.$$

The thickness increases as x to the power 1/2. The turbulent layer grows more rapidly with downstream distance.

Another perspective is provided by examining the variation of C_f with downstream distance in a boundary layer undergoing transition from a laminar to a turbulent state. To avoid particulars of the configuration, this is commonly plotted in nondimensional form as C_f versus R_θ. Momentum thickness, Θ, substitutes for distance downstream. In this form, the data are not sensitive to the upstream origin of the boundary layer. The geometry could be a flat plate with a rounded leading edge, the leading edge could be beveled, or something more complex. R_θ provides a local characterization of the boundary layer.

Figure 6.13 illustrates the development of C_f. The lines are the empirical data correlation (6.27) and the analytical solution for laminar flow (6.26). Symbols are experimental data. The experimentally measured C_f starts at the laminar level when $R_\theta \lesssim 200$. It then peels off the laminar curve, reaches a minimum at $R_\theta \approx 250$ and begins to rise. It reaches the turbulent data correlation at $R_\theta \approx 600$.

This evolution embodies the phenomenon of transition from laminar to turbulent flow. As R_θ increases, the boundary layer eventually becomes unstable and regular, laminar motion is replaced by irregular, turbulent flow. Turbulent eddies greatly increase the rate of momentum mixing. They transport momentum from the upper part of the boundary layer toward the wall, where it effects a frictional force. That is reflected in an increased surface shear stress.

The minimum in the data of C_f versus R_θ can be defined as the transition Reynolds number. The particular value of about 250 in Figure 6.13 is not general. It corresponds to an experiment in which there was 2.5% turbulent intensity in the stream above the boundary layer. The percentage of free-stream turbulence intensity is defined as $\mathrm{Tu} \equiv 100 u'/U_\infty$. The formula

$$R_{\theta_{tr}} = 163 + e^{6.91 - \mathrm{Tu}} \tag{6.31}$$

fits experimental data on the location of transition versus free-stream turbulence intensity. This implies that transition will be inevitable if $R_\theta > 1,165$, corresponding to $\mathrm{Tu} = 0$. Equation (6.31) also implies that the flow will remain laminar, even in the face of large external disturbances, if $R_\theta < 160$.

It should be warned, however, that these numbers are true only for zero pressure gradient boundary layers. Adverse pressure gradient makes the boundary layer more prone to transition; favorable pressure gradient makes it less so. When to compute the flow as laminar, and when to compute it as turbulent, is often uncertain. The state of the boundary layer is quite crucial in many applications, as should be obvious from Figure 6.13. Assuming laminar flow when it should be turbulent could underestimate drag by an order of magnitude. An erroneous assumption could have another, more drastic effect on the overall flow: a turbulent flow might remain attached under conditions where a laminar flow would separate. In fact, promotion of turbulence is a means to avoid separation.

An striking example of the ability of turbulence to delay separation is provided by the "drag crisis." If the boundary layer on a cylinder or sphere is laminar it will separate shortly before the topmost point (§4.1.1.3); but if the boundary layer is turbulent, the flow will remain attached beyond the top and separate further to the rear of the object. The drag on a bluff body is primarily the consequence of the low base pressure on its rear. As the separation point moves back, less surface area lies in the low pressure, separated zone. The integrated effect is that drag decreases. Starting at low Reynolds number with a laminar boundary layer, as the Reynolds number increases the drag coefficient falls, as shown by Figure 6.14. As explained on page 142, the decrease of C_D in laminar flow is a consequence of the way drag coefficient is defined. But, beyond a critical value ($Re \approx 10^5$ for a cylinder), C_d drops abruptly. This occurs because the boundary layer makes a transition to turbulence. Then separation shifts rearward and the size of the base pressure region decreases.

6.3.2 Wall Functions

The law-of-the-wall (6.21) and logarithmic overlap law [(6.23) and (6.24)] provide a theoretical anchor for the description of turbulent boundary layers. It is seen in Figure 6.10, at right, that the logarithmic layer is a distinct region in a log-linear plot. Put in linear–linear coordinates, in Figure 6.10 at left, it is barely discernible. It seems rather remarkable that this law should be such a prominent theoretical feature. But it is, indeed, not only a valuable theoretical anchor, it has very practical implications too.

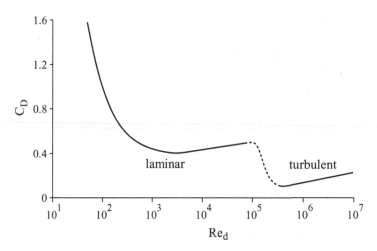

Figure 6.14. Drag crisis on a sphere. At $Re \approx 10^5$ the drag coefficient, C_D, suddenly drops when the boundary layer on the sphere becomes turbulent.

A well-established practice is to fit the log-law to experimental data to measure skin friction. Suppose measurements of u are made at two points, y_1 and y_2 inside the log layer. Then formula (6.24) gives

$$u_2 - u_1 = \frac{u_*}{\kappa} \log(y_2/y_1). \tag{6.32}$$

Given $\kappa = 0.41$, the value of u_* can be solved from the two data points. With more than two measurements, u_* can be obtained from a least-squares fit of the logarithmic formula to the data. Thus skin friction can be found without making measurements very near to the wall. In high Reynolds number flow, it becomes virtually impossible to measure the velocity shear right next to the wall. Fitting data to the log-law is the only means to obtain the skin friction.

For its validity, Eq. (6.32) requires that y_1 and y_2 fall in the layer $0.2\delta_{99} > y > 40\nu/u_*$. That is a consistency check which can be made after u_* is determined.

These same ideas translate into a computational method called the *wall function*. The viscous wall layer is extremely thin, but the velocity increases greatly across it. That can be substantiated by a few estimates. The bottom of the log-layer is at the fraction $(40\nu/u_*)/\delta_{99}$ of the boundary layer thickness. Estimating $U_\infty/u_* \sim 20$, this is $800/R_{\delta_{99}}$ of the thickness. A 99% thickness Reynolds number of 10^4 is quite moderate: the viscous wall layer is then 8% of the overall boundary layer thickness. In Figure 6.10, it is seen that the velocity is 50% of the free-stream value at the top of the viscous layer. Thus, 50% of the velocity profile lies within the first 8% of the distance from the wall, according to this estimate.

Clearly, the law-of-the-wall layer is a region of steep velocity gradient. Steep gradients are demanding of grid resolution. Long, narrow, high aspect cells are needed to compute in this region. That adds significantly to both the number of elements needed to compute the flow and to the computational stiffness. Hence, the viscous layer is deleterious to both cost and convergence.

Theory provides a detour. If the wall law (6.21) is valid, then, in principle, a solution to that region can be obtained once and for all. Then we need only compute the flow above the universal layer, matching to the known solution to provide a boundary condition for the rest of the domain. Most of the geometrical region is then solved by computation, but the thin wall layer is avoided by invoking a known, wall-function solution. The grid requirements can be reduced substantially.

To circumvent the wall layer, the first grid point is put at a y_1 that is somewhere above the surface. It is assumed to be within the universal wall layer. If the velocity, u_1 has been computed at y_1, then the skin friction can be found from Eq. (6.21), written as

$$u_1 = u_* F(y_1 u_*/\nu).$$

This is implicitly a formula for u_*, given u_1. It is a drag law, because $\tau_w = \rho u_*^2$ provides the shear stress on the wall. Thence, the skin friction can be obtained without integrating to the wall.

How is the law-of-the-wall function, F, determined? If $y_+ > 40$ then it is the logarithmic formula $F = 1/\kappa \log(y_+) + B$. This is acceptable if the first grid point is far enough from the wall to satisfy $y_+ > 40$. Sometimes the wall functions are implemented with this constraint.

However, the theory of a universal layer applies below the log-layer, as well. One simply needs a suitable formulation to represent F. One way to view the wall layer is that it is a constant stress layer; in other words, it is a zone of turbulent Couette flow. If the stress is constant, independent of y, then that constant is equal to the wall stress. Above the wall, the stress is part viscous, part turbulent. The sum of these is constant. If the turbulent component is modeled by the eddy viscosity assumption, then within the wall layer this sum is

$$\tau = (\mu + \mu_T)\frac{\partial u}{\partial y} = \tau_w.$$

μ is the molecular viscosity and the eddy viscosity μ_T represents turbulent mixing. Thence, the velocity at y_1 is

$$u_1 = \int_0^{y_1} \frac{u_*^2}{\nu + \nu_T} dy. \tag{6.33}$$

This is the needed function $u_1 = u_* F(y_1 u_*/\nu)$. An assumed distribution of eddy viscosity makes the integral concrete. For instance, the VanDriest formula is

$$\nu_T = \kappa u_* y[1 - \exp(y_+/26)].$$

Alternatively, a boundary layer can be computed, all the way to the wall, and the portion of the solution that lies in the wall layer can be tabulated, or fit by a curve, for future use as a wall function.

We provide no further details on wall functions. It is critical to recognize their limitation. They have a good theoretical basis, but that theory invokes an equilibrium assumption that is not always met. If the boundary layer leaves the wall, then it is clear that the wall function becomes erroneous in the vicinity of separation. It can continue to be used as an operational device, but it cannot be expected to be accurate. Indeed, strong pressure gradients, large surface curvature, and many other substantial perturbations will invalidate the surface layer equilibrium assumption and make wall functions inaccurate.

6.3.3 Free Shear Layers

Laminar free-shear layers were discussed in §4.2. Turbulent shear layers are analogous; they simply spread in consequence of turbulent mixing rather than by molecular diffusion. As might be imagined, that means they grow more rapidly than their laminar counterparts.

Shear layer thickness, δ, characterizes the extent of a distinct, vortical region of the mean flow. As the shear layer proceeds downstream, the thickness of the vortical region grows. Because turbulent stress is responsible for the growth, it is insensitive to Reynolds number once the flow is fully turbulent. This is a general characteristic of Reynolds averaged analysis of turbulent flow; molecular viscosity plays a small role except in the viscous wall layer. Gridding for numerics is largely insensitive to Reynolds number – in stark contrast to laminar flow.

In most engineering applications shear layers are not self-similar; that is, they are not simply characterized by a thickness and a velocity scale. For example, immediately after a round jet exits a nozzle it consists of a *potential core* surrounded by a thin shear layer, the latter being the downstream continuation of the boundary layer inside the nozzle. The free shear layers spread, eventually merging in the center at about 5 jet diameters downstream. Only after about 15 diameters does the flow fully develop into a self-similar jet. That might not happen: before it can become self-similar, the jet might be altered by interactions with geometry; for instance, it might impinge on a wall, as in Figure 4.28 on page 163.

As in Chapter 4, scalings of ideal, fully developed shear flows will be described. Again, these give guidance and estimates to computational analyses.

The integrated momentum flux that is used in quasi-one-dimensional analysis is $J \equiv \rho u^2 A$ [see Eq. (4.8)]. A is the cross-sectional area and u is an average velocity. For a round jet, A is proportional to jet thickness squared, δ^2. If the centerline velocity, U_{cl}, is invoked as a velocity scale, then the momentum flux, J, is proportional to $\rho(U_{cl}\delta)^2$. In the absence of external forces, momentum flux is constant. It follows that

$$\delta U_{cl} = \text{constant} \tag{6.34}$$

in constant density flow. Thus, the centerline velocity is inversely proportional to the thickness: $U_{cl} \propto 1/\delta$. We would like to make an estimate from this of how the thickness grows. That will be done by invoking an eddy viscosity assumption. This is where the phenomenology enters.

Diffusional spreading is assumed to be analogous to the laminar case, as in Eq. (4.7), except with an enhanced viscosity. An important difference is that the eddy viscosity depends on the flow. *Eddy viscosity is not a property of the fluid; it is a property of the flow.* Hence, it, too, must scale on the characteristic length and velocity. Because ν_T has dimensions of ℓ^2/t, eddy viscosity must be proportional to $U_{cl}\,\delta$.

The formula (4.7) for diffusional spreading can be written

$$\frac{D\delta^2}{Dt} = 2\nu_T.$$

On the left, $D\delta^2/Dt \sim U_{cl}d\delta^2/dx$. On the right, $\nu_T \sim U_{cl}\delta$. Hence

$$U_{cl}\delta\,d\delta/dx \sim U_{cl}\delta.$$

Thus, the turbulent jet spreads linearly,

$$\delta \propto x. \tag{6.35}$$

That is the same as in laminar flow. But in turbulent flow the constant of proportionality is independent of Reynolds number.

Let thickness of the round jet be defined as the radius at which the velocity is one-half of the centerline value. Then the constant of proportionality obtained from experiments is $d\delta/dx = 0.094$ with about 10% variability from one measurement to the next.

Spreading rates for other shear layers are summarized below, without derivation. The constants of proportionality are from experimental data; they depend on the precise definition of thickness, so they should be regarded as order of magnitude values. In the formula for the two-dimensional wake, d is a diameter of the wake-generating body, and C_D is its drag coefficient:

$$
\begin{aligned}
&\delta \sim 0.094x &&\text{round jet}\\[2mm]
&\delta \sim 0.21\sqrt{C_D d}\,\sqrt{x} &&\text{two-dimensional wake}\\[2mm]
&\delta \sim 0.084\frac{U_\infty - U_{-\infty}}{U_\infty + U_{-\infty}}x &&\text{mixing layer.}
\end{aligned}
\tag{6.36}
$$

Although there is some similarity between these and laminar spreading rates, the critical distinction is that they do not depend on Reynolds number. Physically, spreading is effected by turbulent stirring, not by laminar diffusion. Growth rates are larger than in laminar flow, and the mechanism is different.

6.3.4 Computed Axisymmetric Jets

An axisymmetric turbulent jet was computed with a two-equation turbulence model. Contours of eddy viscosity are overlaid by streamlines in Figure 6.15. The eddy viscosity starts to become large at the edge of the shear layer. It diffuses into the jet, reaching the axis just before the outflow boundary. A turbulent jet becomes fully developed after about 15 diameters; hence, this domain is not long enough for the jet to become self-similar. The eddy viscosity reaches a level

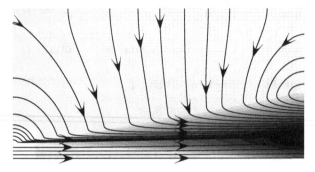

Figure 6.15. An axisymmetric jet. Streamlines over contours of eddy viscosity. The maximum contour level is $\nu_T/\nu = 2,300$.

2,300 times higher than molecular viscosity. Mixing is dominated by turbulence. Streamlines show amient fluid being entrained across the computational boundaries (see §4.2.2).

In the lower part of the figure, swirl has been added. At the inlet the velocity consists of an axial component, U, and an angular component, Ωr, for r inside the jet. The swirl parameter is defined to be the ratio of angular velocity at the radius, a, of the jet to axial velocity: $S = a\Omega/U$. In this computation $S = 2$. Sufficient swirl will provoke a region of backflow shortly downstream of the jet exit. Backflow is caused by the axial pressure gradient. The pressure on the axis is less than ambient in consequence of radial equilibrium between pressure gradient and centrifugal acceleration [Eq. (1.47), page 32]. Near the exit this produces a low pressure on the jet axis. Farther downstream, the swirl has diffused outward. The pressure on the axis is higher than upstream. Thus, the pressure gradient force is directed upstream on the axis. This is the origin of the recirculating velocity seen in Figure 6.16. Eddy viscosity is convected upstream by the backflow, and it can be seen that turbulent mixing is initiated immediately as the jet enters the domain. The spreading rate is greatly elevated by swirl.

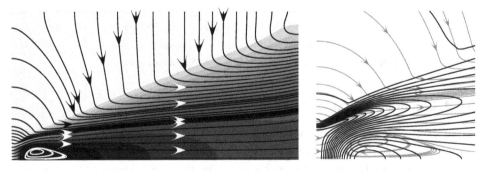

Figure 6.16. An axisymmetric jet with swirl. At left, streamlines plotted over contours of eddy viscosity, the maximum contour being $\nu_T/\nu = 4,200$. At right, zoom of recirculating region. u contours overlaid on streamlines.

It must be remarked, however, that strong swirl can suppress turbulence. This was discussed on page 223. The centrifugal force can suppress the radial component of turbulent velocity fluctuations. That in turn can suppress the production of turbulent energy – at least that can occur in nature. Unfortunately, the two-equation models described in §6.2 are not able to capture swirl stabilization.* In some situations, erroneous predictions could occur. In the present case, the turbulence is not suppressed because the recirculation maintains high levels of k on the axis.

One application of swirling jets is to flame stabilization in combustors. The backflow carries hot gas toward unburned gas, entering in the jet. Thereby, the flame is kept lit.

6.4 Eddy Simulation

The discussion of eddy simulation will be preceded by a quick summary of some classical ideas about the spectrum of turbulent motions. The mode of thought introduced in the classical literature formed the basis for early work on numerical simulation. Were computer power sufficient, there is little doubt that the pioneers of turbulence theory would have delved into eddy simulation. Indeed, a classic article written by Taylor and Green in 1936 (Taylor, 1971) presented a direct numerical simulation performed by series expansion in time. The Taylor–Green study became a milestone in this subject. It was entitled "Mechanism of the Production of Small Eddies from Large Ones." That is a defining problem of the classical theory of homogeneous turbulence.

6.4.1 Spectrum of Turbulence

Turbulence is characterized by a continuous spectrum of eddy sizes. Ideas about characterizing turbulence gel around dividing the spectrum into large, small, and intermediate ranges. Loosely speaking, the spectrum is a plot of energy versus eddy size (Figure 6.17). The energy is zero at very small size, increases to a peak at a large scale, and then falls at a very large scale. The conventional way to display the spectrum is as at the right of Figure 6.17. This is energy density versus wave number, defined as $k = 2\pi/\lambda$. Wavelength, λ, characterizes eddy size. Small wavelength corresponds to large k. Smaller eddies have low energy; hence, the spectrum tapers off rapidly at high k.

A plot of the vorticity spectrum would go the other way: small eddies have high vorticity – at least until some very small size is reached, below which there is little turbulent motion. The rate of energy dissipation goes like vorticity: it is largest at small scales. That is because vorticity and dissipation involve derivatives

---

* Second moment closure models – see Durbin and Reif (2001) or Pope (2000) – do capture effects of swirl.

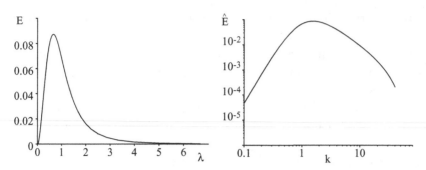

Figure 6.17. Energy spectra in physical and Fourier space. In physical space, at left, the spectrum can be understood as average energy versus eddy size. Energy is normally plotted against one over the size, as at the right. The axes are scaled arbitrarily.

of velocity. (The rate of energy dissipation is $\mu|\nabla \boldsymbol{u}|^2$. Differentiation amplifies small scales.)

This is the picture that emerges: on average, the energy is highest at large scales, whereas the rate of dissipation peaks at small scales. We are led to the concept of an *energy cascade*: to achieve an equilibrium, energy must flow from the large scales, where it is concentrated, to the small scales, where it is being dissipated. That is what is meant by the energy cascade. If the turbulence is maintained by mean shear, the shear produces the large-scale energy. It cascades to small scale as a consequence of the fluid dynamics. The energy and dissipation spectra, described in the previous paragraph, imply these dynamics, but the reasoning is not concrete. What fluid dynamical processes are responsible for energy transfer to small scales?

That is debatable. Large eddies might develop instabilities that produce smaller and smaller scales of motion or straining might stretch and fold large-scale vortices to create smaller-scale structure. Somehow or other, energy is transferred from the large to the small scales: that is a central tenet of turbulence theory. It is also an underpinning of all types of eddy simulation.

The notion of an energy cascade was introduced by L. F. Richardson in his prescient monograph on weather prediction by numerical process – published in 1922, two decades before the invention of the digital electronic computer. Richardson proposed the notion of simulating the evolution of weather patterns by solving discretized conservation equations. His fantastic vision was of the seats of a concert hall filled with technicians at adding machines, the nodes and processors of a computational grid. A conductor on the podium directs the flow of data between the nodes. To describe the role of turbulence in the atmosphere, Richardson paraphrased a poem by Jonathan Swift: in Richardson's version, "Big whorls have little whorls, which feed on their velocity; and little whorls have smaller whorls, and so on to viscosity (in the molecular sense)." This remains our conceptual view: large eddies grind down into smaller and smaller ones, until energy is dissipation by viscous action at the smallest scales.

The smallest scales are supposed to provide dissipation at just the right rate to maintain equilibrium with the grinding down. The philosophy of large eddy simulation (LES) is that the smallest scales do not have to be computed; an equivalent, artificial dissipation can perform their role. This is analogous to the thinking behind wall functions. The small scales are hypothesized to have universal structure. They can be represented by a fixed formula. This saves computational expense.

The large and small scales of the energy spectrum are analogous to the inner and outer zones of the turbulent boundary layer. The analog of the logarithmic overlap law is Kolmogoroff's $-5/3$ law. This law, devised in 1941 by the great mathematician A. N. Kolmogoroff, was originally conceived as the $+2/3$ law,

$$E \propto (\varepsilon r)^{2/3}. \tag{6.37}$$

Kolmogoroff asked the question, "If dissipation removes energy at a rate ε, how should energy increase with the size of the eddies in order to maintain equilibrium?" His answer is a matter of dimensional analysis. Energy, per unit mass, has dimensions of velocity squared ℓ^2/t^2. Dissipation is its rate of change, $\varepsilon \sim \ell^2/t^3$. r is the size of the eddies, $r \sim \ell$. Both sides of $E \propto (\varepsilon r)^{2/3}$ have dimensions of velocity squared. If dissipation controls the size of eddies, the $r^{2/3}$ law follows.* The smallest eddies are also influenced by molecular viscosity. The set of parameters ε, r, and v, combines into the nondimensional variable $r(\varepsilon/v^3)^{1/4}$. Define η to be $(v^3/\varepsilon)^{1/4}$. Then formula (6.37) can be extended over the entire universal range as

$$E = (\varepsilon r)^{2/3} F(r/\eta).$$

The $r^{2/3}$ law obtains when $r \gg \eta$ so that $F(r/\eta) \to F(\infty)$ becomes a constant of proportionality.

Dissipation is in control only at small scales. At large scales, the size and energy of eddies are controlled by how they are produced. If they originate in the wake of a body, the size of the body sets the large scale. If they are boundary layer eddies, they scale with boundary layer thickness. Universality can only be assumed at much smaller scales than the geometry and away from boundaries. If the geometrical scale is δ, Kolmogoroff's law (6.37) is applicable when

$$\eta \ll r \ll \delta.$$

This is called the *inertial subrange*. It is the range in which neither viscosity or geometry are influential.

Near a wall, the size of the largest eddies is confined by the distance to the wall. As the surface is approached, the largest eddies become small and affected by viscosity. Figure 6.18 suggests schematically how size diminishes as the wall is approached. There is no room for eddies much smaller than the largest to exist. Ideas about the energy cascade and small-scale universality have no merit in the near-wall zone.

---

* In Fourier transform space, this becomes $\hat{E} \propto \varepsilon^{2/3} k^{-5/3}$, where the wave number, k, has dimensions of $1/\ell$ and the spectral density $\hat{E}$ has dimensions of ℓ times velocity squared. This is why Kolmogoroff's law is usually cited as the $-5/3$ law.

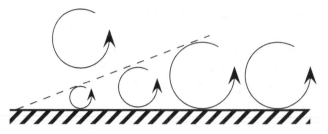

Figure 6.18. The average size of eddies increases with distance from the wall until a maximum size is reached.

6.4.2 Direct Numerical Simulation

Turbulence consists of irregular, eddying motion. It is governed by the Navier–Stokes equations. With sufficient grid and time-step resolution, and with sufficient computer power, eddying can be simulated accurately. Fully resolved simulation is called direct numerical simulation, or simply, DNS.

The simulation of flow in a ribbed channel, shown in Figure 6.1, required 70×10^6 grid points and 10^5 time-steps. How are these values devised? The grid must be fine enough to resolve all scales of the turbulence. For channel flows and boundary layers, this is found to require streamwise and spanwise spacings of $\Delta x_+ \sim 20$ and $\Delta z_+ \sim 6$. Plus units (page 227) are motivated by law-of-the-wall scaling and phenomenology. The phenomenology is that eddies near a wall form a pattern of perturbation jets or *streaks*. These are elongated patterns that can be seen in contours of u velocity, such as Figure 6.19, and in isosurfaces of vorticity, as in Figure 6.20. On average, they are spaced about 100-plus units in z and have about a 20-plus unit width. $\Delta z \sim 6$ is the grid spacing needed to resolve them accurately. The streaks can be quite long, on the

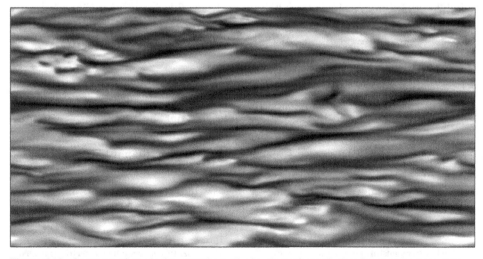

Figure 6.19. Contours of u velocity near the wall of a plane channel, showing streaky appearance. This is a plan view, in a plane at $y_+ = 10$, parallel to the wall.

Figure 6.20. Isosurfaces of x vorticity near a wall in a perspective view. Dark gray surfaces are positive; light gray are negative.

order of 1,000-plus units in the x direction. However, smaller-scale features also are seen in Figure 6.19, and it is found that Δx cannot be much larger than Δz.

Consider flow through a channel of height h and width w. In plus units $h_+ = hu_*/\nu$. This is a Reynolds number. For fully developed turbulence to exist, this must be on the order of 10^3 or more. If the width is comparable to the height, the above estimates tell us that on the order of $w_+/\Delta z_+$ points are needed in the span; for $w_+ = 10^3$ this is about 160 points. If the length of the channel is four or five times its height, then on the order of 250 points are needed in the x direction.

The grid requirements in the y direction are more difficult to estimate. The steep shear near the wall must be well resolved. This demands $\Delta y_+ < 1$. Farther from the surface, the spacing can expand: Δy_+ can increase to values comparable to Δz_+. Hence, the number of points in y will be greater than $h_+/\Delta z_+ \sim 160$; 250 might be a reasonable estimate for the simple channel flow. This gives a total of

$$N_x N_y N_z \sim 250 \times 250 \times 160 = 10^7.$$

The way this was devised makes it clear that an element of art is needed. More correctly, the grid requirements can be estimated, but grid independence must be checked via computer simulations, and the grid must be refined if it is deficient.

Complexities in the geometry require local grid refinement and increase the total count of grid points. Suffice it to say that DNS is very demanding of computer resources – and of human effort.

The time-step is set by stability and accuracy considerations. A simple physical criterion subsumes both. A fluid particle should travel less than one grid point per time-step; that is, $u_i \Delta t < \Delta x_i$. This must be met for convection in any direction, x, y, or z. The values $u_+ \sim 20$ and $\Delta x_+ \sim 20$ are representative. Hence, typically, $\Delta t_+ \sim 1$. If the length of the channel is $\sim 5 \times 10^3$ plus units, then on the order of 10^4 time-steps are needed for a fluid particle to pass through the domain. A full simulation requires about 10 through flow times. That means $\sim 10^5$ time-steps. The number is determined by two factors. About half of the time-steps are needed to pass through a transient stage, in which the turbulent solution is established. The other half is required to compute statistics of the flow.

Unfortunately, statistics converge slowly, like $1/\sqrt{N_{\text{smpl}}}$, the inverse of the square root of the number of independent samples. Samples become independent when they are from different eddies. Many samples from a single eddy will not help convergence. The time average must extend over many eddy timescales. This entails long computing time. It is alleviated a bit if averaging can also be done in directions of spatial homogeneity: for instance, the ribbed channel of Figure 6.1, was averaged in time and in the z direction.

In complex geometries, with no direction of homogeneity, longer averaging time is needed. The number of time-steps required to obtain converged statistics can be exceedingly large. That makes DNS viable only for fundamental studies, not for routine flow computation.

We have introduced the notion of grid requirements by making estimates for flow in a channel. Another starting point is to estimate how of grid requirements depend on Reynolds number. Classic theory is invoked to estimate the number of active modes. The theory is Kolmogoroff's $+2/3$ law. As will be seen in Eq. (6.40), this law implies that the number of spatial modes increases as Reynolds number to the power $3/4$ in each direction. In three dimensions,

$$N \sim N_x N_y N_z \sim Re^{9/4}. \tag{6.38}$$

Similarly, the number of frequency modes increases as Reynolds number to the power $1/2$. The total increases as almost the third power of Reynolds number. If all of these modes must be computed accurately, the expense becomes overwhelming. That limits DNS to fairly low Reynolds numbers.

The estimate (6.38) is appropriate to *isotropic*, or nondescript, turbulence. Isotropy means directional independence. Statistics are unaltered by rotation of the coordinate axes. Rotation by $90°$ about the z axis will convert u to v and v to $-u$. If this is to have no effect on the statistics, it must be true that $\overline{u'^2} = \overline{v'^2}$. Furthermore, because $\overline{u'v'}$ is rotated into $-\overline{u'v'}$, that quantity must vanish. Pursuing this constraint on all components of the Reynolds stress tensor (6.5) leads to the conclusion that isotropy demands

$$\overline{u'u'} = \overline{u'^2}I, \tag{6.39}$$

where I is the identity matrix. All directions are on a par in isotropic turbulence.

Real-world turbulence is not isotropic. One direction or another has a distinctive property. Wall-bounded turbulence is not isotropic because the wall normal is an identifiable direction. However, it remains true in nonisotropic turbulence that grid requirements increase steeply with Reynolds number, approximately as given by Eq. (6.38).

6.4.3 DNS of Isotropic Turbulence

The earliest direct simulations were of isotropic turbulence. DNS is a tool for basic studies. It is the computer analog to a basic experiment, undertaken to measure fundamental quantities, or to gain understanding of isolated phenomena. Homogeneous,

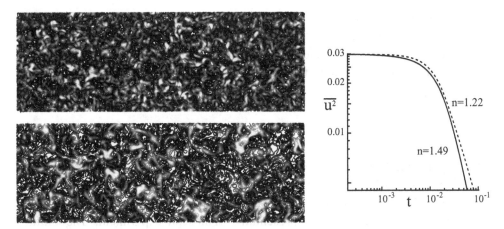

Figure 6.21. Contours of velocity magnitude in two simulations of isotropic turbulence. The representative eddy size is three times larger in the lower case. The evolution of variance with time is plotted at right.

isotropic turbulence is an example of precisely that concept of isolating fundamental aspects of turbulence. The effort to understand isotropic turbulence occupied many researchers in the early days of turbulence theory and again in the early days of computer simulation.

Very efficient, pseudospectral methods are applicable to this case. The simulation starts by summing a large number of sine waves to represent the initial velocity field. These have wavelengths, λ, in the x, y, and z directions. The wave number vector, k, is formed from the inverse of the wavelengths

$$k = \left(\frac{2\pi}{\lambda_x}, \frac{2\pi}{\lambda_y}, \frac{2\pi}{\lambda_z} \right).$$

This is a vector in three dimensions. It has a magnitude and two angular directions, its latitude and longitude. If the turbulence is isotropic, its averaged energy depends only on the magnitude, not on the angles. The sine waves of the initial velocity field are given amplitudes that are consistent with isotropy, they also are given random phases – and the simulation begins.

The simulation consists of marching the Navier–Stokes equations forward in time. As time evolves, nonlinearity activates the energy cascade, and the turbulence develops into a realistic state. Two examples of the instantaneous field of velocity magnitude are shown in Figure 6.21. The property of isotropy is that the contours show no directional preference; some are elongated more toward the horizontal, some more toward the vertical. Because of the directional independence, all three components of turbulent intensity are equal; $\overline{u'^2} = \overline{v'^2} = \overline{w'^2}$.

The notion of *homogeneity* is also illustrated by Figure 6.21. These instantaneous contours show the spatial variation of velocity. They have a random appearance. There is no coherent pattern to this randomness – contrast it to Figure 6.1, where small eddies are seen near the wall, and larger ones are seen in the middle of the channel. At any point, the same level of turbulent fluctuation is as likely as at any other point. The intensity, $\overline{u'^2}$, is not a function of position. Homogeneity means that statistics do not depend on position in space. Figure 6.21 is an instantaneous view of a statistically homogeneous field.

Given homogeneity and isotropy, the turbulent energy can depend only on time. Its evolution is plotted in Figure 6.21 on log-log axes. After an initial period, it decays exponentially, according to

$$\overline{u'^2} \propto t^{-n}.$$

The exponent n decreases with increasing turbulent Reynolds number. For the two fields in Figure 6.21 it has the values of $n = 1.49$ and $n = 1.22$. Theoretically, it approaches 2.5 as $Re \to 0$, and it is speculated that $n \to 1$ as $Re \to \infty$. These extremes are never seen in practice; 1.2 is representative of most lab experiments. The solution (6.15), to two-equation models, agrees in form with the DNS result.

6.4.4 Approximate Simulation of Large Eddies

Efforts have been made to shortcut the expense of DNS while continuing to compute the random eddying motion by Navier–Stokes simulation. The estimate (6.38) is based on the ratio of largest to smallest scales of the eddies – the allusion to largest and smallest means representative dimensions; of course, there is no largest or smallest eddy per se. The largest scale is set by some aspect of the flow geometry or computational domain. Let it be L. The smallest scale, η is set by viscosity; it is the scale at which viscous dissipation of vorticity dominates over stretching. For fixed L, η decreases with Reynolds number and the ratio L/η increases.

η is the scale at which universality applies. Recall Kolmogoroff's reasoning that these scales are controlled by ε. Viscosity and dissipation combine to produce

$$\eta = (\nu^3/\varepsilon)^{1/4} \qquad (6.40)$$

with the dimension of length. This is called either the dissipation length or the Kolmogoroff length. The appearance of $\nu^{3/4}$ is why $L/\eta \sim Re^{3/4}$. This leads to the scaling (6.38).

Do we really need to simulate all scales or can something less costly be done? Motions at scale L are the most energetic. It might be hoped that to a good approximation, only these scales need be simulated. That is the idea of large eddy simulation (LES). If the range of eddy size were cut off at a fixed scale, L_{min}, then the grid requirement L/L_{min} would be insensitive to Reynolds number. Great reduction in grid requirements might be obtained.

In practice that does not work. The problem stems from focusing on the most energetic eddies. If only they were needed, then LES would be very effective. However, if instead of energy, the focus were on vorticity – mean-squared vorticity is called enstrophy – then the opposite picture emerges: the smallest scales dominate. That is the character of hydrodynamic turbulence; energy dominates at large scale; enstrophy dominates at small scale. Fortunately, the primary role of small-scale vorticity often is to dissipate turbulent energy. The theory that those scales are universal has been mentioned previously: maybe they need not be simulated; maybe they can be replaced by a model that plays a similar role to theirs. The model need only produce an appropriate rate of dissipation. That is the rationale for LES: the large scales are simulated, the small scales are represented by a *subgrid* scale model.

Unfortunately, that simplistic idea is not satisfactory near to boundaries. The idea of small-scale universality is invalid in the vicinity of surfaces. The shortcuts behind LES are best justified for free-shear flows, away from walls.

This last remark is a caveat about accuracy. No accuracy criterion for LES exists at present. In particular, the idea of grid independence (Chapter 2) is not applicable to LES. That is because in current practice, the cutoff $L_{\min}$ is defined to be the grid spacing. Hence, the notion of "large" means large relative to the particular grid.

In this brief introduction to the field of large eddy simulation, we mention the concepts but will not dwell on the mechanics of how simulations are performed. Suffice it to say that a good deal of attention must be paid to adjusting the grid and inflow conditions to capture particular scales. A good deal of computing is required to obtain quantitatively accurate statistics. With due care accurate simulations can be performed.

6.4.4.1 Filtering and Subgrid Models

The separation into large eddies and subgrid components has a formal definition. Notionally, the flow field is passed through a filter that lets scales larger than the grid size pass. These are the eddies that will be simulated. The scales that do not pass through the filter are the subgrid scales. These must be modeled. To illustrate, the box of turbulence in Figure 6.21 is passed through a low pass filter, defined as follows. The velocity is represented by a sum of Fourier modes in each of the three directions:

$$u(x) = \sum_{-N_x}^{N_x} \sum_{-N_y}^{N_y} \sum_{-N_z}^{N_z} \hat{u}(n_x, n_y, n_z) e^{i\pi n_x (x/L_x)} e^{i\pi n_y (y/L_y)} e^{i\pi n_z (z/L_z)}. \qquad (6.41)$$

$\hat{u}$ is the amplitude of the Fourier component. Filtering means multiplying each amplitude by a number less than unity. The simplest filter is a cutoff. The cutoff filter equates $\hat{u}$ to zero after some value Nc of n. Thus, the filtered velocity is

$$\langle u(x) \rangle = \sum_{-Nc_x}^{Nc_x} \sum_{-Nc_y}^{Nc_y} \sum_{-Nc_z}^{Nc_z} \hat{u}(n_x, n_y, n_z) e^{i\pi n_x (x/L_x)} e^{i\pi n_y (y/L_y)} e^{i\pi n_z (z/L_z)}.$$

The angle brackets signify the operation of filtering. Figure 6.22 compares a fully resolved DNS field to the same field, filtered with $Nc = N/8$. The 1/8 filter isolates large scales. A 1/4 filter would look very similar to the full field. To see a difference in that case, vorticity is contour plotted in Figure 6.23.

The reason the velocity filter must be set to $N_c = 1/8N$ for these particular data is because the energy, $|\hat{u}|^2$ of the small scales is low. Removing even 3/4 of the modes has little visable effect; when 7/8 of them are removed in each direction, the

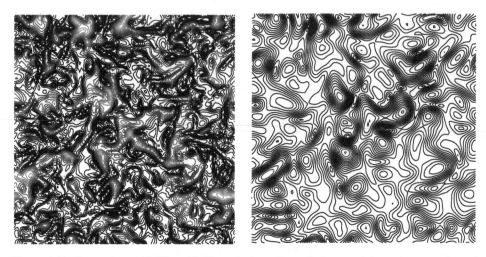

Figure 6.22. Comparison of DNS and LES resolutions. The turbulence at left was low pass filtered to produce that at right. Contours of velocity magnitude are plotted.

large features become predominant. A corollary is that this DNS is well resolved. It contains all scales of motion, down to those with extremely small energy.

It is not a rule that LES resolution is 1/8 that of DNS. It should be significantly more coarse than DNS, but there is no rule how much more. An LES grid for engineering flows might typically be a factor of 5 coarser in each direction than a corresponding DNS grid. The total reduction in grid count is this cubed, or 125: LES grids might be two orders of magnitude smaller than DNS grids. With a commensurate increase in time-step, a simulation could be more than two orders of magnitude less expensive than a fully resolved DNS. That makes eddy simulation feasible, albeit often only as a qualitative tool.

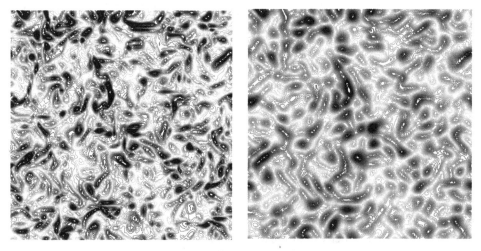

Figure 6.23. Comparison of DNS and LES resolution of vorticity magnitude. The data at left were filtered to produce those at right.

The vorticity is the curl of the velocity. Differentiation enhances small scales. The magnitude the vorticity in a Fourier mode is $|k\hat{u}|$, where k is the wave number vector, $(n_x, n_y, n_z)\pi/L$. Because enstrophy is larger at small scale, a noticeable difference between the original and filtered fields is seen with $N_c = 1/4N$ in Figure 6.23. The reader should be able to identify patterns in the right panels of Figures 6.22 and 6.23 that can be seen in the left panels and smaller features in the left panels that are not apparent on the right.

Formally, filtering leads again to Eq. (6.4). But now U is the *filtered* velocity and u' is the *subgrid* scale fluctuation. The filtered velocity is not ensemble averaged; it is still a random field from which only the small scales have been removed. The stress $\langle u'u' \rangle$ represents the effect of the small, subgrid eddies upon the larger, resolved scales. The two concepts central to LES are the filter and the subgrid model. In practice, the act of filtering need not be made explicit; but it is essential to realize that it only removes the small scales, leaving the randomness at large scales intact. The latter is to be simulated.

The subgrid scale model must be made explicit.* This is usually done with an eddy viscosity – the Smagorinsky model being the most common. Smagorinsky was a meteorologist. The seminal work in LES was done in the meteorological community in the late 1960s. If the largest scales are atmospheric, it is self-evident that simulating the whole spectrum, down to the smallest scales, is too demanding. It is not surprising that this is where LES was born. LES was subsequently applied to engineering flows, starting in the early 1970s.

Smagorinsky's model is an eddy viscosity. It becomes a subgrid model by making it a function of the grid spacing. The grid spacing is treated as a *mixing length* from which an effective viscosity is formed. The intent is to mimic the dissipative effect of small-scale eddies. The mixing length idea was introduced by Prandtl by an analogy to the mean-free path of molecules in the kinetic theory of gases. A contaminant is carried a distance ℓ_{mix} and then mixes with its environment. If starts at $x - \ell_{mix}$ and ends up at x, its concentration is $c(x - \ell_{mix})$. The concentration of the environment is $c(x)$. The fluctuation before mixing takes place is

$$c' = c(x - \ell_{mix}) - c(x) \approx \ell_{mix}\frac{\partial c}{\partial x}.$$

The same reasoning applied to velocity gives

$$u' = u(x - \ell_{mix}) - u(x) \approx \ell_{mix}\frac{\partial u}{\partial x}.$$

The flux $\overline{u'c'}$ is

$$\overline{[u(x - \ell_{mix}) - u(x)][c(x - \ell_{mix}) - c(x)]} \approx \left(\ell_{mix}^2\frac{\partial U}{\partial x}\right)\frac{\partial C}{\partial x}.$$

The bracketed term is the mixing length formula for eddy diffusion: $\nu_T = \ell_{mix}^2|\partial U/\partial x|$. In LES, the grid spacing plays the role of mixing length. The velocity gradient is

---

* The term *implicit subgrid scale modeling* exists in the literature. It means the numerical dissipation serves as the subgrid model.

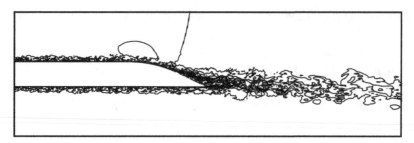

Figure 6.24. Large eddy simulation of flow past the trailing edge of a hydrofoil. Contours of velocity. Figure courtesy of Dr. Meng Wang.

represented by the rate of strain. This leads to the Smagorinsky formula

$$\nu_{sg} = C_s \Delta^2 |S|, \tag{6.42}$$

where Δ is the grid spacing and $|S|$ is the magnitude of the rate of strain. C_s is a constant that has a value of about 0.2. The simulation is effected by solving the full, unsteady Navier–Stokes equations, with the subgrid contribution (6.42) added to the molecular viscosity.

6.4.4.2 Numerical Illustrations

When discussing discretization on page 85 we noted that upwind biasing of the convective derivative produced a dissipative numerical error. The leading error of a centered discretization is dispersive. On a nonuniform grid the error may not be entirely dispersive, but, generally, centered schemes are less dissipative than upwind schemes. For that reason central differences are often preferred for LES. Dissipative numerical errors can overwhelm the eddying, causing the turbulence to decay when it should not. This is especially a problem for LES because of the marginal resolution of small eddies. Dispersive errors are not innocuous, but they can be masked by the eddying.

A related guideline for LES is that the numerical discretization should be energy conserving in the absence of viscous dissipation. The numerics should cause fluctuations to neither grow or decay; that should be left to the physics. Again, this concept derives from the meteorological community. If the fluctuations are weather systems, the numerics should neither cause or suppress storms. Energy conservation also ensures a type of stability: errors cannot grow explosively if fluctuating energy is conserved.

An instantaneous field of turbulent flow past a hydrofoil trailing edge illustrates the nature of large eddy simulation. The chord Reynolds number is 2×10^6. Contours of the streamwise velocity, scaled by free-stream speed, are plotted in the range 1.27 to -0.24 in Figure 6.24. They demonstrate the wide range of spatial scales resolved by the LES. Approximately 7 million grid points are used in the simulation, performed on a body-fitted structured grid. To capture the full range of scales, down to the dissipation range, would require the order of 100 million grid points: DNS would be more than an order of magnitude more expensive than LES.

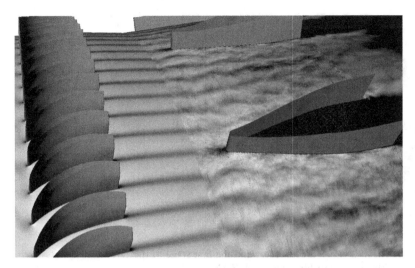

Figure 6.25. Turbulent flow in a compressor and the diffusing section of a gas turbine. The compressor is computed with RANS, the diffusing section with LES.

The maximum near-wall grid spacings in plus units [Eq. (6.19)] are 62, 55, and 2 in the streamwise, spanwise, and wall-normal directions, respectively. The subgrid-scale eddy viscosity varies strongly with local mesh spacing and flow features. Its time-averaged value has a maximum of approximately 3 times molecular viscosity in the boundary layers and 20 times molecular viscosity in the wake. The ratio ν_{sg}/ν is a measure of how well the eddies are resolved. If it is small compared to unity, the resolution is close to DNS. The values of 3 and 20 indicate that this is in the LES regime but still very well resolved. The subgrid viscosity plays a role, but it is still comparable to molecular viscosity.

The simulation of Figure 6.24 was motivated by the problem of noise generation as eddies pass a trailing edge. This is an area in which LES holds promise: unsteadiness of large eddies is essential to generating acoustic waves. Aeroacoustic noise is an application for which LES has many merits.

A second illustration of LES comes from a simulation of flow inside a gas turbine engine. Figure 6.25 illustrates the flow in the compressor and the diffuser, upstream of a combustor chamber. This figure is a contour plot of the instantaneous axial velocity at the midspan of the compressor blades and the diffuser vanes. On the left, the flow was computed with RANS; on the right, the flow was computed with LES. The eddying motion is resolved in the LES portion; however, the mean velocity would look similar to that, presented on the left.

As discussed previously, the resolution requirements for LES at high Reynolds numbers are severe, especially in the boundary layers. LES computations like these challenge the capabilities of the largest of modern-day computers. The RANS part of this simulation required 10^6 grid points within the domain shown. That zone consists of 14-blade passages. The two blade passages in the diffuser section contain 1.3×10^6 grid points. However, the latter is under resolved and would require about an order of magnitude more points to obtain suitable resolution of the large eddies.

Figure 6.26. Turbulent flow the in a combustor and the first stage of the turbine. The combustor jets are computed with LES, the turbine stage with RANS.

The second figure from the gas turbine simulation, Figure 6.26, shows combustor jets upstream of the stator and rotor of the first stage of the turbine. The jets are computed with LES and the turbine with RANS. This is a more attractive application of LES than Figure 6.26 because LES is better suited to free-shear flows than to boundary layers. The "large eddies" are larger in a free-shear layer. They are produced by shear layer instabilities (§6.5). The instabilities develop into large vortices, characteristic of detached shear layers.

Because these large, energetic scales of turbulence are resolved by LES, a more accurate description of scalar mixing is obtained. In this application, that leads to improved predictions when combustion takes place in the jets. RANS computations of turbulent jets in the combustor often fail to produce sufficient mixing.

However, as the flow reaches the turbine blades, it becomes dominated by smaller-scale eddies and resolution requirements for LES become demanding. Accurate prediction of the flow also requires a precise description of the turbulent boundary layers around the rotor and stator blades. The cost of resolving these with LES becomes prohibitive. Hence, the turbine rotor and stator are simulated with RANS in Figure 6.26.

As was mentioned in connection with the previous example, a way to assess the role of the subgrid model is to compare the subgrid viscosity to the molecular viscosity, ν_{sg}/ν. If the ratio is small the simulation is nearly DNS; if it is not small, the simulation is a genuine LES; if it is quite large, the simulation is underresolved. For the combustor, the ratio of subgrid to molecular viscosities is about 10–20 in the most of the chamber, but ν_{sg}/ν reaches values of 100–200 within the fuel jets. The latter indicates that the simulation is not adequately resolved.

As the wall is approached, LES becomes increasingly inaccurate unless a quite fine mesh is used. The size of energetic eddies becomes smaller and so must the grid (see Figure 6.18). The log-layer theory can be interpreted to say that eddy size decreases linearly with distance from the wall within that layer. The desire to avoid

DES LES

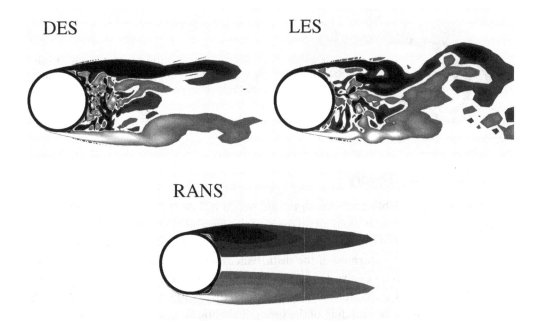

RANS

Figure 6.27. Vorticity contours from DES, LES, and RANS in flow around a sphere. Figure courtesy of G. Constantinescu.

this demanding, near wall region, led to the development of detached eddy simulation (DES).

Detached eddy simulation is a hybrid of RANS and LES. Near to boundaries it becomes RANS, far from boundaries it becomes eddy simulation; hence, the allusion to "detached eddies." RANS and LES are merged by an eddy viscosity model. Near boundaries, the full eddy viscosity is in effect. Proceeding away from the surface the eddy viscosity is reduced to the point that instabilities inherent in the Navier–Stokes equations can become operative. Irregular eddying then develops. Far from the surface, the grid spacing provides the length scale, making the model behave like an LES subgrid model. For instance, the k-ω viscosity can be written as $\nu_T = C_\mu \sqrt{k}\ell$, where $\ell = \sqrt{k}/\omega$. The DES formulation is

$$\ell = \min[\sqrt{k}/\omega, C_{\text{des}}\Delta],$$

where C_{des} is a constant that determines when the grid will clip the eddy viscosity. As a wall is approached the first factor tends to zero; hence, there will be no clipping. Moving out from the wall, the first factor increases and eventually becomes larger than the second; then the grid spacing becomes the length scale.

The LES and DES approaches are compared by simulations of flow round a sphere in Figure 6.27. The detaching shear layer develops large-scale unsteadiness. In DES, that happens farther from the surface than in LES. In the LES and DES approaches, it is felt that large-scale unsteadiness is far more important than the fine details.

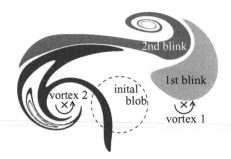

Figure 6.28. Convection by blinking vortices. Evolution of a dye blob is displayed at three times, shown by the three shades of gray.

6.5 Instability Theory

There are no reliable methods to predict when a flow should be turbulent. Most theoretical efforts start from an analysis of the instability of laminar flow. Instability theory addresses the question of whether small disturbances will grow. It does not go so far as to determine if the disturbances will evolve into turbulent motion. Sometimes they only evolve to another laminar state. However, as this is the concluding section of a chapter on turbulence, the discussion here is of ideas the have illuminated our understanding of the onset of chaotic flow. Texts such as Drazin and Reid (1995) and White (1991) can be consulted for a comprehensive development of the theory of hydrodynamic instability.

The ability of the Navier–Stokes equations to produce turbulence is quite remarkable and quite perplexing. Turbulence is a highly nonlinear phenomenon. However, the constant density, incompressible Navier–Stokes equations contain only one nonlinear term, $u \cdot \nabla u$. This is simply the convection of momentum. It seems that the ultimate source of chaotic flow simply is convection of momentum.

Convection by blinking vortices is a toy that illustrates how convection can generate complexity. Two vortices are situated as in Figure 6.28. A circular blob of dye is initially situated between them. Vortex 1 is allowed to convect the blob for some period of time; vortex two is inactive during this time. Then vortex 1 is switched off and vortex 2 is allowed to convect the blob. Call this one "blink." Figure 6.28 shows the dye distribution after one, two, and three blinks. Clearly, the dye blob is becoming increasingly complex. This does not illustrate turbulence: the blinking vortices have laminar streamlines; however, it is suggestive of the extent to which seemingly simple convection can generate chaotic mixing.

The blinking vortex is a kinematic perspective on chaos. Dynamics, and in particular dynamical instability, play a causal role. Turbulence is usually considered to be the end product of a succession instabilities. Initially, small perturbations grow into large disturbances. These undergo secondary instabilites and evolve into a more complex state. For instance, the first instability might produce regular wavy motion; the next might contort the wave crests. Through further instabilities erratic, turbulent motion emerges. Those further instabilities might not be clear-cut; they might be described succintly as a rapid breakdown into turbulence. That is the gist of Figure 6.29. Waviness is seen in a short region after the nozzle exit; then corrugations on the waves become noticeable. Shortly thereafter the fluctuations becomes chaotic.

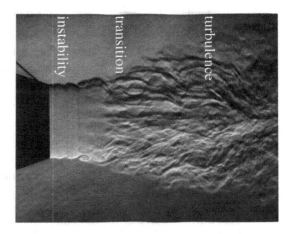

Figure 6.29. Instability, transition, and turbulence in a jet. Photograph courtesy of Professor P. Bradshaw.

A system is *stable* if any small perturbation decays. If some perturbations decay, but others grow, the system is *unstable*. Figure 6.30 is a standard illustration. A ball sitting at the bottom of a bowl is at a stable equilibrium. A small push to one side or the other will cause the ball to oscillate about the bottom, eventually settling back to its equilibrium position. Any perturbation decays, so the system is stable. But the apex of a dome is an unstable equilibrium position. Place the ball at the apex and give it a small push to one side or the other. The ball will not return to its rest position, it will roll off the dome. Small perturbations grow, so the system is unstable.

Mathematically, the two equilibria are at $\theta = \pm 90°$. Newton's second law gives the dynamical equation

$$\ddot{\theta} = -\frac{g}{R}\cos\theta.$$

The equilibria are where $\ddot{\theta} = 0$; hence, $\theta_{eq} = \pm 90°$. Near equilibrium, $\theta = \theta_{eq} + \varepsilon\theta'$, where ε represents the very small push. Then, by Taylor series approximation, $\cos\theta \approx -\varepsilon\theta'\sin\theta_{eq}$ and

$$\ddot{\theta}' = \sin\theta_{eq}\frac{g}{R}\theta'.$$

If the coefficient of θ' on the right is positive, the equilibrium is unstable because the perturbation grows in time: indeed, the solution for θ grows exponentially with time. For the case $\sin\theta_{eq} = 1$, the equilibrium is unstable. That corresponds to the ball sitting at the top of the dome in the left pane of Figure 6.30. When the ball sits at the bottom of a bowl, $\sin\theta_{eq} = -1$. Then the perturbation does not grow; θ oscillates about $-90°$. A small amount of friction will cause θ' to damp to zero.

Figure 6.30. Basic notion of stability.

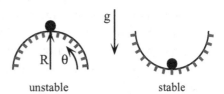

unstable   stable

The fluid dynamical version of instability consists of growing waves. Figure 6.29 is an example. Just beyond the nozzle exit, a short region of laminar flow, with growing waves is visible. They are characterized as instability waves because they grow in amplitude with downstream distance. By analogy to the ball sitting on the inverted bowl, a small perturbation must have triggered this instability wave. It then grows exponentially as it propagates downstream. This is termed *spatial instability*. In fluid flows, one is sometimes concerned with development of an instability in time or *temporal instability*. The classical example is a fluid heated from below. As the bottom gets increasingly hot, the fluid becomes buoyantly unstable and is set into motion. The amplitude of the motion grows with time and can become turbulent.

6.5.1 Inflection Point Theorem

The spatially growing instability seen in Figure 6.29 takes the form of a sine wave times an exponential growth: $\sin(\alpha x)e^{\beta x}$. Instability corresponds to $\beta > 0$. Stability theory predicts the value of β and identifies the factors that determine its sign. A very important result of stability theory is Rayleigh's *inflection point theorem*. This theorem says that in shear flows the sign of β cannot be positive unless an inflection point exists in the velocity profile. Instabilities of the sort seen in Figure 6.29 are attributed to the existence of inflection points in the velocity profile. Many other examples of shear flow instability can be interpreted as inflectional instability. This very important theorm should be appreciated by any student of fluid dynamics. Its proof can be found in many texts, such as White (1991) or Panton (1997).

Consider a parallel shear flow, with velocity profile $u(y)$. An inflection point is the value of y at which the slope du/dy has a maximum; thus, it is where $d^2u/dy^2 = 0$. For the sake of concreteness, if $u = 1 - (1 + y)e^{-y}$, then $d^2u/dy^2 = 0$ at $y = 1$. An inflection point is shown in Figure 6.31. It was remarked in Chapter 5 that the u profile in a boundary layer subjected to adverse pressure gradient must contain an inflection point. Jets, mixing layers, and wakes all contain inflection points. For that reason, this class of flows tends to become turbulent at fairly low Reynolds numbers.

Rayleigh's considerations on shear flow instability are presented in his famous book *The Theory of Sound*, published in 1877. It had been observed that flames were sensitive to sound. When a tuning fork was brought near to a burner flame, it was observed to flicker. Rayleigh hypothesized that the flame was simply providing a visual indicator; the flickering was due to an instability of the jet of combustible gas. He showed by concrete analysis how instabilities can grow on a shear flow. He also proved his famous *inflection point theorem*. It says that a parallel shear flow is stable if there is no inflection point in the velocity profile. The common supposition is that a profile with an inflection point is unstable. That is not what the theorem says, but it is true that an inflection point is almost always destabilizing.

Figure 6.31 at left shows two velocity profiles, the dashed curve with and the solid curve without an inflection point. The inflection point is where the slope of velocity versus y changes from increasing with y to decreasing with y. In this case, the inflectional profile corresponds to a mixing layer. A cat's-eye pattern of streamlines

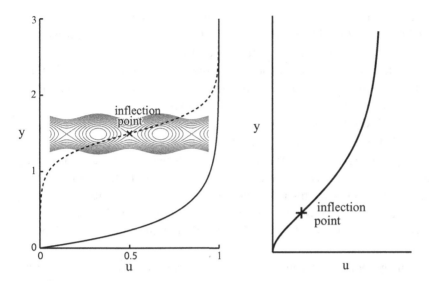

Figure 6.31. Velocity profiles with (– – – –) and without (——) inflection point. A cat's-eye pattern, associated with inflection point instability, is superposed. At right, an inflectional velocity that might correspond to a boundary layer subjected to adverse pressure gradient.

is superposed on the figure to suggest the form of instability. It consists of a sinusoidal vertical velocity added to the mean shear. This is called a Kelvin–Helmholtz instability. In Figure 6.29, a cat's-eye pattern can be seen at the edge of the jet.

Rayleigh's theorem, that an inflection point is a necessary condition for inviscid instability, might be explained qualitatively as follows. A uniform gradient of vorticity is perturbed by displacement of fluid elements. The displacement is in the form of a wave, as shown at the left of Figure 6.32. Vorticity is conserved under displacement. Subtracting the perturbed from the unperturbed distribution of vorticity gives the perturbation. Suppose that the vorticity increases monotonically with y. Then, where the vorticity has been displaced upward the perturbation is negative: at the given y, the perturbed vorticity is less than it was initially. Similarly, a downward displacement causes a positive vorticity perturbation.

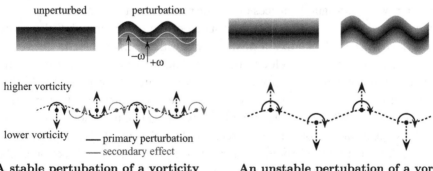

A stable pertubation of a vorticity gradient.

An unstable pertubation of a vorticity extremum.

Figure 6.32. Physical mechanism of Kelvin–Helmholtz instability.

Schematically, the perturbation creates a row of alternating signed vortices, as suggested in Figure 6.32 by the dark circles, labeled "primary perturbation." The induced velocity of the positive vortices is downward on their right side and upward on their left. The negative vortices have the opposite sense of rotation: up on their right and down on their left. Consequently, the induced flows reinforce. They create a secondary displacement of the unperturbed vorticity, which draws higher vorticity down and lower vorticity up, alternatively – as shown by the lighter circles in Figure 6.32, labeled "secondary effect." The induced flow of these secondary vortices opposes the original displacement. Where the initial displacement was downward, the induced secondary velocity is upward; where the initial displacement was up, the induced velocity is down. This is the sign of stability: a perturbation produces a disturbance that acts to oppose the perturbation. It is like a displacement of the ball at the bottom of the bowl in Figure 6.30: a displacement to the left produces a restoring force to the right.

This reasoning about displacement of vorticity is predicated on a nonvanishing vorticity gradient. Without a vorticity extremum, we have stability that is the statement of the inflection point theorem.

The origin of inflection point instability is less obvious. In a parallel shear layer, the vorticity equals $-du/dy$. If $du/dy > 0$, the vorticity is negative. The gradient of vorticity vanishes at an inflection point, $d^2u/dy^2 = 0$, where the vorticity has a minimum. The vorticity magnitude decreases in both directions from the inflection point. The stability argument recited in the previous paragraph for the case of a monotonic vorticity distribution is no longer applicable. Rayleigh's theorem now allows the possibility of inviscid instability. Whether it occurs is a matter for concrete analysis; however, numerous stability analyses have shown that inflection points almost always lead to instability.

Given that observation, we might argue qualitatively as follows. Consider a displacement of the vorticity in the y direction. Let the amplitude of the displacement vary sinusoidally in x. Where the minimum vorticity is displaced upward, the perturbation will consist of a local decrease of vorticity, below the unperturbed level. Where it is displaced downward, the perturbation is again a decrease. This is shown schematically at the right side of Figure 6.32. The perturbation can be thought of either as a row of like-signed vortices or as a displaced sheet of vorticity.

In a frame of reference moving at the speed of the inflection point, the undisturbed velocity profile looks like a mixing layer with positive velocity above the inflection point and negative velocity below it. The lift force on a vortex is proportional to the oncoming velocity times minus the circulation [Eq. (1.64)]. The perturbation vorticity is negative. Hence, where the oncoming velocity is positive, the force is upward. Where the oncoming velocity is negative lift is downward. Thus, the vortices experience a force in the direction of displacement. This is the sign of instability; the perturbation produces a force that acts in the same direction as the displacement. It is analogous to the ball atop the dome in Figure 6.30. A displacement to the left produces a force to the left.

Inflectional profiles are found in jets, wakes, mixing layers, as well as in adverse pressure gradient boundary layers. If they are all unstable, they should not exist – as least not in a steady, laminar state. The instability is counteracted by viscous damping. The instability is suppressed when viscous forces are dominant. The steady flow remains stable until a critical Reynolds number is reached. Indeed, beyond that, the flow is either unsteady or turbulent. It has already been seen on page 144, in Figure 4.13, that unsteadiness develops in a wake when the Reynolds number is moderate. This can be interpreted as the steady flow becoming unstable.

6.5.2 Instability in Boundary Layers

Now we come to a rather unusual result of viscous instability theory. Rayleigh's theorem is a result of inviscid stability theory. It says that without an inflection point, the flow is inviscidly stable. Viscosity should make it only more so. Oddly enough, that is wrong. One of the most remarkable, and counterintuitive, results of theoretical fluid dynamics is that viscosity can destabilize a boundary layer. This was first surmised by Heisenberg. Tolmein obtained explicit instability results. That was in the 1920s. The theory involved subtle mathematics. For years, it stood as a fascinating theory, unsubstantiated by experiment. Finally, in the late 1940s, Schubauer and Skramstead produced Tollmien–Schlichting waves in the laboratory, verifying theoretical predictions of viscous instability. The instability is called a Tollmien–Schlichting (T-S) wave after the seminal work of Tollmien and the extensive researches of Schlichting.

In the years between their theoretical discovery and their observation in the lab, some skepticism was expressed over whether T-S waves played a role in transition to turbulence. If they were its precursor, why were they so hard to detect? To see T-S waves, experiments are conducted in very low-noise wind tunnels. Indeed, to bring out the instability, coherent forcing usually is introduced, either by sound or by a vibrating ribbon. Then, disturbances are seen that propagate and grow in excellent agreement with theory. Why were they so hard to detect in early experiments?

An instability is characterized by its growth rate. If it grows as it propagates downstream, its amplitude varies as $e^{\beta x}$, where β is the growth rate. Rayeigh's theorem says that $\beta = 0$ if there is no inflection points, and if viscosity is negligible. With viscous damping $\beta < 0$ becomes a decay rate. When viscosity is destabilizing it causes β to become positive.

If the Reynolds number is sufficiently low, the boundary layer is stable and $\beta < 0$. Tolmein's result is that $\beta > 0$ can occur in a boundary layer, even without an inflectional profile. But Rayleigh's result is that $\beta = 0$ in the inviscid, infinite Reynolds number limit. In combination, these results say that $\beta \leq 0$ below some critical Re, $\beta > 0$ in some range of Re (Tolmein) and $\beta \leq 0$ above some other Re (Rayleigh). That is indeed the theoretical result for any given disturbance. A sinusoidal disturbance, varying as $\sin(2\pi x/\lambda)$, has a lower and an upper critical Reynolds number. There is a pair for each wavelength λ. The upper and lower values form curves $Re_l(\lambda)$ and $Re_u(\lambda)$, which are called the lower and upper branches. Figure 6.33 shows the well-known, banana-shaped curve of T-S instability. The upper and lower branches join at the minimum critical Reynolds number – near $R_{\delta*} = 520$ in Figure 6.33, which is the critical value for the Blasius profile. In between the upper and lower branch

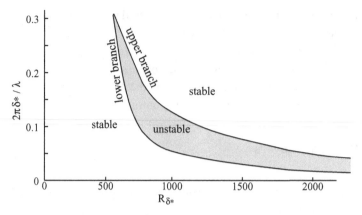

Figure 6.33. Region of Tolmein–Schlichting instability.

the boundary layer is unstable. Above the critical Re a zone of growing disturbances exists.

The instability mechanism does not have a simple physical explanation. Viscous friction causes a phase change across a thin layer, very close to the wall, a bit like that in the oscillatory solution of §1.5.2, but now the oscillation is superposed on the mean shear. The effect of the viscous phase change is to cause Tolmein–Schlichting waves to become unstable.

Via this subtle mechanism, a very weak instability is caused. The maximum growth rate in the Blasius boundary is only $\beta\delta_* = 3.8 \times 10^{-3}$. That mode has a wavelength of $45\delta_*$. In one wavelength, the amplitude of the T-S wave grows by less than $e^{0.171} = 1.19$. Other wavelengths have a lower growth rate. That explains why T-S waves were so difficult to discover: they grow very slowly and are easily masked by other disturbances.

A corollary can be drawn from the stringent lab conditions that are needed to see T-S waves: in the natural environment, viscous instablity is rarely the precursor to turbulence. Inflectional instabilities have an order of magnitude larger growth rate. Any disturbance that causes an inflectional velocity profile will provide a more likely transition route. These include cross flow on swept back wings, adverse pressure gradients in decelerating boundary layers and skewing of the velocity profile on three-dimensional bodies. Any of these can cause transition via inviscid, inflection point instability.

Even in the absence of an inflectional profile, it is found that as little as 0.5% intensity of turbulence in the free stream will induce disturbances in the boundary layer that bypass the viscous instability. Theory and computer simulations show that such perturbations amplify in a shear flow by displacing its mean vorticity. Displacement leads to long, jetlike perturbations that lift up from the wall and become the seat of instability and, ultimately, of transition to turbulence.

Although this description of hydrodynamic instability is couched in terms of transition from laminar to turbulent flow, *instability* should not be confused with *transition*. The former simply means that some perturbations will amplify. The flow

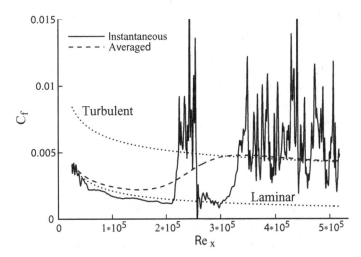

Figure 6.34. Instantaneous and averaged skin friction in a boundary layer as it transitions from the laminar to the turbulent state.

remains laminar. It may evolve into a different laminar flow, say, one containing two-dimensional vortices: vortex shedding behind a circular cylinder can be explained as the end product of in instability of the steady flow. For transition to occur, the perturbation must develop strong three dimensionality. That, in turn, leads to irregular velocity fluctuations and a cascade of energy to small scales.

The start of transition per se, is somewhat nebulous. It does not occur at a sharply defined position. A transition Reynolds number might be defined through the behavior of the curve of averaged skin friction as described on page 232. Transition begins where C_f hits its minimum. However, this is an averaged viewpoint. The actual process is intermittent in space and time. Figure 6.34 shows how $C_f(x)$ appears at an instant: the transition region is intermittently turbulent; sometimes C_f is at the turbulent level, sometimes at the laminar level. Its average lies in between.

Spatial features that are the source of intermittency are illustrated by Figure 6.35. At left, the flow is laminar; at right, it is turbulent. A patch of turbulence is seen just to the left of middle. Above the patch the boundary layer is laminar. This is called a *turbulent spot*. Spots form intermittantly in the transitional region. From top to bottom within the transition region of Figure 6.35, the flow alternates from

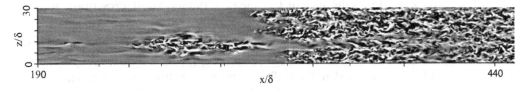

Figure 6.35. A turbulent spot visualized by v contours. This is a plan view through a boundary layer, seen from above. At left, the boundary layer is laminar; at right, it is turbulent; in the middle, it contains a turbulent patch in a laminar background.

laminar to turbulent and back to laminar. This figure is from a computer simulation. Laboratory experiments show the same type of turbulent spots.

This section has endeavored to give some flavor of instability theory and of the nature of the onset of turbulence. Extensive results exist on instability (Drazin and Reid 1995); very little analysis is available for transition. The transitional Reynolds number cannot be predicted. Transition is susceptible to computer simulation, but that is very costly because of the need for a fine grid and temporal accuracy. One is left with empirical formulas, like Eq. (6.31), to estimate where transition occurs.

EXERCISES

6.1 *Log-law.* A turbulent boundary layer has a measured thickness $\delta_{99} = 3.2$ cm. It is not possible to measure very near to the wall. The velocity U has been measured at three points above the wall. It is found that $U = 40.1$ m/s at $y = 2.05$ cm, $U = 34.7$ m/s at $y = 0.59$ cm and $U = 30.7$ m/s at $y = 0.22$ cm. Estimate the friction velocity u_*. Assess the validity of this estimate.

6.2 *Shear layer spreading.* How does U_{cl} vary with x in a two-dimensional, turbulent plane jet? Compute a plane jet with the k-ε model, starting from plug flow. Determine how far from the entrance the self-similar x scaling is reached.

6.3 *Eddy viscosity.* Use the VanDriest formula, cited below Eq. (6.33), to calculate the law of the wall, u_+ versus y_+. What value do you obtain for the additive constant B in the log-law? Explain how this provides a wall function for computational analysis.

6.4 *Turbulent drag.* A long, thin, broad plate is dropped edge-on in a fluid. It falls under gravity and reaches a terminal velocity, V_T. Ignore leading and trailing edge effects. Assume that the boundary layer is turbulent throughout.

Formulate a simple estimate of V_T, given the mass (m), length (L), and width (w) of the plate, and whatever fluid properties you need.

6.5 *Two equation models.* Write the equations of the k-ω model for a homogeneous shear flow. In Eq. (6.14), use $\varepsilon = k\omega$.

The condition of moving equilibrium is $d\omega/dt = 0$. What is the ratio of production to dissipation, $\mathcal{P}/\varepsilon$ under this condition? What is the corresponding ratio for the k-ε model?

6.6 *Compute a t.b.l.* Compute the turbulent boundary layer over a flat plate at Reynolds number $Re_L = 10^6$. Introduce a symmetry plane upstream of the plate, extending a distance of $L/4$ ahead of the solid wall. Prescribe a uniform, low intensity of turbulence at the inflow.

Use the k-ω model. Compute solutions on both a fine grid that resolves the near-wall region and on a coarse grid by applying a wall-function boundary condition. Ensure that for wall integration the first cell center stays below $y^+ = 1$. For the wall function calculation, use grids with the first cell center in the range $30 \leq y_1^+ \leq 300$. In the case of wall integration, do not exceed 30,000 grid cells; for the wall function, do not use more than 10,000 cells.

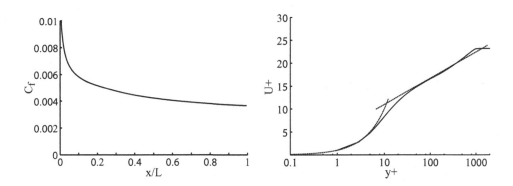

Compare results with the friction coefficient C_f and the velocity profile at $x = L$ presented in the accompanying figures; they were computed on a fine grid, using a two-layer k-ε model.

6.7 *Universal range.* Apply Kolmogoroff's dimensional reasoning to determine how mean-squared vorticity (enstrophy) varies with eddy size in the inertial sub-range. Write a functional form that extends to the entire universal range. Let the unknown function in this formula be $Ae^{-\alpha\eta/r}$, where A and α are constants. At what eddy size is the enstrophy maximum? Explain why most of the kinetic energy dissipation occurs at subgrid scales in large eddy simulations.

6.8 *DNS of a boundary layer.* Estimate the number of grid points required for DNS of a turbulent boundary layer. The momentum thickness Reynolds number is $R_\theta = 1,500$. Let the height of the domain be 20Θ and the length be 300Θ. The width should be about six times the streak spacing. This is a flat plate boundary layer, so Eq. (6.27) can be used in making the estimate.

6.9 *Inflection point theorem.* Comment on the inviscid stability of the following velocity profiles:
(1) $U = \alpha y + e^{-y^2/2\sigma^2}$;
(2) $U = \alpha y + (1 - e^{-y/\delta})$, $y > 0$, $\delta > 0$.

6.10 *Free-stream turbulence.* A laminar, flat plate boundary layer is subjected to 3% free-stream turbulence. The mean free-stream air velocity is $U_\infty = 10$ m/s. At what distance from the leading edge will transition to turbulence occur?

7 Compressible Flow

Two principles distinguish compressible flow: gases heat when compressed and cool when expanded; disturbances propagate at the speed of sound. The first alludes to thermodynamics. The second alludes to gas dynamics.

7.1 Thermodynamics

Heating by compression converts work into thermal energy. This is a reversible conversion in the sense that the thermal energy can be converted back into work. Heating also occurs by frictional dissipation of fluid kinetic energy into thermal energy. That is an irreversible process; viscosity cannot convert the thermal energy back into ordered flow. Friction increases entropy.

Compression and expansion occur in the course of the motion of a gas. For instance, on approaching a blunt body, the flow will slow, and fluid elements will be compressed. That is the ultimate motive for reviewing basic thermodynamics: the governing equations of compressible flow must be consistent with thermodynamics, extended to a spatially distributed system. However, we start with the thermodynamic description of compression and expansion of a homogeneous gas and then proceed to discuss compressible fluid dynamics. Comprehensive texts (Saad, 1997) can be consulted if the reader desires a thorough treatment of thermodynamics. The following is an informal treatment that provides background to compressible flow analysis.

Define a fluid element as a fixed mass, $\mathcal{M}$, of gas. This occupies a volume element, $\mathcal{V}$, which contains that mass. The volume defined in this way is termed *specific volume* – *specific* properties are those associated with a given quantity of mass. The mass of the fluid element is invariant, because that is how the element is defined: its volume can change. Indeed, compressibility is the property of volume change in consequence of pressure variations.

Pressure does work when it causes these volume changes. Work is force times displacement. Force is pressure times area and is directed normal to the surface acted upon. The net force is $F = P \cdot A$, if this is summed over the area of the fluid element. For simplicity, consider an analog: a piston is a cylindrical chamber of gas with a single movable surface. The work that must be done to compress the piston a distance dx is Fdx. The energy acquired by the gas is minus this: it is $-Fdx$. If the force is exerted by pressure, $-Fdx = -P \cdot Adx$. Adx is simply the decrease of the fluid volume, $d\mathcal{V}$. Hence, energy acquired by the gas equals $-Pd\mathcal{V}$. This is positive for a volume decrease and negative for a volume increase.

The fixed mass of gas, $\mathcal{M}$, occupying a volume $\mathcal{V}$ defines the density $\rho = \mathcal{M}/\mathcal{V}$ within the fluid element. Hence, pressure work also can be written as $-Pd\mathcal{V} = \mathcal{M}Pd\rho/\rho^2$. The expression of work in terms of density will be useful when we move on to discuss compressible flow.

7.1.1 First Law

Energy conservation, the first law of thermodynamics, requires that the work appears as heat energy or the internal energy of the gas. Heat energy per unit mass, or specific heat energy, is denoted as $\mathcal{E}$. Thus, energy conservation under compression is stated as

$$d\mathcal{E} = Pd\rho/\rho^2. \tag{7.1}$$

In a simple gas, $\mathcal{E}$ is a state variable that depends on the gas temperature. The relation of internal energy to temperature $\mathcal{E}$ is expressed by a coefficient of heat capacity, c_v. This is simply the derivative of internal energy with respect to temperature at constant volume:

$$c_v = \partial \mathcal{E}/\partial T \mid_v .$$

It might seem inconsistent to invoke the heat capacity at constant volume, when we are considering heating by volumetric compression; that is not so. c_v simply characterizes the thermal energy stored in the fluid. The condition of constant volume refers to how it would be measured: a given amount of heat, δq, is added, without allowing the gas to expand or contract and the temperature rise, δT, is measured. For instance, the gas in a closed cylinder could be heated by flowing electric current through a wire. If the cylinder cannot expand, all the heat goes into internal energy: $\delta q = \delta \mathcal{E}$. The heat added and the temperature rise are measured. The ratio of $\delta q/\delta T$ gives c_v.

The internal energy is stored in the random motion of molecules. Diatomic gases are particularly important, because they are the major component of dry air (N_2, O_2). At room temperature, diatomic molecules can be thought of as rigid dumbbells. Energy is stored primarily in their five degrees of freedom, two rotational and three translational. Kinetic theory gives $c_v = 5/2R$, where R is the gas constant. (At high temperature, vibrational degrees of freedom contribute and c_v approaches $7/2R$, although only at several thousand degrees Kelvin. At low temperature, the rotational contribution becomes small, and c_v approaches $3/2$. The value $c_v = 5/2R$ is commonly used in computations for air.)

The first law of thermodynamics (7.1) now has the simple, explicit form

$$\rho c_v dT = P\frac{d\rho}{\rho}, \tag{7.2}$$

which relates work to temperature change. This is the form when there is no heat addition. If an amount of heat, ρdq is added, the first law reads

$$\rho c_v dT = P \frac{d\rho}{\rho} + \rho dq. \tag{7.3}$$

In addition to compression, gas temperature can rise because heat is added by external sources. External sources could be conduction through walls, radiation, or chemical reaction. In the absence of these effects, the flow is called *adiabatic*, and Eq. (7.2) applies. In a flowing gas, heating can also occur internally, through friction.

7.1.2 Equation of State

The energy Eq. (7.2) is a relation between three variables P, T, and ρ. A second relation is provided by the equation of state. An equation of state is a connection between thermodynamic variables, which exists independently of the process taking place – for instance, whether the fluid is flowing, being compressed, or being heated. It represents a statistical balance, on the molecular scale. Pressure is a consequence of momentum transfer by molecular collisions. The frequency and intensity of those collisions is determined by thermal motion; hence, the equilibrium pressure is a function of temperature. The equation of state expresses the idea that gas pressure is not an independent quantity; rather, it is a macroscopic variable related to temperature via the thermal motion of molecules.

The equation of state of an ideal gas is written as

$$P = \frac{n \mathcal{R} T}{\mathcal{V}}.$$

Here $\mathcal{R}$ is the universal gas constant and n is the number of molecules in a volume $\mathcal{V}$. For gas dynamics, this better stated in terms of the density. The density is related to the number of molecules per unit volume by $\rho = n \mathcal{M}_w / \mathcal{V}$, where $\mathcal{M}_w$ is the molecular weight of the gas. Hence

$$P = \rho RT, \tag{7.4}$$

where $R = \mathcal{R}/\mathcal{M}_w$. This form of the gas law is suited to flow computations in which density is a dependent variable.

It is common to refer to R as the gas constant and $\mathcal{R}$ as the *universal* gas constant: $\mathcal{R} = 8.312 \, J/(\text{mole} \cdot {}^\circ K)$. The gas constant depends on the molecular constitution of the gas: for air, $\mathcal{M}_w = 29 \, \text{g}/\text{mole}$ and $R = 287 \, J/(\text{kg} \cdot {}^\circ K)$.

7.1.3 Entropy and Irreversibility

Basic thermodynamics cannot be summarized without some mention of entropy – although, to some extent, it presently is of interest only as an indicator. Changes of entropy indicate that heat has been added, or removed, by something other than reversible work. Any concrete computation must represent those others processes; no reference to entropy per se is needed. In particular, heat can be added where the fluid passes a hot wall; it can be added by the action of viscous friction; it can radiate into the flow and be absorbed. In each case, the mechanism is represented by the equations or by boundary conditions. Entropy can be evaluated in a postprocessing step. Nevertheless, discussion of entropy is warranted because of its role as an indicator.

Essentially, there are two types of entropy source, diabatic and irreversible. This is embodied in the second law of thermodynamics

$$dS = \frac{dq}{T} + dS_{\text{irrev}}, \tag{7.5}$$

where entropy is denoted by S. The first term is the diabatic heating; the second is irreversible entropy production. Diabatic processes are reversible sources of entropy. A flow with no diabatic entropy sources is *adiabatic*; a flow with neither diabatic or irreversible sources is *isentropic*.

Diabatic heating refers to processes external to the system that add entropy. These include chemical reaction or radiation. We are not concerned with such phenomena here; for our purposes, heat transfer through boundaries is the only diabatic source of concern. This entropy source enters computational analysis via boundary conditions. Walls might be insulated, constant temperature, or have a prescribed heat flux. Insulated walls are adiabatic: mathematically, the normal derivative of temperature is zero at the surface.

Calculations with adiabatic walls are an element of heat transfer analysis. Suppose one wants to know whether walls at a particular temperature will transfer heat to the fluid or extract heat from the fluid. The wall temperature can be compared to the fluid temperature that would exist if the walls were insulated. To determine whether heat will flow from a wall to the fluid, a first computation is done with adiabatic walls. The temperature of the fluid adjacent to the surface is called the adiabatic wall temperature, T_{adiab}. If the actual wall temperature is greater than the adiabatic wall temperature, $T_w > T_{\text{adiab}}$, heat will flow from the wall to the fluid. In compressible flow of a gas, it is necessary to define heat transfer relative to adiabatic conditions because the compression and expansion of the gas as it flows and dissipation of energy by viscosity will produce temperature variations, even in the absence of heat addition.

Consider the example of flow through a turbine stator passage, as in Figure 7.1. The flow is transonic, with the inflow Mach number $M_{\text{in}} = 0.25$ and and total temperature $T_{0\,\text{in}} = 400°\text{K}$. The exit Mach number is $M_e = 0.92$. A turbulence model was invoked so that the boundary layers on the vane surface remain attached.

As the flow accelerates within the center of the passage, the temperature falls as a consequence of energy conservation, Eq. (7.8): thermal energy is converted to kinetic energy. In the center of the passage, the contours are progressing down in temperature from left to right.

In the case at the left of Figure 7.1, the vane wall is adiabatic. Temperature contours approach the surface perpendicularly. They do not quite reach it because of dissipative heating within the thin boundary layer. There is a small temperature rise just next to the wall, causing the contours to divert downstream.

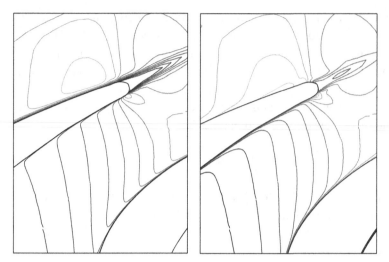

Figure 7.1. Temperature contours in a turbine passage. On the left, the blade is adiabatic; on the right, it is at constant temperature.

At the right of Figure 7.1, the vane surface was kept at the temperature $T_w = 298°$ K. This is lower than any temperature internal to the flow; thus, it corresponds to a cooled vane. It is seen that the temperature contours turn upstream, parallel to the wall as the vane surface is approached. Hence, the temperature gradient is from cool at the wall to hot in the flow. Heat flows down that gradient toward the surface.

Figure 7.1 is the same geometry as Figure 1.6 on page 25. Although Figure 1.6 was used to illustrate loss of total pressure in a turbine cascade, total pressure and entropy are closely related – for instance, see Eq. (E7.2). The region where total pressure has been lost is the region where entropy has been created.

In gas flows of present interest, irreversible entropy addition [Eq. (7.5)] is caused viscous friction. The rate of frictional working is stress times rate of strain: $\tau{:}s$. For a Newtonian fluid, this equals $2\mu|s|^2$, which is nonnegative. Hence, viscous dissipation can only increase entropy. Irreversible entropy can be thought of as a placeholder for viscous dissipation. It becomes important in high speed boundary layers and in shock waves. From Eq. (7.5), in an adiabatic process $dS = dS_{\text{irrev}}$.

7.1.4 First Law with Flow

Returning to the first law of thermodynamics (7.2): at constant pressure, it can be written

$$d\left(\mathcal{E} + \frac{P}{\rho}\right)\bigg|_p = 0. \tag{7.6}$$

The quantity $h = \mathcal{E} + P/\rho$ defines the specific enthalpy or the enthalpy per unit mass. An interpretation is that P/ρ is potential energy of a fluid element and $\mathcal{E}$ is its thermal energy. Then the above expresses conservation of the total energy, in the absence of flow. It is readily seen that

$$dh = c_v dT + RdT = c_p dT$$

for an ideal gas, having used $P/\rho = RT$. The heat capacity c_p is defined as

$$c_p = c_v + R.$$

The flowing fluid element also has kinetic energy, which, per unit mass, is $1/2|\boldsymbol{u}|^2$. Including it, the conservation of total specific energy (7.6) must be extended to

$$\mathcal{E} + P/\rho + 1/2|\boldsymbol{u}|^2 \equiv h + 1/2|\boldsymbol{u}|^2 = \text{constant}. \tag{7.7}$$

Although a condition of constant pressure was mentioned when enthalphy, h, was defined, that is not a constraint on this equation. For instance, if temperature and density are constant, Eq. (7.7) corresponds to Bernoulli's Eq. (1.38), on page 25. The extra contribution, dP/ρ that was added to Eq. (7.2) in going from energy conservation without flow, can be regarded as potential energy stored in the pressure.

If the heat capacities are independent of temperature, then $h = c_p T$. The constant on the right side of Eq. (7.7) is stated as $c_p T_0$. Then conservation of total energy is expressed as

$$T + \frac{|\boldsymbol{u}|^2}{2c_p} = T_0. \tag{7.8}$$

T_0 is called the *stagnation temperature*. (Stagnation temperature is also called total temperature.) The actual temperature, or static temperature, varies as the gas flows, but it does so such that the stagnation temperature is unchanged. The stagnation temperature is the temperature that a fluid element would have were it brought to rest by converting its kinetic energy into thermal energy. It might not be obvious, but this notion includes slowing the element by viscosity; conservation of stagnation temperature is not restricted to isentropic flow. For instance, stagnation temperature is conserved across shock waves, even though they add entropy to the flow.

However, total temperature conservation does assume adiabatic flow. Departures from adabiatic conditions cause changes in total temperature, T_0. If a fluid element passes a heated or cooled wall, heat will be added or removed by conduction. Nonadiabatic walls are often the primary cause of total temperature variations.

Conservation of T_0 provides a rule of thumb for computational analysis. If the stagnation temperature is uniform at the inlet, then all fluid elements carry the same T_0 as they flow downstream. Their temperature and velocity will change, but they should do so such that the the total temperature is conserved. That is strictly true only for steady, adiabatic flow with unit Prandtl number. However, it is often a very good insight into compressible flow.

7.1.5 Isentropic Relations

If the flow is adiabatic, dq is zero in Eq. (7.3). However, in a flowing fluid heat can be created by dissipation of kinetic energy. Eliminating pressure from Eq. (7.3) via the equation of state and representing dissipative processes as entropy creation gives

$$\rho c_v dT = RTd\rho + \rho Td\mathcal{S}_{\text{irrev}} \tag{7.9}$$

or, integrating between two states

$$\Delta \mathcal{S}_{\text{irrev}} = c_v \log(T_2/T_1) - R\log(\rho_2/\rho_1). \tag{7.10}$$

For instance, states 1 and 2 can be upstream and downstream points in a flow.

$Td\mathcal{S}_{\text{irrev}}$ represents heat created by friction. If friction can be ignored, and this term set to zero, useful formulas emerge. This is the assumption of *isentropic flow*.

Invoke the definition $c_p = c_v + R$ and denote the ratio of heat capacities as $c_p/c_v = \gamma$. Then $R = c_v(\gamma - 1)$. The relation (7.9) between density and temperature becomes

$$\rho dT = (\gamma - 1)Td\rho.$$

Integrating this between two states gives

$$\log(T_1/T_2) = (\gamma - 1)\log(\rho_1/\rho_2). \tag{7.11}$$

Exponentiating both sides of this equation shows that $T/\rho^{\gamma-1}$ is constant between the two states. If a gas is compressed isentropically, temperature and density both increase, so as to maintain this ratio constant. For a diatomic gas, under normal conditions, $\gamma = 1.4$. If the initial temperature and density are T_1 and ρ_1, and the final density is ρ_2 then the final temperature is

$$T_2 = T_1(\rho_2/\rho_1)^{0.4}.$$

From the equation of state (7.4) and the isentropic relation (7.11), it follows that

$$T\rho^{1-\gamma} = \text{constant},$$
$$P\rho^{-\gamma} = \text{constant}, \tag{7.12}$$
$$PT^{-\gamma/\gamma-1} = \text{constant}.$$

With $\gamma = 1.4$, pressure is proportional to $\rho^{1.4}$ and to $T^{3.5}$.

Corresponding one-dimensional, gas dynamics relations are obtained after rearranging Eq. (7.8). Anticipating the next section, we cite the formula

$$a^2 = \gamma RT \tag{7.13}$$

for sound speed. For an ideal gas $R = c_p - c_v$. Then the formula (7.13) allows (7.8) to be rewritten

$$\frac{T_0}{T} = 1 + (\gamma - 1)\frac{|\boldsymbol{u}|^2}{2a^2} = 1 + \frac{\gamma - 1}{2}M^2. \tag{7.14a}$$

If a gas flows isentropically, Eqs. (7.12) continue to apply. Let state 1 be the ambient conditions and state 2 be stagnation conditions. The constants in Eqs. (7.11) and (7.12) define the stagnation properties: for instance,

$$T\rho^{1-\gamma} = T_0\rho_0^{1-\gamma}.$$

Thereby the isentropic flow formulas

$$\frac{P_0}{P} = \left(1 + \frac{\gamma - 1}{2}M^2\right)^{\gamma/(\gamma-1)},$$

$$\frac{\rho_0}{\rho} = \left(1 + \frac{\gamma - 1}{2}M^2\right)^{1/(\gamma-1)}$$

$$(7.14b)$$

derive from (7.14a). These give an idea of how pressure, temperature, and density vary with speed in compressible, isentropic flow. Unlike stagnation temperature, P_0 and ρ_0 are not conserved in every adiabatic flow – only if it is also isentropic. They change across shock waves (§7.3).

7.2 Mach Waves

Disturbances propagate at the speed of sound. In the incompressible approximation, the speed of sound is infinite; action at a distance is recovered. A fluid element approaching an obstruction knows of its presence through a pressure field that the element feels before it reaches the object. In the incompressible limit, that pressure field is established instantaneously. A distant boundary is sensed immediately by the fluid element.

But in compressible flow, a sound wave must propagate from the fluid element, to the surface and back in order for its presence to be felt. The pressure field is established at the speed of sound, not instantaneously. It follows that an element moving faster than the speed of sound will not know of the presence of an obstruction. Before a sound wave can propagate to the object and back, the fluid element has reached the object. That is not quite correct; rather, the element discovers the presence of an obstruction abruptly at a shock wave. Despite that qualification, the principle remains: the object cannot produce an upstream influence into a supersonic stream. That perspective is essential to understanding compressible flow. It is embodied in the *Mach cone*.

A pebble dropped in a pond produces a circular ripple that spreads radially. It propagates at the relevant wave speed (in this case a combination of gravity and surface tension sets the speed). So, too, a point impulse in a gas produces a spherical ripple that propagates radially at the wave speed of sound. A sequence of pebbles will produce a nested set of circles, as in Figure 7.2, at left. If the pebbles are dropped into a moving stream, the set of circles will be displaced. If the speed of the stream is u, then at time t the center of the circle will be displaced to $x = ut$. Meanwhile, it has grown to a radius at, where a is the wave speed. If $u < a$, the set of circles will look like the left side of Figure 7.2. All circles will lie inside that generated by the first pebble.

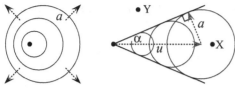

Subsonic   **Supersonic**

Figure 7.2. Acoustic waves from subsonic and supersonic and moving sources. The envelope of waves from a supersonic source creates a Mach wave.

If $u > a$, the set of circles will look like the diagram at the right of Figure 7.2. They will no longer be nested. A line drawn tangent to the envelope of circles makes an angle α to the x axis, with

$$\sin \alpha = a/u = 1/M. \tag{7.15}$$

This defines the *Mach wave*. Note that when $M < 1$ the Mach wave does not form: mathematically, $\sin \alpha$ would have to be greater than unity. $M < 1$ corresponds to the case at left in Figure 7.2, in which there is no envelope of waves.

The physical significance of the Mach wave follows from noting that a point source can influence other points in the domain only if those points can be reached by sound waves. In three dimensions, the circles of Figure 7.2 become spheres, and their envelope is a cone, the Mach cone. Only objects lying inside the Mach cone know of the presence of the source. An object at point X in the figure is influenced by the source; an object at point Y is oblivious to it.

By a change of reference frame, the right-hand schematic in Figure 7.2 applies to a stationary gas and a point source moving at speed u to the left. The Mach cone then is a wake, within which the fluid has been disturbed by the source; outside the Mach cone the fluid remains quiescent.

These basic notions set the stage for discussing compressible flow phenomena. The Mach cone defines a region of influence, within which objects know of each other's presence. In supersonic flow, the region of influence lies in the downstream direction. Disturbances cannot propagate upstream against the supersonic flow. The supersonic region is influenced by inlet conditions but not by downstream surfaces or exit conditions. Mathematically, the unidirectional influence is described as *hyperbolic* behavior.

In subsonic, but still compressible, flow the Mach cone does not exist. The region of influence is a spherically expanding wave. If one waits long enough, all points of the fluid enter that sphere; alternatively, in steady state all points of the fluid can communicate. Mathematically, the omnidirectional influence is described as *elliptic* behavior.

7.2.1 Speed of Sound

The idea that sound has a speed is quite intuitive. A disturbance at a distance will be heard somewhat later than it is seen. The distance to the disturbance divided by the time delay is the speed of sound. Sound propagates by elastic compression and expansion of gas. Compression raises the pressure locally, producing a pressure excess that accelerates and compresses the gas ahead of the wave, causing it to propagate. Behind the compression is an expansion that decelerates the gas back to rest. The elasticity of air is represented by the increase of pressure with density,

$$\partial p / \partial \rho. \tag{7.16}$$

Sound waves are very small amplitude perturbations. Viscous dissipation can be ignored, making the above the derivative at constant entropy. Then from Eqs. (7.12) and (7.4) the derivative is found equal to γRT.

The acceleration caused by pressure perturbations is in accord with the momentum equation

$$\frac{\partial \rho u'}{\partial t} = -\frac{\partial p'}{\partial x} = -\left(\frac{\partial p}{\partial \rho}\right) \frac{\partial \rho'}{\partial x}.$$

Density variations caused by compression and expansion can be seen as a product of mass conservation,

$$\frac{\partial \rho'}{\partial t} = -\frac{\partial \rho u'}{\partial x}.$$

If the mass flow increases in the direction of u, more mass will exit a control volume than enters, and the density will fall.

Because sound waves are tiny perturbations of density, pressure, and velocity, they are denoted by a prime. Formally, the equations have been linearized by neglecting any products of primed variables that arise in the full momentum and continuity equations. Combining the last two equations gives

$$\frac{\partial^2 \rho'}{\partial t^2} = \frac{\partial}{\partial x} \left(a^2 \frac{\partial \rho'}{\partial x}\right), \tag{7.17}$$

where $a^2 = \partial p / \partial \rho = \gamma RT$. This is a wave equation. It describes the propagation of disturbances at speed a – which, hence, is the speed of sound.

Equation (7.17) is simply an approximation to the full equations governing compressible flow. It follows that sound propagation is implicit in any compressible flow computation. There is no need to distinguish sound from flow, unless one is concerned with acoustics. Acoustics is treated distinctly from fluid dynamics because it is the limit of very small, unsteady, often high-frequency disturbances. These are not usually resolved in a flow simulation. Indeed, in steady flow, sound is not produced; there must be some unsteadiness to determine the frequencies of the sound waves.

7.2.2 Occurrence of Shock Waves

Even though acoustics are not of concern, the speed of sound is quite relevant to compressible flow. Aspects of compressible flow are understood in terms of the limit placed on the propagation speed of pressure disturbances: they cannot propagate faster than the speed of sound. A fluid element approaching an obstruction at supersonic speed cannot be influenced by the object because the Mach cone of the fluid

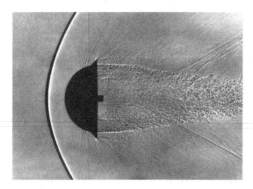

Figure 7.3. Shadowgraph showing the bow shock in front of a hemisphere in supersonic flow. Courtesy of NASA Ames: photograph by L. Jones (1958).

element lies behind the element. The Mach cone defines the region that can influence its trajectory. The fluid in front of a particle moving supersonically lies outside its Mach cone. This presents a dilemma: ultimately, the fluid element will be diverted around an body in its path. If it were to collide with the body, its velocity would go to zero. Before it became zero, it would become subsonic, its Mach cone would cease to exist, and the obstacle could affect its motion. So it does not collide with the body. Prior to reaching the surface, the velocity becomes subsonic so that the path of the fluid element can anticipate the surface and divert smoothly around it.

This has not resolved the dilemma: a supersonic element cannot be aware of the body ahead of it; hence, it cannot have the information to adjust it smoothly to subsonic velocity. The problem has been moved away from the surface to somewhere in the fluid, but it remains a paradox: the element must anticipate what is ahead of it. The resolution of this quandary is quite remarkable. Transition from supersonic to subsonic speed occurs abruptly at a *shock wave*, that is, the entity with which the fluid particle collides.

Shock waves are an intriguing natural phenomenon. They are nature's method for circumventing the impossibility of pressure propagating upstream into supersonic flow. The lesson is that when a supersonic flow cannot meet downstream conditions, shock waves are likely to form to adapt the flow. Figure 7.3 is an example. A supersonic flow approaches a hemisphere. To the left of the curved shock wave, the flow proceeds in the x direction at supersonic speed. It must divert around the obstacle. The shock wave forms and turns the flow before it reaches the surface. We will discuss how this is done below. Loosely, the shock wave is an abrupt conversion from supersonic to subsonic flow. After crossing the shock, upstream influence can occur, and the gas flows smoothly around the front of the hemisphere.

This picture might be reminiscent of surface ripples seen when a stream flows around a rock in its path. They are similar: the ripples in the stream have the same cause as those that radiate away from a pebble dropped into a pond. In the latter case, they propagate; in the former case, they are held stationary in the oncoming flow. To hold the ripples stationary, the oncoming flow speed must counter the propagation speed of the ripple. Returning to compressible flow, if the shock wave were a sound wave, the oncoming flow must have the speed of sound to hold it in place. The wave in Figure 7.3 wants to propagate to the left, but the oncoming flow balances the

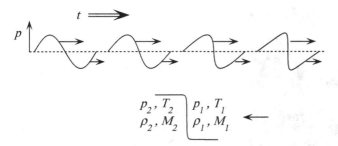

Figure 7.4. Steepening of a wave by the crest overtaking the trough. At bottom, the shock as a jump.

propagation speed to hold the wave in place. It follows that shock waves can form only in supersonic flow.

That last remark needs qualification because the reasoning invoked the speed of sound, whereas the flow is supersonic. Recall that sound is the limit of tiny perturbations. As such, they cannot represent a true shock. As the magnitude of pressure disturbances becomes appreciable, the wave speed increases. A finite amplitude wave has a speed greater than sound in the ambient medium. Where the gas is compressed, its temperature rises, and so does the sound speed. Relative to the approach flow the shock speed is supersonic.

That dependence of sound speed on compression explains how shocks form. Consider a sinusoidal pressure wave, like that at the top of Figure 7.4. As the pressure rises so does the sound speed; where it falls the speed decreases. Hence, the peak of the wave moves fastest and begins to catch up with the trough. As the wave propagates, it steepens. That is shown along the top of Figure 7.4 by a time sequence. The crest cannot overtake the trough because there can be only one value of pressure at any point. When the slope becomes vertical, a discontinuity forms. This is the shock wave.

Wave steepening explains how shocks can occur. The actual shock is a discontinuous jump between two states, supersonic approaching the jump and subsonic leaving it. Instead of the sine wave illustrated in Figure 7.4, the disturbance can be a ramp (exercise 7.4) that also steepens into a discontinuity. Then the fully formed shock is a step between two states, as illustrated at the bottom of Figure 7.4. In the frame of the wave, the flow approaches from the right, so that the shock is held stationary by the supersonic flow. That corresponds to Figure 7.3.

By assuming that the shock can be held stationary in supersonic flow, we have assumed that the wave speed is greater than the speed of sound. Indeed, these finite amplitude compression waves do propagate faster than the ambient speed of sound. Their speed increases with the size of the jump. For a given approach flow, only a particular strength of shock can be held stationary. Hence, the shock strength can be determined as a function of the approach Mach number.

Why do these nominally discontinuous jumps occur in supersonic flow? They are not always an inevitability. Shock waves are needed when the flow cannot adapt to downstream conditions by isentropic expansion or compression. Shocks then fit

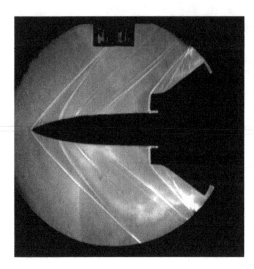

Figure 7.5. Oblique shocks in flow round an airplane. Shlieren photograph courtesy of NASA Ames.

themselves into the flow in a manner that will allow the flow to adjust. Shocks are described either as *normal* or *oblique*, depending on whether they are perpendicular or angled to the approach flow. The curved shock in Figure 7.3 is normal at the center but oblique on the upper and lower sides. Figure 7.5 shows an object with a sharper nose from which oblique shocks emanate.

Usually computational schemes are constructed to automatically capture shocks. Shocks are not actually jumps; they have some finite width, which is determined by molecular diffusion. They will form only in the presence of dissipation. However, their actual thickness is extremely small. To capture them, numerical diffusion must be added to a computation. That enables the solution algorithm to produce the relatively sharp changes that define shock waves. If the grid resolution is inadequate, the solution will be overly diffused; crisply capturing shocks is a challenge. However, with proper numerics the jumps will occur correctly and automatically in the solution.

The theory of shock waves provides an understanding of such solutions. Some basic elements of compressible flow theory are reviewed in the next sections; more comprehensive developments can be found in specialized texts such as Anderson (1982).

7.3 One-Dimensional Gas Dynamics

7.3.1 Normal Shock

Consider a discontinuity between two flow states, as at the bottom of Figure 7.4. The gas flows from right to left through a stationary discontinuity. In crossing the jump, mass, momentum, and energy must be conserved. These are stated as

$$\rho_1 u_1 = \rho_2 u_2,$$
$$p_1 + \rho_1 u_1^2 = p_2 + \rho_2 u_2^2, \qquad (7.18)$$
$$h_1 + 1/2u_1^2 = h_2 + 1/2u_2^2.$$

Subscripts 1 and 2 denote the two states. The equation of state (7.4) applies on both sides of the jump. For a perfect gas enthalpy is proportional to temperature, $h = c_p T$. Then energy conservation can be written as

$$T_1 + \frac{u_1^2}{2c_p} = T_2 + \frac{u_2^2}{2c_p} = T_0. \tag{7.19}$$

The total temperature is unchanged across the shock; T_0 is the stagnation temperature.

It is not obvious, but Eqs. (7.18) can be solved for state 2 as a function of state 1. For instance,

$$M_2^2 = \frac{2 + (\gamma - 1)M_1^2}{2\gamma M_1^2 - (\gamma - 1)} \tag{7.20a}$$

(Anderson, 1982). If $M_1 > 1$, this shows that $M_2 < 1$. On physical grounds, the flow into a shock must be supersonic ($M_1 > 1$) – otherwise, there would be no shock. Then the flow out of the normal shock will be subsonic. The presure, density, and temperature rise across the shock and the velocity falls:

$$\frac{\rho_2}{\rho_1} = \frac{u_1}{u_2} = \frac{(\gamma + 1)M_1^2}{2 + (\gamma - 1)M_1^2},$$

$$\frac{p_2}{p_1} = 1 + \frac{2\gamma}{\gamma + 1}(M_1^2 - 1), \tag{7.20b}$$

$$\frac{T_2}{T_1} = \frac{p_2}{p_1}\frac{\rho_1}{\rho_2}.$$

Equations (7.20a) and (7.20b) are the jump conditions that should be observed across a normal shock. Computational data should be in close agreement with these formulas if numerical resolution is sufficient.

Entropy rises across a shock. That is a consequence of the requirement that the Mach number must decrease, and it follows diagnostically from Eq. (7.20b). Entropy is an indicator that irreversible dissipation of kinetic energy into heat has occurred. A detailed analysis would require dissipative processes to be represented explicitly. It is impressive that the one-dimensional analysis produces the shock jump without this detail. A corollary is that accurate representation of the dissipative process is unnecessary. In computations, numerical dissipation suffices to capture shocks. Dissipation must be present, but the shock structure need not be computed accurately. The one-dimensional analysis tells us that consistency with inviscid conservation laws will suffice to produce the correct jump.

The magnitude of entropy creation is a measure of shock strength. Equation (7.3) can be written as

$$dS = c_v \frac{dp}{p} - c_p \frac{d\rho}{\rho}$$

after invoking the gas law,

$$p = \rho R T, \tag{7.21}$$

and $c_p = c_v + R$. Integrating across the shock

$$\Delta \mathcal{S} = c_v \log \left[\frac{p_2}{p_1}\right] - c_p \log \left[\frac{\rho_2}{\rho_1}\right]$$

$$= c_v \left\{ \log \left[1 + \frac{2\gamma}{\gamma - 1}(M_1^2 - 1)\right] + \gamma \log \left[\frac{2 + (\gamma - 1)M_1^2}{(\gamma + 1)M_1^2}\right] \right\} > 0. \tag{7.22}$$

This is positive because $M_1 > 1$. However, when M_1 is just slightly supersonic, the entropy change is very small. A Taylor series expansion of Eq. (7.22) shows that $\Delta \mathcal{S} \sim (M_1 - 1)^3$ when $M_1 \to 1$. Low supersonic Mach number shocks are nearly isentropic. An instance will be encountered in Figure 7.9, where a flow is compressed through a weak shock, then reexpands to nearly the conditions upstream of the shock, with almost no increase of entropy. The entropy rise can also be described as a loss of total pressure – see exercise 7.6.

7.3.2 Supersonic Flow over a Hemisphere

The jump conditions (7.20b) are referred to as Rankine–Hugoniot relations. Anderson (1982) contains enjoyable sections on the historical persona of compressible flow theory. Rankine was an engineer and thermodynamicist, known for the Rankine cycle and the Rankine scale of absolute temperature. He first devised the equations relating the states on either side of a normal shock in 1870. Rankine's equations were independently rediscovered by Hugoniot 17 years later.

A criterion imposed on CFD codes meant for supersonic flow is that they should preserve Rankine–Hugoniot relations. Some aspects of compressible computation will be described in §7.5. The essential numerical properties are obedience to local conservation of mass, momentum, and energy and preservation of the physics implied by the Mach cone.

The geometry of Figure 7.6 was chosen to produce a bow shock. The bluntness of a hemisphere ensures that the shock will stand in front of the body, giving a region near the symmetry line in which a normal shock forms.

Supersonic flow of air at atmospheric pressure and ambient temperature is considered. The inflow is at $M = u_\infty/a = 1.84$. To better resolve shocks, the grid was successively refined using mesh adaptation, with pressure gradient as a criterion for refinement. A sample adapted mesh is provided later in Figure 7.10.

The right portion of Figure 7.6 shows the velocity and pressure along the symmetry line. The velocity falls and the pressure rises across the shock, lying at $x/D = -0.2$. The dashed, horizontal lines show the prediction of Eq. (7.20b) for state 2. At $M_1 = 1.84$, the downstream pressure, p_2 should equal $3.78p_1$. With $p_1 = 1.013 \times 10^5$ Pa, this gives $p_2 = 3.83 \times 10^5$ Pa. The axial velocity downstream of the shock is $u_2 = u_1/2.42 = 264$ m/s from the Rankine–Hugoniot equations (7.20b): this corresponds to $M_2 = 0.61$. The numerics agree with these jump relations.

Between the nose of the shock and the nose of the body, the flow is subsonic. Pressure rises on a curve between $x/D = -0.2$ and the stagnation point, $x/D = 0$. The velocity decreases smoothly to zero at the stagnation point.

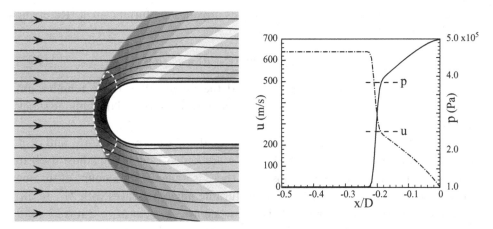

Figure 7.6. Supersonic flow past a hemisphere. Contours of Mach number. The sonic line is dashed. At right, velocity (—·—) and pressure (———) on the stagnation line.

This is rather different from the supersonic region. There the velocity is constant until the shock is reached, at which it changes abruptly. That is true over the whole region upstream of the bow shock, as can be seen in the Mach number contours in the left pane of Figure 7.6. This embodies the principle of no upstream influence into supersonic approach flow. The streamlines remain straight, oblivious of the approaching surface until they encounter the bow shock.

In general, shocks will not be exactly perpendicular to the oncoming flow. If they lie at an angle, they are termed *oblique*. The shock wave in Figure 7.6 is oblique and curved. Its role can be seen in streamlines: the bow wave turns the direction of the flow, so that it passes around the surface. Supersonic flow cannot turn in anticipation of the surface. Were it to meet the front of the body, it would have to turn abruptly and follow the surface tangentially. Instead, it is turned away from the surface by an oblique wave. It might initially seem counter intuitive, but the flow turns *toward* the shock, as seen in the velocity vectors of Figure 7.7. Another curiosity is that the flow remains supersonic after crossing the shock, except in a zone near the nose of the body. The white, dashed curve is the sonic line. Inside it, the Mach number is subsonic; outside it, the flow remains supersonic. Flow out of the subsonic zone reexpands to supersonic speeds at it rounds the hemisphere.

The turning of streamlines toward the oblique shock has a simple explanation: the component of velocity parallel to the shock is unaltered across the discontinuity. The perpendicular component decreases, just as across a normal shock.

Figure 7.7. Definition of directions at an oblique shock. The gray line is the shock wave.

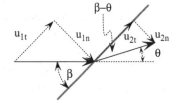

Trigonometry applied to the velocity triangles provides the turning angle. Referring to Figure 7.7,

$$\tan \beta = u_n/u_t|_1; \quad \tan(\beta - \theta) = u_n/u_t|_2. \tag{7.23}$$

Because the normal velocity decreases and the tangential velocity remains unchanged across the discontinuity, $\theta > 0$, and the flow turns toward the oblique shock wave.

The turning angle, θ is a function of the incident Mach number, M_1, and of β. Given a geometry, as in Figure 7.5, M_1 and θ are known, and β will adjust itself to produce the required turning – if possible. There is a maximum amount of turning that can occur in this way. The maximum turning angle increases with M_1. In air, the absolute maximum turning angle is 46°, when $M_1 \to \infty$; at $M_1 = 2$ it is 23°. If θ is larger than the maximum, the shock must move ahead of the body and become curved, as in Figure 7.6. The reader may wish to look forward to Figure 7.9 for a computed example of flow turned by an oblique wave.

Let us pursue these points analytically. Elementary considerations provide a formula for the shock angle, β, that is needed to produce a given turning angle, θ. The normal component of velocity satisfies the jump conditions (7.18), whereas the tangential component is unaltered:

$$u_{1t} = u_{2t}.$$

The solution [(7.20a) and (7.20b)] is modified only by replacing M_1 by M_{1n} or, by Figure 7.7, $M_1 \sin \beta$. Then, according to expressions (7.23),

$$\frac{\tan(\beta - \theta)}{\tan \beta} = \frac{u_{n2}}{u_{n1}} = \frac{\rho_1}{\rho_2},$$

where the last step follows from mass conservation. Hence, the turning angle can be found from expression (7.20b) for ρ:

$$\frac{\tan \beta}{\tan(\beta - \theta)} = \frac{(\gamma + 1)M_1^2 \sin \beta^2}{2 + (\gamma - 1)M_1^2 \sin \beta^2}. \tag{7.24}$$

A few solutions to this for turning angle versus shock angle, with given incident Mach number, are shown in Figure 7.8.

Consider specified values of M_1 and θ. These are set by the incident flow and geometry. Figure 7.8 provides two possibilities for the shock angle, β. These are called weak and strong cases. For the most part, the weak shock allows the flow to remain supersonic after it is turned ($M_2 > 1$), whereas the strong shock is a transition from super to subsonic flow. (There is a small range in which both solutions have $M_2 < 1$.) The weak case is commonly seen in flows like Figure 7.5. Either can occur, depending on downstream conditions, but the strong solution is rare.

In the limit of no turning, $\theta = 0$, the two solutions are a normal shock, $\beta = 90°$, and a Mach wave, $\beta = \sin^{-1}(1/M_1)$. These are the logical limits of "strong" and "weak" discontinuities: the weak case limit is a sound wave; the strong is a jump, perpendicular to the stream.

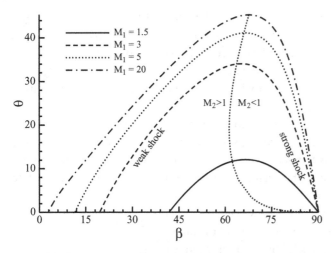

Figure 7.8. Shock and turning angles for a few Mach numbers. The dash-dot curve separates the strong ($M_2 < 1$) and weak ($M_2 > 1$) shock solutions.

It can be seen in Figure 7.8 that on the weak branch the wave angle, β, increases with decreasing M_1 for a given deflection, θ. Lower speed requires a larger wave angle to turn the flow. Each curve of constant Mach number has a maximum. The corresponding θ is the maximum turing angle that can be accomplished at this speed. For instance, at $M = 1.5$, an oblique shock cannot turn the flow angle more than $12°$; at $M = 3$, this angle is $34°$. If θ is larger than this, Eq. (7.24) has no solution. What happens if the turning angle is too large?

Nature resolves this difficulty by moving the shock forward of the surface, as a *detached* shock. The detached shock is normal to the oncoming flow ahead of the nose, and oblique at the sides of the body, as in Figures 7.3 and 7.6. Downstream of the curved front, there will be super- and subsonic regions. Where the flow is subsonic, the upstream influence of the body will turn the flow direction; in the supersonic regions, the obliqueness of the shock front will turn the flow.

7.3.3 Expansion Fan

Is it possible for θ to be negative in Figure 7.7? Yes. θ is the angle that the flow must turn to follow the surface of a wedge. If $\theta > 0$, the wedge slopes up into the oncoming flow. If $\theta < 0$, the surface falls away from the flow. In Figure 7.5, portions of the airplane do turn away from the flow. The bow shock turns the flow upward to follow the top surface and downward to follow the bottom. Subsequently, the surface turns away from the flow: eventually, it turns back to the horizontal. A series of waves is observed where the surface slope is changing back toward horizontal. Clearly, something happens where $\theta < 0$.

Figure 7.5 is a Schlieren photograph. The Schlieren technique makes density gradients visible. The index of refraction of light increases with density. As light waves traverse the wind tunnel test section, they are bent by density variations.

After it exits through the test section window, the beam of light is intersected by a knife edge. If the density variation has deflected light rays below the edge, a dark area appears. If light rays that would have struck the knife have been deflected up, above the edge, a bright area appears. A variant is to replace the knife by a grating of colored bands. Deflected white light will produce a colored image of the density gradients.

Figure 7.5 is a grayscale image of a color Schlieren. The bright lines show where steep density gradients occur. Some are at shocks, where the surface of the airplane turns into the flow. The density rises as the gas is compressed across the shock front. Other bright lines occur where the surface turns away from the flow. These must be expansions, where the density falls.

If θ is negative in Figure 7.7, the flow accelerates across the wave front. This is a consequence of trigonometry and invariance of the tangential velocity: $u_{1t} = u_{2t} = u_t$. The magnitude of the velocities before and after the wave are

$$|u_1| = u_t/\cos\beta; \quad |u_2| = u_t/\cos(\beta - \theta). \tag{7.25}$$

If $\theta < 0$, then $|u_2| > |u_1|$. The velocity increases, so the flow remains supersonic.

This is called an *expansion wave* – the terminology will become evident in §7.4. It is quite different from a shock wave. The latter is a discontinuity that is produced by wave steepening. The expansion wave does not steepen and is not a discontinuity. It causes a smooth, isentropic turning of the flow. In fact, the flow turns through a series of expansion waves or an expansion fan. This is also called a Prandtl–Meyer expansion fan, after early analyses by Prandtl and his student.

The analysis of oblique shocks was completed by asserting that the normal component of velocity satisfies the jump conditions, which led to Eq. (7.23). There are no jump conditions to complete the analysis of expansion waves. Instead, the expansion wave is asserted to be a Mach wave. Then the unknown angle, β, is specified as $\beta = \alpha$ with α given by Eq. (7.15). Hence, we find

$$\frac{|u_2|}{|u_1|} = \frac{\cos\alpha}{\cos(\alpha - \theta)}$$

from Eq. (7.25). For small turning angles

$$\frac{\cos\alpha}{\cos(\alpha - \theta)} = \frac{1}{\cos\theta + \tan\alpha\sin\theta} \approx 1 - \frac{\theta}{\sqrt{M^2 - 1}},$$

where $\sin\alpha = 1/M$, implying $\tan\alpha = 1/\sqrt{M^2 - 1}$, was sustituted. Across an infinitesimal expansion wave, the velocity changes by

$$\frac{\Delta|u|}{|u|} = \frac{|u_2| - |u_1|}{|u|} = \frac{-\Delta\theta}{\sqrt{M^2 - 1}}. \tag{7.26}$$

The turning through an expansion is a sum of these small turns, as seen in between the dashed lines in Figure 7.9.

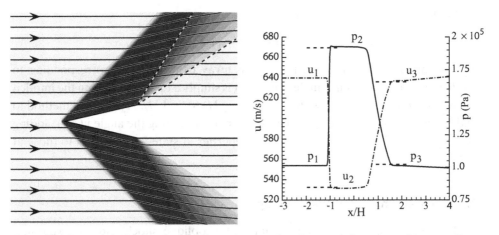

Figure 7.9. Supersonic flow over a two-dimensional wedge. Contours of pressure with streamlines overlaid. At right, the curves are values along a streamline. The horizontal dashed lines show predictions of oblique shock and expansion fan theory.

The Mach number ratio is

$$\frac{M_2}{M_1} = \frac{|u_2|}{|u_1|}\sqrt{\frac{T_1}{T_2}}$$

because sound speed is proportional to the square root of temperature. The expansion is isentropic; hence, temperature satisfies Eq. (7.14a), giving

$$\frac{M_2}{M_1} = \frac{|u_2|}{|u_1|}\sqrt{\frac{1 + (\gamma - 1)M_2^2/2}{1 + (\gamma - 1)M_1^2/2}}.$$

For small changes, $M_2 = M_1 + \Delta M$, $\Delta M \ll 1$, this is approximately

$$\frac{\Delta M}{M} = \frac{-\Delta\theta}{\sqrt{M^2 - 1}} + \frac{1/2(\gamma - 1)M\Delta M}{1 + 1/2(\gamma - 1)M^2}$$

or

$$-\Delta\theta = \frac{\sqrt{M^2 - 1}}{(1 + 1/2(\gamma - 1)M^2)}\frac{\Delta M}{M}.$$

The Prandtl–Meyer analysis gives the Mach number increase through an expansion fan as the integral of this: given θ and M_1,

$$-\theta = \int_{M_1}^{M_2} \frac{\sqrt{M^2 - 1}}{1 + 1/2(\gamma - 1)M^2}\frac{dM}{M}. \tag{7.27a}$$

Clearly, if $\theta < 0$, then $M_2 > M_1$; the flow accelerates through the expansion fan. The integral (7.27a) has the closed form

$$\nu(M) = \sqrt{\frac{\gamma + 1}{\gamma - 1}}\tan^{-1}\sqrt{\frac{\gamma - 1}{\gamma + 1}(M^2 - 1)} - \tan^{-1}\sqrt{M^2 - 1} \tag{7.27b}$$

so that Eq. (7.27a) can be written

$$-\theta = v(M_2) - v(M_1).$$

This is only valid when the surface opens away from the flow, corresponding to $\theta < 0$. Given a valid turning angle, and M_1, it is simply a matter of solving the implicit relation $v(M_2) = v(M_1) - \theta$ with (7.27b) for $v(M)$ to find M_2. A graphical method to find M_2 consists in plotting the function v versus M, finding the angle $v(M_1)$, adding $-\theta$ and looking on the axis to find M_2. Properties at state 2 are related to those at state 1 by isentropic flow.

7.3.4 Supersonic Flow over a Wedge

A slab with a wedge fore-body combines the oblique shock and the expansion fan. In Figure 7.9, the incident Mach number is $M_\infty = u_\infty/c = 1.84$, with the velocity $u_\infty = 640$ m/s. The ambient pressure is 1 atm. The half-angle of the wedge, $\theta = 12.5°$, defines the turning angle. The angle of the oblique shock is computed to be $\beta = 46.2°$, from the oblique shock relation (7.24). This shock is relatively weak, thus the flow downstream of the shock remains supersonic. Recall that only the normal component of velocity jumps across the shock. The tangential component is unaltered and the Mach number can remain greater than unity.

The flow velocity u_2 and pressure p_2 can be computed from the normal shock relations (7.20a) and (7.20b). First M_{n2} is found from $M_{n1} = M_1 \sin \beta = 1.33$ in Eq. (7.20a), giving $M_{n2} = 0.77$. At $M_{n1} = 1.33$, the jump relations (7.20b) give $p_2 = 1.89 p_1$ and $u_{n2} = u_{n1}/1.57$. The normal velocity $u_{n2} = 294$ m/s follows from $u_{n1} = u_1 \sin \beta = 462$ m/s; and the pressure $p_2 = 1.91 \times 10^5$ Pa follows from the atmospheric pressure $p_1 = 1.01 \times 10^5$ Pa. The tangential velocity velocity $u_{t2} = u_{t1} = u_1 \cos \beta = 443$ m/s is unaltered.

The velocity magnitude downstream of the shock is then

$$u_2 = \sqrt{443^2 + 294^2} = 532 \text{ m/s}.$$

The Mach number downstream of the shock is $M_2 = M_{n2}u_2/u_{2n} = 1.39$.

Continuing along the surface, the flow rounds the corner, where the wedge joins the slab, through an expansion fan. Flow variables downstream of the expansion fan can be computed directly from the Prandtl–Meyer relation (7.27a) and the isentropic formulas (7.14a) and (7.14b). We start with the values for the flow after the oblique shock: from $M_2 = 1.39$ and turning angle $\theta = -12.5$ of the expansion, the Mach number after the expansion fan can be computed from (7.27a) to be $M_3 = 1.82$. This is very close to the incident Mach number of 1.84; the flow expands isentropically back to conditions almost identical to those upstream. There is very little entropy gain across the oblique shock.

The pressure p_3 is computed from p_2, $M_2 = 1.39$, $M_3 = 1.82$ and Eq. (7.14b):

$$p_3 = p_2 \left[\frac{1 + 1/2(\gamma - 1)M_2^2}{1 + 1/2(\gamma - 1)M_3^2} \right]^{\gamma/(\gamma-1)} = 1.005 p_1.$$

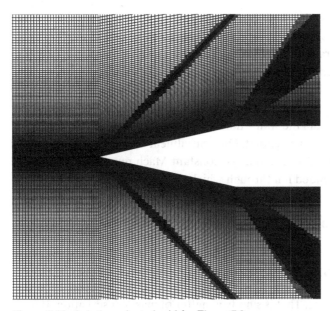

Figure 7.10. Solution adapted grid for Figure 7.9.

The net effect of the oblique shock followed by the expansion fan is only a 1/2% pressure rise. The velocity magnitude u_3 is found to be 636 m/s.

The Mach wave at the start of the fan makes an angle $\sin^{-1}(1/1.39) = 46°$ to the incident flow, or an angle of this plus the wedge angle to the horizontal, that is 58.5°. The end of the fan is a wave at angle $\sin^{-1}(1/1.82) = 33.3°$ to the horizontal. Contour lines of constant pressure show the expansion fan in Figure 7.9. The theoretical angles for its beginning and end are marked on the figure with dashed lines. Theory is extremely accurate.

An initial structured grid with 244×190 cells was used in the computations. This grid was then successively refined to better resolve shocks, using mesh adaptation (Chapter 2) with pressure gradient as a criterion for refinement. The final grid, Figure 7.10, had 68,493 nodes and 67,123 cells. The adapted grid is able to crisply capture the oblique shocks and expansion fans.

Streamlines overlay pressure contours in Figure 7.9. The streamlines show the abrupt turning to the direction of the wedge through an oblique shock. This is followed by curved trajectories that reorient the flow to the direction of the after body.

This two-dimensional, illustrative computation is in excellent agreement with the one-dimensional analyses. The plot at the right of the figure shows how the pressure and velocity vary along a streamline. First, they are constant at their incident levels, until the shock is met, at $x/H = -1$. Then the pressure rises and velocity falls. The levels remain constant until the expansion fan is encountered, at $x/H = 0.6$, where the velocity increases and the pressure falls smoothly as the flow expands. The horizontal dashed lines show the levels computed above from the one-dimensional formulas. They are in excellent agreement with the computation.

This two-dimensional wedge flow can be contrasted to the hemisphere (Figure 7.6). Away from the nose of the hemisphere, the shock wave is oblique. However, it does not turn the flow parallel to the surface. It turns it away from the surface, but the flow direction continues to change after passing through the shock wave. The differences are that the hemisphere is axisymmetric and that it is curved. Generally, supersonic flow will follow curved streamlines. This can be imagined to be directed by Mach waves emanating from the wall – the expansion fan is a particular example, in which the Mach waves all start at a point. On a curvilinear body, the starting points are spread along the surface. The contours of constant Mach number in Figure 7.6 can be thought of as an extended fan through which the flow turns isentropically.

7.4 Quasi-One-Dimensional Gas Dynamics

Another important compressible flow phenomenon is "choking." Consider flow through a duct with varying cross-sectional area – for instance, a circular tube with radius varying with distance, $r(x)$. In subsonic flow, as the cross-sectional area decreases, the flow will accelerate. That is simply a matter of mass conservation. It may seem peculiar, but in supersonic flow, as the cross section decreases, the flow decelerates. That is the result of a simple, one-dimensional analysis, described below. It does not violate mass conservation because the density rises in the contraction. The gas is compressed as it is forced through the decreasing area. This is loosely like the flow of traffic where two lanes merge into one: the density goes up and the velocity goes down. That analogy cannot be taken too far: collisions between automobiles are not elastic; but the point to be made is that density rising and velocity falling is not implausible when one thinks in terms of a stream of gas molecules.

The opposite behaviors of subsonic and supersonic contractions raises the question of what happens if a supersonic flow enters a large contraction. If the contraction is large enough, at some point the Mach number will drop to unity. Beyond that, the flow would be subsonic; but then further contraction would cause the speed to rise, so it could not remain subsonic. Clearly, this is a paradox. The resolution is that the Mach number cannot drop below unity in a supersonic contraction. In the absence of a shock, the Mach number will either remain above unity or will reach unity where the cross-sectional area reaches a minimum. The latter is called *choked* flow. It is a law of physics – more correctly, of quasi-one-dimensional gas dynamics – that the Mach number cannot fall below the value of unity, which is reached at the minimum area – also called the "throat." Making the throat smaller constricts the flow. The mass flux decreases to maintain sonic conditions at the throat. That is a rather useful observation. It means that flow can be metered by passing it through a sonic throat. The mass flow is independent of downstream conditions, as long as the throat remains choked.

From what has just been said, it might seem impossible to bring a supersonic duct flow to subsonic conditions. Contracting the area brings it to sonic speed but not lower. The transition can be accomplished by a converging–diverging nozzle. When the flow is supersonic, it decelerates to Mach one in a converging section. A

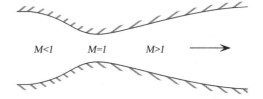

Figure 7.11. Converging–diverging nozzle. $M<1$ $M=1$ $M>1$

subsonic flow decelerates if the cross section increases in the direction of flow. If the duct expands after the throat, the Mach number will continue to decrease to subsonic levels. Hence, the duct area must decrease to a minimum, at which the Mach number reaches unity, then increase.

The reasoning just given can be reversed. If the flow enters the nozzle subsonically, it can accelerate only to Mach one in the converging section. The nozzle must then diverge if the flow is to become supersonic. That is the basic concept of a rocket nozzle. It is illustrated by Figure 7.11.

The converging–diverging, or de Laval, nozzle is a device for converting from subsonic to supersonic flow or vice versa. Suppose the flow enters subsonically. It accelerates in the contracting section. If the contraction is sufficient, the minimum area becomes choked. The mass flow then is unaffected by what happens downstream of the throat: the expanding section is supersonic, so disturbances cannot propagate upstream. Another consequence of choking is that, if there are no shocks, the flow downstream of the throat is uniquely determined by the shape of the duct. That is simply a consequence of the condition that $M = 1$ at the throat. A sonic throat regulates the flow.

In particular, Eqs. (7.14a) and (7.14b) on pages 270–271 can be invoked. These formulas evaluated at $M = 1$ are denoted by asterisks:

$$\frac{T_0}{T_*} = \frac{\gamma+1}{2}; \ \frac{\rho_0}{\rho_*} = \left[\frac{\gamma+1}{2}\right]^{1/(\gamma-1)}; \ \frac{P_0}{P*} = \left[\frac{\gamma+1}{2}\right]^{\gamma/(\gamma-1)}. \tag{7.28}$$

If the flow remains isentropic after passing through a choked throat, it can be expressed in terms of conditions at the throat. This is done by rewriting (7.14a) and (7.14b) as

$$\frac{T_*}{T} = \frac{T_*}{T_0}\frac{T_0}{T} = \frac{2+(\gamma-1)M^2}{\gamma+1} \tag{7.29a}$$

and similarly

$$\frac{P^*}{P} = \left[\frac{2+(\gamma-1)M^2}{\gamma+1}\right]^{\gamma/(\gamma-1)} = \left(\frac{\rho*}{\rho}\right)^{\gamma} \tag{7.29b}$$

for temperature, pressure, and density. Hence, conditions downstream of a choked throat are determined entirely by the asterisked values, at the sonic point.

These isentropic formulas are converted to a one-dimensional model of flow through a duct by relating the Mach number to the cross-sectional area. An expression for $M(A)$ is found from mass continuity.

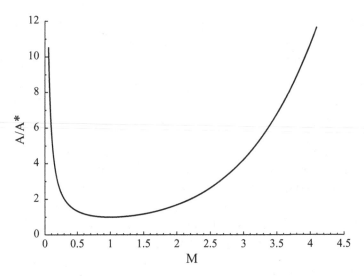

Figure 7.12. Area ratio versus Mach number in choked flow.

The condition of mass conservation, in a quasi-one-dimensional approximation, is that $\rho u A$ is constant. This can be restated in terms of Mach number as

$$\rho\sqrt{\gamma R T}\,MA = \text{constant}. \tag{7.30}$$

Then equating mass flux at any position in the duct to that at the sonic throat

$$\rho T^{1/2}MA = \rho_* T_*^{1/2}A_*.$$

With Eqs. (7.29a) and (7.29b) for T_*/T and ρ_*/ρ, we find that

$$\frac{A}{A_*} = \frac{1}{M}\left[\frac{2+(\gamma-1)M^2}{\gamma+1}\right]^{\frac{\gamma+1}{2(\gamma-1)}}. \tag{7.31}$$

With $\gamma = 1.4$, the exponent on the far right is 3.

Suppose that the duct area is known. The Mach number at any point is determined by the ratio of the local cross-sectional area to the minimum area, A_*. Formula (7.31) is solved for M, given area ratio A/A_*. That equation is found to have two solutions, one less than unity and one greater than unity. For instance, if $A/A_* \gg 1$ either M is small and Eq. (7.31) with $\gamma = 1.4$ is approximately

$$M = \frac{A_*}{A}\left(\frac{2}{\gamma+1}\right)^{\frac{\gamma+1}{2(\gamma-1)}} = \frac{A_*}{1.2^3 A} < 1;$$

or M is large, and (7.31) gives approximately

$$M = \left(\frac{A}{A_*}\right)^{\frac{\gamma-1}{2}}\left[\frac{(\gamma+1)}{(\gamma-1)}\right]^{\frac{\gamma+1}{4}} = \left(\frac{6^3 A}{A_*}\right)^{0.2} > 1.$$

The full shape of the curve defined by Eq. (7.31) is plotted in Figure 7.12. In the case that a subsonic flow passes through a sonic choke, this curve determines the Mach number on either side of the throat solely from the area ratio A/A_*. The flow downstream is known independently of conditions upstream. Similarly, the flow upstream of the throat is unaffected by the downstream portion. The latter is not surprising because the downstream portion is supersonic, so disturbances cannot propagate upstream against the flow.

The phenomenon of choking can be understood by returning to Eq. (7.30). Writing the mass flux at the throat in terms of stagnation properties

$$\dot{m} = \rho_* \sqrt{\gamma R T_*} A_* = \rho_0 \sqrt{\gamma R T_0} A_* \left[\frac{2}{\gamma+1}\right]^{(\gamma+1)/2(\gamma-1)} \tag{7.32}$$

having substituted relations (7.28). The last factor equals 0.579 in diatomic gases – for which $\gamma = 1.4$. Thus, the mass flow is determined by stagnation properties, independently of downstream conditions. A choked nozzle regulates the flow, in the sense that if it is supplied by a tank with a particular temperature and pressure, the mass flow rate is unaffected by the exhaust pressure. To be choked, the exhaust pressure must be less than 0.528 times the stagnation pressure. For instance, a $1\,\mathrm{cm}^2$ nozzle, supplied by $P_0 = 2\,\mathrm{atm}$ and $T_0 = 300°\mathrm{K}$, exhausting to 1 atm produces a mass flux of $23.6\,\mathrm{g/s}$. If it exhausts to $1/2$ atm, the mass flux is still $23.6\,\mathrm{g/s}$.

The simple ideas covered to this point provide a basis for understanding many aspects of compressible CFD. They must be tempered by an intuition for how to resolve paradoxes. This has already arisen in the discussion of oblique shocks (page 281): in that case, when the geometrical turning angle is too large no shock can produce it. The resolution in that case was for the shock to detach from the surface and become curved.

A similar paradox arises in the flow through a nozzle, exiting to an ambient pressure. If A_e is the exit area, then Eq. (7.31) or Figure 7.12 provides an exit Mach number, M_e. If the flow is isentropic, Eq. (7.14b) provides the exit pressure P_e, given the stagnation pressure:

$$\frac{P_e}{P_0} = \left(1 + \frac{\gamma-1}{2} M_e^2\right)^{-\gamma/(\gamma-1)}. \tag{7.33}$$

Suppose the flow exhausts to the atmosphere. The exit pressure can be equal to, greater than, or less than atmospheric pressure. In the first case, the flow is said to be perfectly expanded; in the second, it is *underexpanded*; in the third, it is *overexpanded*. Over- and underexpanded conditions only occur if the exit Mach number is supersonic. If the flow is underexpanded, it will expand further after exiting the nozzle. This might be observed in contours of density or pressure as an expansion fan emanating from the nozzle exit into the external flow.

If the flow is overexpanded, a paradox exists. The flow that reaches the end of the nozzle has a pressure lower than ambient. It must proceed from the low pressure in the nozzle into the higher-pressure ambient fluid. As it exits the nozzle, the area expands and the pressure should fall; but then it fails to match to ambient. The resolution, again, is to introduce a discontinuity. Because the gas is flowing supersonically, a shock wave can form in the expanding section of the nozzle. The flow then adapts to the external conditions by a pressure rise through the shock wave. Let us pursue this step by step.

The role of exit pressure can be understood by considering a nozzle that is fed from a large tank with a fixed total pressure, P_0, or, in a computational analysis, with a total pressure that is prescribed at the inlet. The nozzle exits into a chamber with controlled ambient pressure, P_a. In a computation, this is a satisfactory exit boundary condition, provided the outflow is subsonic, as is explained in §7.5. We want to describe the flow inside the nozzle as the ambient pressure, P_a, is decreased, starting from the stagnation pressure, P_0.

When $P_a = P_0$, there is no flow. If P_a is slightly less than P_0, the gas will begin to flow, remaining subsonic throughout the nozzle, exiting at ambient pressure. The corresponding exit Mach number is given by Eq. (7.14b):

$$M_e^2 = \frac{2}{\gamma - 1}\left[\left(\frac{P_0}{P_a}\right)^{(\gamma-1)/\gamma} - 1\right].$$

Mass conservation (7.30) and the isentropic formulas (7.14b) relate the Mach number at any other section to the area ratio A_e/A and the exit conditions:

$$\rho_e M_e \sqrt{T_e} A_e = \rho M \sqrt{T} A. \tag{7.34}$$

A unique quasi-one-dimensional solution exists.

The velocity will be greatest at the point of minimum area or the throat. Now let P_a drop farther. At some point, the velocity at the throat will reach the speed of sound. Just before that, the flow is subsonic, and hence it slows down in the expansion following the throat. That means that the pressure rises in that section, and P_a is greater than the pressure at the throat. When sonic conditions are just reached, the throat pressure is

$$P_* = P_0\left(\frac{\gamma + 1}{2}\right)^{-\gamma/(\gamma-1)} = 0.528 P_0$$

for $\gamma = 1.4$. The ambient pressure is higher than $0.528 P_0$ when sonic conditions are barely obtained at the throat and the rest of the duct is subsonic. It is given by Eq. (7.33) with the subsonic root of Eq. (7.31) for M_e.

If the ambient pressure is dropped slightly more, the flow should be supersonic after the throat. However, were it to expand supersonically from the throat all the way to the exit, then the pressure would fall *below* the throat value of $0.528 P_0$. But the exit pressure is *higher* than $0.528 P_0$. Now there is a dilemma: the flow cannot remain subsonic; it must expand supersonically after the throat, but a supersonic expansion will not bring the pressure up to the ambient level.

To recover the ambient pressure, the flow must be subsonic before the exit is reached so that the pressure can rise above the value at the throat. What happens is that the flow starts to expand supersonically, but a normal shock wave forms in the nozzle. The pressure rises across the shock and rises further in the subsonic expansion downstream of it, such that the ambient pressure is reached at the exit; see Figure 7.13.

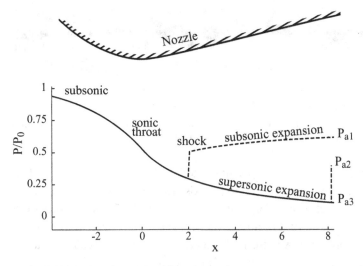

Figure 7.13. Transonic nozzle with various exit pressures.

A shock must exist in the expanding section for a range of ambient pressures. When P_a is just below that for sonic conditions to be reached, the shock is right at the throat and has vanishing strength. As the exit pressure is reduced, the shock blows out of the throat and moves to a location in the downstream nozzle. It now has a finite strength, characterized by the pressure jump. A case in this range is shown as the curve labeled P_{a1} in Figure 7.13. As P_a drops farther, the shock moves farther and farther along the expanding nozzle toward the exit.

Eventually, the ambient pressure falls to the level at which the shock lies in the exit plane. This is case P_{a2} in Figure 7.13. From the shock relations (7.20b), the ratio of the ambient pressure to the exit pressure is simply that across a shock at the exit Mach number:

$$\frac{P_a}{P_e} = 1 + \frac{2\gamma}{\gamma+1}(M_e^2 - 1).$$

The exit Mach number is determined by the supersonic root of Eq. (7.31) and the ratio of the exit to the throat area. Because $M_e > 1$, $P_a > P_e$. The jet is not perfectly expanded until the ambient pressure drops to P_{a3} in Figure 7.13; that is the pressure created by an isentropic, supersonic expansion. There is a range of pressures between P_{a2} and P_{a3} for which the shock has reached the end of the nozzle, but the jet that exits the nozzle is still overexpanded. In that range of pressures, the final rise from P_e to P_a occurs outside the nozzle, through oblique shocks. In fact a series of oblique shocks usually occurs, crisscrossing the jet, as seen in Figure 7.14. The series of shocks occurs because the oblique shock turns the flow direction toward the jet centerline, causing the two sides to flow toward each other, creating another oblique shock that turns the flow back, away from the axis.

When the ambient pressure falls to P_{a3} the flow is perfectly expanded. Further decrease of the ambient pressure leads to an underexpanded jet exiting the nozzle. It

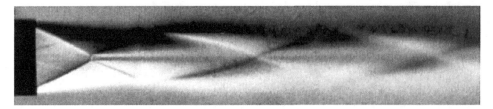

Figure 7.14. Overexpanded jet showing oblique shocks emanating from the exit plane. Jeronimo and Van Der Haegen, Schlieren technique, lab notes VKI (2002).

expands farther through a fan at the exit, as illustrated in Figure 7.15. Again, the flow direction is diverted, this time away from the jet axis. That overexpands the jet near the axis and a series of shock waves form, as seen in Figure 7.15. That photograph shows a normal shock in the central region of the jet. The jet has overexpanded and shocks up to the higher pressure. A pattern of pressure waves is observed, extending downstream. Such cellular patterns are commonly seen when jets are not perfectly expanded. A delightful collection of Schlieren photographs of various phenomena related to shock waves can be found in VanDyke (1982).

7.4.1 Shock Patterns and Shock-Induced Separation

The quasi-one-dimensional theory is quite elegant. Is it an accurate portrayal of real flow though converging–diverging ducts? If not, when is it wrong; when is it accurate?

Let us examine a computed flow in an axisymmetric convergent–divergent nozzle and see how it compares to quasi-one-dimensional theory. The geometry is that seen in Figure 7.16. This is a radial section of an axisymmetric converging–diverging pipe, followed by a large cylindrical plenum chamber. The three-dimensional geometry is this section, rotated about the centerline.

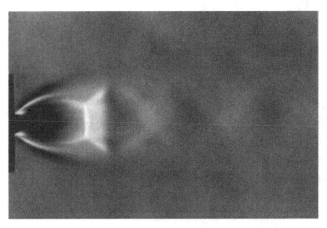

Figure 7.15. Underexpanded jet flame, showing expansion at the exit plane, followed by shock cells. Devaud et al., Shock Waves, **12**, 241–249 (2002).

Figure 7.16. Mach number contours in subsonic flow through the nozzle.

Initially, a multiblock, structured grid consisting of 300×60 cells in the nozzle and 80×120 cells in the plenum was created. It was addressed in the node, cell, and connectivity format of an unstructured mesh. That was so that an unstructured solver, with adaptive capability could be used. Mesh adaptation (§2.1.5) was performed for the cases with supersonic flow, refining in regions with strong pressure gradients.

The boundaries of the plenum, situated at the exit of the nozzle, are maintained at atmospheric pressure, P_{atm}. The boundaries of the duct are adiabatic, no-slip surfaces. Various flow regimes were achieved by altering the total pressure, P_0, at the inlet. The total temperature, T_0, was held fixed at 300°K. A turbulence model was activated; it only influences the boundary layer on the nozzle wall, in most cases. It will be seen shortly why the boundary layers were treated as turbulent.

For a choked, convergent–divergent nozzle with a given exit, A_e, and throat, A_*, areas, the pressure ratio P_0/P_e can be calculated for shock-free, isentropic flow. First, the exit Mach number M_e is found from Eq. (7.31) (or Figure 7.12) and the area ratio. Then the pressure ratio is obtained from Eq. (7.14b)

$$\frac{P_0}{P_e} = \left(1 + \frac{1-\gamma}{2}M_e^2\right)^{\frac{\gamma}{\gamma-1}}.$$

Suppose the exit pressure equals the pressure in the exit plenum chamber, which is 1 atm: $P_e = P_{atm}$. For the nozzle in this example, $A_e/A^* = 4$; then Figure 7.12 shows the exit Mach number to be $M_e = 2.94$. The above equation gives $P_0/P_e = 34.0$: the total pressure at the inflow has to be approximately 34 atm.

What happens if the total pressure at the inflow is lower than that? If P_0 is just barely larger than the exit pressure P_e, the flow is subsonic everywhere. That case is shown by the uppermost pair of curves in Figure 7.17, labeled 1. These are for $P_0/P_{atm} = 1.01$. This is the only case in which the flow was not regulated by a choked throat. For that reason, the theoretical curve was calculated using the mass flux obtained by the CFD. The isentropic formula does not acknowledge viscous forces on the walls. At the low pressure ratio of this case, they cannot be entirely ignored: the computed mass flux is about 30% lower than the inviscid, isentropic formula implies. The extreme limit would be a straight duct, for which the isentropic pressure difference would be 0; the actual pressure drop would be entirely viscous. One-dimensional theory can be modified for frictional losses (see Anderson, 1982) but that was not done here.

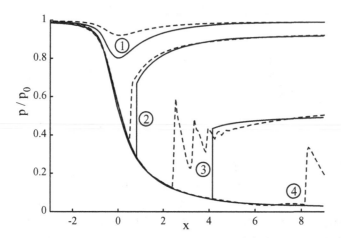

Figure 7.17. Pressure distributions down the axis of the nozzle. (———) One-dimensional analysis; (- - - -) CFD.

As P_0 is increased, the flow in the divergent part of the nozzle becomes super-sonic. The pressure in the divergent part falls below atmospheric pressure, as was explained in connection with Figure 7.13. The only way the ambient condition in the plenum can be met is for a normal shock to form in the diffusing section.

The pressure ratio is set to $P_0/P_{atm} = 1.1$ for the curve labeled 2 in Figure 7.17. The shock is not entirely normal; it is slightly curved in the vicinity of the walls because of the flow nonuniformity and the boundary layers. But the agreement between CFD and theory in case 2 of Figure 7.17 is acceptable. The most notable discrepancy is that the computed shock location is slightly before the theoretical position. Mach number contours in Figure 7.18 show a small cross-stream variation in the upstream region and a larger variation in the downstream, diffusing section. The one-dimensional theory would be very accurate were the contours vertical.

Further increase of P_0 results in the shock being pushed increasingly downstream and at the same time becoming stronger, as it sits in a higher Mach number section. The reader may be feeling uncomfortable in light of material in previous chapters. The pressure rises across the shock, imposing an adverse pressure gradient on the boundary layers. How long can they be assumed to be innocuous?

A sufficiently strong shock will cause the boundary layers to separate shortly downstream of its location. This separation calls the quasi-one-dimensional analysis

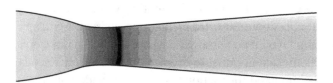

Figure 7.18. Mach number contours in choked flow through the nozzle, with a weak shock. The darkest contour is the maximum Mach number of 1.3. This is case 2 of Figure 7.17.

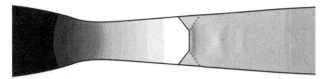

Figure 7.19. Pressure contours illustrating the λ shock pattern. The flow expands through the choked throat to the minimum pressure, then shocks up. This is case 3 of Figure 7.17.

into question. The separated wall layer will produce Mach number variations across the duct. It has a potential to dramatically change the structure of the shock.

Indeed, it does: there is a strong interaction between the core flow and the wall layer. The shock induces separation; the separation modifies the Mach number distribution; the shock structure changes. What is seen in computations and experiments is even more intriguing than could be anticipated. Shock-induced separation leads to a rich class of phenomena.

The view of simple, one-dimensional flow must be severely revised. Near the wall the separation bubble will force the flow away from the surface. The supersonic approach flow must turn through an oblique shock and flow over the recirculation zone. Having turned away from the wall to skirt the backflow, it must subsequently return to the downstream direction and proceed through the nozzle. This is accomplished by a second oblique shock. This behavior is shown by the pressure contours in Figure 7.19.

The full shock structure consists of the main, normal shock in the middle of the nozzle and a pair of oblique shocks near the wall, forming a λ shape. Lines have been superposed on the pressure contours of Figure 7.19 to bring out the λ structure. The point where the three legs of the λ meet is the *triple point*. The normal shock in the middle of the channel is called the Mach stem and the feet are called the incident and reflected oblique shocks. The last term arises from considering the first leg to be a wave that starts at the wall, propagates up to the base of the normal shock and is reflected back to the wall along the other leg. It is as if the triple point were replaced by a plane, reflective wall. Indeed, oblique shocks can crisscross a duct, reflecting from wall to wall. In that case, there is no Mach stem; the upper and lower legs of the λ's in Figure 7.19 join to form a cross. The downstream crisscross pattern is created by the reflections.

We see from Figure 7.17, case 3, that the normal shock appears even farther upstream of the predicted location than in the previous case. Also the computed pressure trace exhibits peculiar, sharp oscillations that are not seen in the simple theory; are these real or a numerical error? They are real; they have been accurately resolved by adapting the computational mesh. The oscillations are the signature of a pattern of shocks: the flow decelerates through a normal shock, reexpands to higher speed, shocks down, reexpands, and so on. Gyrations between super- and subsonic flow are displayed in Figure 7.20.

The pressure ratio here is $P_0/P_{\text{atm}} = 2.0$. Streamlines and Mach number contours, corresponding to Figure 7.19, are provided in Figure 7.21. The top inset zooms

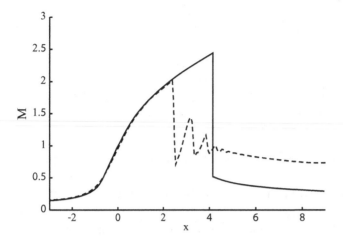

Figure 7.20. Mach number trace along the centerline for case 3. (———) One-dimensional analysis; (– – – –) CFD.

in to show how the flow deflects around the separated zone. An oblique shock turns the flow away from the wall and then another turns it back toward the axial direction. The full view shows an extensive separated zone in the diverging section.

The lower inset zooms in on the expanding section. A ridge in the Mach number isolines creates the impression of a line emanating form the triple point on its downstream side. This is a *slip line*. The flow that passes through the normal shock will exit subsonically: in this case, its Mach number is about 0.7. Flow through the

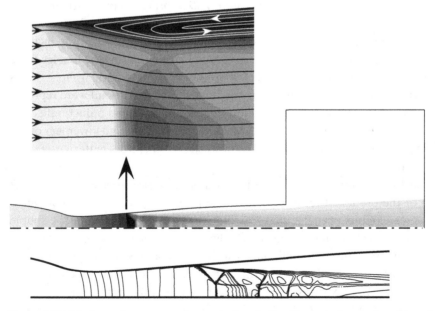

Figure 7.21. Mach number contours in supersubsonic flow through the nozzle, with a shock cell pattern. Upper inset shows streamlines near the shock-induced separation. Lower inset plots isolines of Mach number in the expanding section.

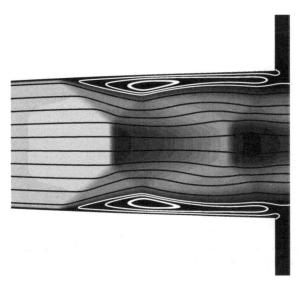

Figure 7.22. Strong shock near the exit. A full cross-sectional view, zoomed on region of shock-induced separation.

oblique legs of the λ can emerge supersonically: in this case its Mach number is about 1.04. The slip line is the consequent discontinuity. Two other slip lines are seen further down the diffuser. They originate at abrupt changes in the angle of the shock.

The separation zone and oblique shocks near the wall are able to expand the core flow back to supersonic in between normal shocks. A close view of this is provided in Figure 7.22. That computation is at $P_0/P_{atm} = 5.0$. The shock now is just upstream of the end of the nozzle and is stronger. The shock pattern creates a recirculation next to the wall. The separated zone acts like an adaptable nozzle, allowing the core flow to compress and expand twice within the region portrayed.

In physical experiments the separated flow is unstable and the shock pattern may be unsteady and it may be asymmetric. The present computations invoke a turbulence model to produce a steady solution. This RANS solution must be understood to represent the ensemble averaged flow (see Chapter 6). Eddies are often strong enough to be seen in Schlieren photographs. They provide mixing that keeps the separated zone relatively thin.

Eventually, when the inlet total pressure is sufficiently high, the shock stands just at the exit plane. In case 4 of Figure 7.17, P_0/P_{atm} is 8. The normal shock is almost at the end of the nozzle. A slightly higher P_0 will blow it out the exit. How does this pressure ratio compare to theory? Given the exit isentropic Mach number of $M = 2.94$ for a perfect, isentropic expansion, the total pressure that drives the shock out of the nozzle can be computed. Setting $M = 2.94$ in the normal jump Eq. (7.20b), and assuming the pressure after the shock to be equal to P_{atm}, gives $P_0/P_{atm} = 3.38$. Because of shock-induced separation, the computational value is over twice this. The flow structure is far more complex than the single, normal shock assumed in the one-dimensional analysis.

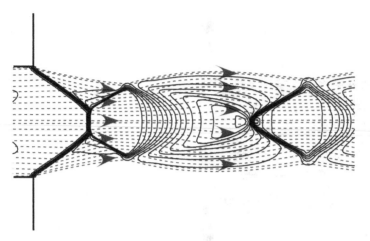

Figure 7.23. Oblique shocks extending from an overexpanded nozzle into the jet, demonstrated via pressure contours. Streamlines are superposed as dotted lines with arrows.

If the value of P_0 is increased a bit beyond 8, the flow becomes supersonic throughout the diffusing portion of the nozzle. However, the pressure at the exit is still lower than the surrounding atmosphere. The jet exiting the nozzle is *overexpanded*. The rise to ambient pressure takes place via a series of oblique shocks, attached to the nozzle exit and extending into the plenum. For an example with $P_0/P_{atm} = 11.0$, see Figure 7.23; an experimental visualization is provided in Figure 7.14. Streamlines are superposed on the pressure contours in Figure 7.23 to show how the shock waves redirect the velocity within the jet.

Further increase of P_0 will finally lead to the situation where $P_e = P_{atm}$, and the flow will be perfectly expanded. Theoretically, for one-dimensional, isentropic flow $P_0/P_{atm} = 33.6$ provides perfect expansion when $A_e/A_* = 4$. The results for $P_0/P_{atm} = 35.0$ are presented in Figure 7.24. It is obvious why a properly designed rocket nozzle should be either perfectly or underexpanded: now there are no shocks or no separation and the velocity varies smoothly in the duct. One-dimensional theory is quite accurate: this corresponds to case 4 of Figure 7.17 but without the shock. As soon as P_0 becomes high enough to force the shock out of the nozzle, the one-dimensional, isentropic theory comes into excellent agreement with computations.

Figure 7.24. A properly expanded, supersonic nozzle. Mach number contours.

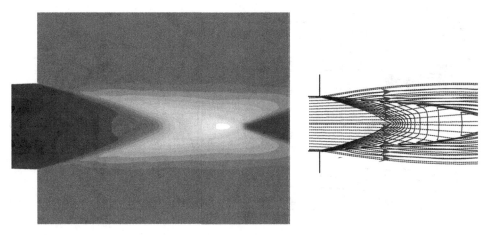

Figure 7.25. An underexpanded jet from a supersonic nozzle. Pressure contours in the plenum. Streamlines and pressure isolines are shown at right.

If the inflow total pressure is increased even further, the pressure at the exit will be higher than atmospheric. The jet exiting the nozzle is *underexpanded*. It equilibrates to ambient conditions across expansion waves, outside the duct. To illustrate, computations for $P_0/P_{atm} = 45$ are presented in Figure 7.25. The high pressure at the nozzle lip is relieved through an expansion fan. The streamlines bow out. Along the axis, the Mach number rises beyond 3, as the velocity overexpands. It is accompanied by a pressure that falls to a low level. An oblique shock recompresses the flow, raising the pressure to meet the ambient conditions. The jet exits the computational domain supersonically. As will be explained in the next section, this means that no boundary condition can be applied: in particular, the pressure at the right boundary of the plenum is not P_{atm}; it is set by extrapolating from the interior.

The de Laval, converging–diverging nozzle explains how a smooth transition from subsonic to supersonic flow can occur; but it need not occur inside a duct. Figure 7.26 is an example of transonic flow over an airfoil. The Mach number is 0.734 in the incident flow. It accelerates to a maximum of $M = 1.3$ at midchord. The convergence and divergence of streamlines as the flow rounds the leading edge accomplishes the transition to supersonic speed. The white curve is the contour $M = 1$. The deceleration toward the trailing edge forces the flow to shock down to subsonic speed. A normal shock sits on the upper surface of the wing at the downstream edge of the sonic line.

The plot of pressure coefficient shows the pressure jump. Note that the ordinate decreases upward: that is so that the upper portion of the plot corresponds to the upper surface of the wing. At one time, it was uncertain whether a transonic airfoil could be shock free. The question was answered in the affirmative: it is possible to diffuse the flow back to subsonic on the rear of the airfoil without a shock. A well-designed, transonic wing will have only a weak shock at design conditions. The pressure distribution will resemble that in Figure 7.26.

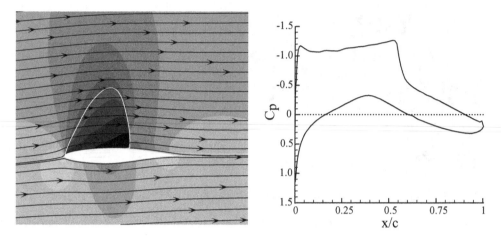

Figure 7.26. Transonic flow over an airfoil. Mach number is contoured, with the sonic line super-posed in white. Figure courtesy of G. Kalitzin.

7.5 Computation of Compressible Flows

A great many computational algorithms have been developed for compressible flow. In some, the motive is consistency with the Mach cone. Physically, this requires that a given grid point should only be influenced by upstream points, lying inside its Mach cone, when the flow is supersonic. In broad terms, this argues for some form of upwind discretization (§2.2). In general purpose CFD codes, the Mach cone is not treated rigorously. Early efforts did so via Riemann solvers. However, this field evolved in the direction of approximate Riemann solvers, flux difference splitting, and characteristic-based dissipation schemes (Tannehill et al., 1997). Any but exact Riemann solvers produce some degree of numerical dissipation. In a sense, that is a consequence of the physics, which produce an upwind bias. Numerical dissipation can provide stability and robustness, without significantly deteriorating accuracy.

Dissipation schemes are divided broadly into matrix and scalar numerical viscosities. Matrix dissipation provides a different level of viscosity for different components of the solution. The components are certain combinations of velocity, pressure, and density. Viscosity has the dimensions of velocity times length times density: $\mu_{\text{numerics}} = v \times \Delta \times \rho$. The length is grid spacing. In matrix dissipation each component has a different velocity, as will be described. Thus, each can have a different value of numerical viscosity.

In the simplest form of scalar dissipation, numerical viscosity is proportional to the maximum of the values of v. In practice, an adjustable constant might multiply the scalar viscosity, $\mu_{\text{numerics}} = c_\mu v_{\text{max}} \times \Delta \times \rho$, so that numerical dissipation can be minimized, subject to stability needs.

It should be noted that $\rho v \Delta$ is the effective viscosity of a first-order method. A second- or third-order method may produce fourth-order diffusion, with viscosity of the form $\rho v \Delta^3$. Numerical diffusion can be made quite small without excessive demand on grid resolution.

Shock waves introduce a new issue. Some readers may be familiar with Gibbs phenomenon in Fourier analysis. An attempt to produce a discontinuous function with a finite number of Fourier modes results in overshoots and oscillations in the vicinity of the jump. The same can occur in numerical computations of shock waves. The overshoot and oscillations are unphysical and can prevent a computation from converging. Pressure should rise monotonically across a shock, without overshooting the downstream level. Various schemes have been developed to preserve monotonicity and avoid oscillations. The successful methods place limits on the numerical convective flux. These limiters are designed to prevent the pressure from rising above the level downstream of the shock or from dropping below the level upstream of it. Near the shock, limiters cause the discretization to become a first-order, upwind approximation. Away from the shock, the flux limitation is removed, and the numerical accuracy becomes second order. Terms like *essentially nonoscillatory*, *monotone upwind*, or, as a class, *total variation diminishing* are applied to various methods for capturing shock jumps via numerical dissipation. The holy grail is to obtain a sharp jump, extending over as few grid points as possible.

The components of the solution variables, mentioned above in connection with matrix dissipation, are characteristic variables. Characteristics also determine what boundary conditions are allowed.

Consider an impulsive force applied at a point in the gas, in one dimension. It will produce sound waves that propagate both to the left and to the right. They move with velocities a and $-a$. Now consider the same impulse applied in a moving gas. The sound will be convected by velocity u, in addition to propagating. It will consist of two waves, moving with velocities $u + a$ and $u - a$. Disturbances other than sound can be carried by the fluid. For instance, sound waves are irrotational, so they carry no vorticity. Vorticity is convected at the fluid velocity, u. There are three components of vorticity, each of which moves with velocity u. Thus, there are five propagation speeds, $u, u, u, u + a, u - a$, three for the components of vorticity and two for sound waves. These are the *characteristic* velocities. When the equations for the five quantities, u, v, w, p, and ρ are solved, five inflow and five outflow conditions are needed. The five characteristic velocities provide an insight into what conditions are permitted.

At a subsonic inflow, $u > 0$ and $u + a > 0$. Four of the five propagation speeds are directed into the domain. That means that four conditions can be specified. The fifth characteristic velocity, $u - a$, is negative: it carries information out of the domain at the inlet. The fifth boundary value must be obtained by extrapolating some quantity from the interior to the inlet. That extrapolation is done internal to the CFD code. (Often, a quantity called a Riemann invariant is what is extrapolated. For one-dimensional, simple waves, this quantity is $u - 2a/(\gamma - 1)$.) At a supersonic inlet, all characteristic speeds are directed into the domain, and five quantities can be specified.

The four quantities specified at a subsonic inlet are usually the total temperature and pressure, T_0 and P_0, and two flow angles. The magnitude of the velocity is inferred. For instance, if the fifth quantity, which is extrapolated from the interior, is the Mach number, then the inlet temperature is

$$T_{in} = \frac{T_0}{1 + \frac{\gamma-1}{2}M^2},$$

where T_0 and M are known. Thus $|u|_{in} = M\sqrt{\gamma R T_{in}}$ is also known. The motive for specifying T_0 is that total temperature is conserved (see page 269). Hence, sensitivity to placement of the inflow boundary is minimized by this choice. Although total pressure is not conserved, as long as no significant dissipation occurs upstream, it is very nearly so. One might conceive of an apparatus that is fed from a large, pressurized tank. The inlet to the computational domain is set by prescribing the total pressure and temperature to be the pressure and temperature of the tank.

At a supersonic exit, all characteristics are directed out of the domain, and all boundary values must be extrapolated from the interior. We have met this in the case of Figure 7.25. It was not possible to specify ambient pressure on the exit plane, but it is not necessary to do so, because a supersonic exit plane has no influence upstream.

At a subsonic exit, one speed, $u - a$, is directed into the domain; therefore, one quantity must be specified. Usually, it is the exit static pressure. The remaining variables are extrapolated to the exit plane from inside the domain.

The origin of the characteristic velocities, whether for purposes of flux splitting or for defining boundary conditions, can be illustrated in two dimensions. The time and x derivatives in the convective derivative of the momentum and mass conservation equations are

$$\frac{\partial \rho u}{\partial t} + \frac{\partial (u\rho u)}{\partial x} + a^2 \frac{\partial \rho}{\partial x} \cdots$$

$$\frac{\partial \rho v}{\partial t} + \frac{\partial (u\rho v)}{\partial x} \cdots \qquad (7.35)$$

$$\frac{\partial \rho}{\partial t} + \frac{\partial \rho u}{\partial x} \cdots .$$

The first two correspond to the convective and pressure gradient terms of Eq. (1.14) after mass conservation is invoked in the form

$$\frac{\partial \rho}{\partial t} + \nabla \cdot (\rho \boldsymbol{u}) = 0 \qquad (7.36)$$

for a variable density fluid. Equation (7.36) follows from Eq. (1.15): $\nabla \cdot \boldsymbol{u}$ was derived as the fractional rate change of volume, $D_t \mathcal{V} = \mathcal{V} \nabla \cdot \boldsymbol{u}$. Equation (7.36) is obtained on replacing $\mathcal{V}$ by $1/\rho$.

In Eq. (7.35), the pressure gradient was converted to density gradient using the formula $dp = a^2 d\rho$. That follows from $p \propto \rho^\gamma$ for an isentropic compression. Differentiating and using the gas law (7.21) gives

$$\frac{\partial p}{\partial x} = \gamma RT \frac{\partial \rho}{\partial x} = a^2 \frac{\partial \rho}{\partial x}$$

(see page 273).

It is natural to regard x and y momenta, ρu and ρv, as the dependent variables, along with density, ρ. If the x derivatives are rewritten for these variables, the equations become

$$\frac{\partial \rho u}{\partial t} + 2u \frac{\partial \rho u}{\partial x} + (a^2 - u^2) \frac{\partial \rho}{\partial x} \dots$$

$$\frac{\partial \rho v}{\partial t} + u \frac{\partial \rho v}{\partial x} + v \frac{\partial \rho u}{\partial x} - uv \frac{\partial \rho}{\partial x} \dots$$

$$\frac{\partial \rho}{\partial t} + \frac{\partial \rho u}{\partial x} \dots .$$

Characteristics are introduced by writing this in vector form, as

$$\frac{\partial}{\partial t} \begin{bmatrix} \rho u \\ \rho v \\ \rho \end{bmatrix} + \mathbf{A} \cdot \frac{\partial}{\partial x} \begin{bmatrix} \rho u \\ \rho v \\ \rho \end{bmatrix} \dots \qquad (7.37)$$

with

$$\mathbf{A} = \begin{bmatrix} 2u & 0 & a^2 - u^2 \\ v & u & -uv \\ 1 & 0 & 0 \end{bmatrix}.$$

The reader can verify that the three eigenvalues of this matrix are $u, u + a$, and $u - a$.

The convection term of Eq. (7.37) can be considered to carry information at these three velocities: the sound speed to the left, relative to the convective velocity; the sound speed to the right; and the convective velocity itself. In subsonic flow, $a > u$, two of these have the same sign as u and one is opposite; in supersonic flow, $a < u$, all have the sign of u. The idea of flux splitting is to define the "upwind" direction based on the sign of the characteristic velocity. In subsonic flow, with $u > 0$, two components are upwinded to the left. For the other, the upwind direction is to the right.

EXERCISES

7.1 *Isentropic flow.* A large tank of air is pressurized to 10 atm. Its temperature is allowed to settle to the room temperature of $300°$K. The air exhausts isentropically through a carefully designed nozzle, expanding to 1 atm. What is the temperature of the air, as it exits the nozzle, in $°$C?

7.2 *Choked flow.* The carefully designed nozzle in exercise 7.1 is replaced by a simple contracting pipe. What is the velocity of the air at the exit? Sketch the development of the jet after it exits the pipe.

7.3 *Rockets.* Show that the thrust produced by a rocket engine is

$$\text{Thrust} = U_e \dot{m} + A_e(P_e - P_{\text{atm}}), \tag{E7.1}$$

where U_e is the velocity at its exit plane and $\dot{m}$ is the mass flux. This can be derived from a momentum balance applied to the following control volume:

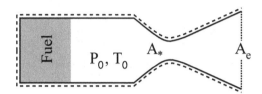

In terms of Mach number and throat conditions, show that Eq. (E7.1) becomes

$$\text{Thrust} = M_e \gamma R \sqrt{T_e T_*} \rho_* A_* + A_e(P_e - P_{\text{atm}}).$$

Then, from the isentropic relations, derive

$$\text{Thrust} = P_0 A_* \left[\gamma \sqrt{\frac{2}{\gamma - 1}} \sqrt{1 - \left(\frac{P_e}{P_0}\right)^{\frac{\gamma-1}{\gamma}}} \left(\frac{2}{\gamma + 1}\right)^{\frac{\gamma+1}{2\gamma-2}} \right. $$
$$\left. + \frac{A_e}{A_*} \left(\frac{P_e - P_{\text{atm}}}{P_0}\right) \right].$$

Let the exit to throat area ratio be 6. At first, the rocket is at ground level, where the ambient pressure is 1 atm. If the flow is perfectly expanded, what is the stagnation pressure?

Suppose the stagnation properties are held constant, whereas the rocket climbs to an altitude where the ambient pressure is 1/2 atm. What is the ratio of thrust at altitude to thrust at ground level?

7.4 *Development into a shock.* Sketch how a smooth ramp down in pressure evolves with time into a sharp discontinuity. The initial condition is shown in the figure at the right.

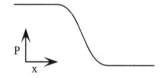

Burger's equation is an analytical model for nonlinear wave propagation. That equation is

$$\frac{\partial u}{\partial t} + u \frac{\partial u}{\partial x} = 0.$$

It has an exact solution

$$u(x, t) = g[x - u(x, t)t],$$

where the function $u = g(x)$ is the initial condition. Unfortunately, this solution is in an implicit form. For each value of x and t, this formula can be solved by Newton's method (page 86) to obtain $u(x, t)$. First let

$$g(x) = 2/3 - 1/3 \tanh(2x).$$

Let the x locations be $x_i = -2 + .05 \times i$ for $i = 0, 1, 2, 3, \ldots 80$. For times running from 0 to 1.5 by 0.15, plot curves of $u(x)$. Repeat for

$$g(x) = 2/3 + 1/3 \tanh(2x).$$

These two cases correspond to shock and expansion waves.

7.5 *Oblique shocks and expansion fans.* A flat plate of finite length is placed at an angle θ to an oncoming supersonic flow. Sketch the pattern of compressible waves that forms at the ends of the plate, noting the difference between the upper and lower sides.

Determine the pressure difference between the upper and lower surfaces if the incident Mach number is 3 and the plate is at an incidence angle of 10°. Assume a weak, oblique shock. Note that the turning angle is $\theta = -10°$ on the upper side.

Determine the lift force per unit width on a plate of length ℓ if the the approach flow is at a pressure P_a. What is the drag force?

7.6 *Normal shock.* Derive a formula for the total pressure ratio, P_{01}/P_{02}, across a normal shock in terms of the Mach number M_1 entering the shock. Deduce that total pressure decreases from the approach to the departing flow. From Eq. (7.10), derive the relation

$$\Delta S = R \log(P_{01}/P_{02}) \qquad \text{(E7.2)}$$

for the entropy change entropy across the shock.

7.7 *Infer the geometry.* Supersonic flow over a two-dimensional wedge was computed. The solution contained an oblique shock. The pressure increase along a streamline crossing the shock was $p_2/p_1 = 4.175$. The inflow was air at atmospheric pressure, $p = 1$ atm, with $T_\infty = 300°$K and $u_\infty = 870$ m/s. What is the half-angle of the wedge?

7.8 *Compression ramp.* Compute the supersonic flow over a compression corner with the angle of 25° as in the figure at right. Set the inflow Mach number to $M_\infty = 2.95$, with $u_\infty = 1,025$ m/s, at $p = 1$ atm and $T_\infty = 300°$K. The downstream boundary, as well as the top boundary, should be set to supersonic outflow.

First, compute the inviscid flow using a coarse, structured grid, with quadrilateral cells. The grid should not exceed 10,000 cells.

The domain extends 2 m upstream of the ramp, and the exit plane is 1 m beyond its start. Make the domain be 2 m high, so that the shock does not intersect the upper boundary. That prevents spurious reflection of the shock from the top of the domain.

Now refine the grid to resolve the boundary layers. The latter should be treated as fully turbulent, invoking a scalar eddy viscosity model. Compute the turbulent flow over the same compression corner. Prescribe a low level of turbulence at the inflow.

Finally, use the computed solution to adapt the grid in the regions of large gradient of Mach number to more accurately capture the shock and the boundary layer. Repeat the process until successfully resolving the recirculation region induced by shock–boundary layer interaction. Suggestion: base adaptive mesh refinement on Mach number gradient.

8 Interfaces

When two fluids occupy the domain, with a sharp boundary between them, we speak of fluid–fluid interfaces or just interfaces. To the extent that the fluids are immiscible, their interface is a type of boundary. The governing laws are unchanged; the new features are boundary conditions. They are of a different nature from those at fixed, solid walls. They depend on the flow on either side of the interface; indeed, the position of the interface is itself a variable. Interface conditions are alternatively described as matching conditions: velocities and stresses on either side must properly match at the interface. Despite this complicating aspect, the view that only boundary conditions are at issue provides some clarity.

The interface may be between liquid and gas – say, water and air. Often the matching conditions are simplified in this case. The density of air is three orders of magnitude smaller than that of water. For many purposes, the forces exerted by the air on the water can be neglected; then the interface is a force-free surface, insofar as the hydrodynamics are concerned. It nevertheless is a moveable surface, whose position must be solved as part of the analysis.

Or the interface could be between two viscous fluids – say, oil and water. The viscosity jumps across their common boundary. Conditions of stress continuity then determine the interaction between the fluid motions.

Oil and water might be placed in a vertical tube. The interface then curves in consequence of surface tension and the angle of contact with the tube. The line of contact is a three-phase boundary, among water, oil, and solid wall. The configuration of the meniscus is determined by properties of the contact line.

An interface is convected by the normal component of fluid velocity. This may take the form of waves that are produced by gravitational acceleration of a density contrast: fluids in containers slosh, boats make undulations in the surface of a lake, ducks make ripples in a pond. Even when a steady deformation of the fluid–fluid interface is being computed it can be understood as a stationary wave pattern. An example is the bow wave of a boat, which can be regarded as a pattern of surface gravity waves that are held stationary by the forward velocity of the boat.

The range of interfacial phenomena is immense. This chapter provides an introduction to some of them and to methods of computational analysis. Given the enormity of the phenomena, it is comforting to fall back on the observation that it is largely a matter of interfacial boundary conditions. The complexities arise when problems are solved, especially when they are solved by computer simulation.

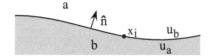

Figure 8.1. Defining sketch for interface conditions.

8.1 Interface Conditions

The velocity must be continuous at the interface between two viscous fluids. Schematically, the configuration is like Figure 8.1. Denote the velocities in each fluid by $u_a(x)$ and $u_b(x)$ and a point on the interface by x_i. Then

$$u_a = u_b \quad \text{at } x_i. \tag{8.1}$$

If that were not true, then an infinite shear would occur as the fluids move relative to one another. The shear times viscosity would cause an infinite stress. Another notation is to represent the jump $u_b(x_i) - u_a(x_i)$ by a bracket $[u]$. Then $[u] = 0$ means that velocity is continuous.

   If one, or both, of the fluids can be treated as inviscid, only the normal component of velocity need be continuous. Without accounting for viscosity, there will be no infinite stress if the tangential velocity has a jump. But the normal velocity must still be continuous. If it were not, a void would open where the fluids moved away from one another. The inviscid approximation may be acceptable for large-scale phenomena like sloshing or surface waves. In these cases, the normal component of fluid velocities on either side of the interface must equal each other, and both must equal the velocity of the surface:

$$\hat{n} \cdot u_a = \hat{n} \cdot u_b = v_{\text{surface}}, \tag{8.2}$$

where $\hat{n}$ is the normal to the interface. This condition provides an equation for the motion of the interface.

   Consider the force balance on a very thin slab of fluid containing the interface. The difference between the stresses on the top and bottom of the slab accelerates it. The mass within the slab is proportional to its thickness. As the thickness goes to zero, so does its mass. If the stress difference did not also vanish, the slab would experience infinite acceleration. Hence, the stress projected on the the interface must be continuous:

$$[\sigma \cdot \hat{n}] = 0. \tag{8.3}$$

As has been explained in Chapter 1, this projection provides the components of force per unit area; condition (8.3) states that the force on the interface is the same whether it is approached from above or from below. In Newtonian fluids, viscous stress is proportional to rate of strain times viscosity, as in Eq. (1.13). If the viscosity differs between the fluids, the rate of strain must be discontinuous for the stress to be continuous.

Consider the simple case of two fluids, occupying layers between plane, parallel walls. The upper wall is moving with velocity U_w and the lower is stationary. This is Couette flow (§1.5). The Navier–Stokes equations show that, in this parallel shear flow, the stress is constant everywhere. The condition (8.3) of stress continuity is automatically met. If the constant value of stress is denoted τ, then in the lower fluid $\mu_b \partial u/\partial y = \tau$ and in the upper $\mu_a \partial u/\partial y = \tau$. If $\mu_a \neq \mu_b$, the shear differs in the two fluids and is discontinuous at the interface. In the lower fluid $\partial u/\partial y = \tau/\mu_b$ and in the upper $\partial u/\partial y = \tau/\mu_a$.

The flow is driven by the moving upper wall. It creates viscous stress that is transmitted through the upper to the lower fluid. The shear in each fluid adapts to produce the requisite stress. Let the upper wall be at $y = H$, and let the interface be at $y = h$. In the lower fluid

$$u = \frac{\tau}{\mu_b} y, \quad y < h$$

and in the upper

$$u = \frac{\tau}{\mu_a} y + \text{constant}, \quad y > h.$$

To match the velocity of the upper wall, the constant is chosen to give

$$u = \frac{\tau}{\mu_a}(y - H) + U_w, \quad y > h.$$

Continuity of velocity (8.1) must be met at $y = h$. Equating the velocities determines the magnitude of the stress, provided that h is specified:

$$\tau = \frac{\mu_a \mu_b U_w}{\mu_a h + \mu_b (H - h)};$$

or the height of the interface is determined if τ is specified:

$$h = \frac{\mu_a \mu_b U_w - \mu_b \tau H}{(\mu_a - \mu_b)\tau}.$$

If the upper fluid is more viscous, $\mu_a > \mu_b$, then a nonnegative h is possible only if

$$\tau < \frac{\mu_a U_w}{H}.$$

The right side is the stress in Couette flow of a homogeneous fluid. Hence, the less viscous, lower layer reduces the shear stress below that of a homogeneous fluid. Exercise 8.1 explores the possibility of using this observation to reduce pumping pressure in a pipe. A very viscous liquid can be lubricated by surrounding it with a less viscous one.

Another simple illustration is provided by Poiseuille flow down an inclined plane. Viscous fluid is fed at a volumetric rate of Q per unit width at the top of a ramp. The ramp is at an angle θ to the horizontal and gravity is directed downward, in the vertical. It will be assumed that the flow develops into a layer of unknown thickness, h, on the ramp. Its surface is stress-free and parallel to the wall. The layer thickness is to be found under the simplifying assumption of parallel flow.

The absence of inertia in parallel flow converts the Navier–Stokes equations (1.14) to

$$0 = -\frac{\partial p}{\partial x} + \mu \frac{\partial^2 u}{\partial y^2},$$

$$0 = -\frac{\partial p}{\partial y} + \mu \frac{\partial^2 v}{\partial y^2} - \rho g. \tag{8.4}$$

Gravitational acceleration has been included in the y component. If these equations are converted into components along and perpendicular to the incline, the perpendicular velocity vanishes. Thus, let

$$u_t = u \cos \theta - v \sin \theta \quad \text{and} \quad u_n = v \cos \theta + u \sin \theta$$

be the components tangential and normal to the ramp. Equations (8.4) become

$$0 = -\frac{\partial p}{\partial x_t} + \mu \frac{\partial^2 u_t}{\partial x_n^2} + \rho g \sin \theta,$$

$$0 = -\frac{\partial p}{\partial x_n} - \rho g \cos \theta.$$

From the normal momentum equation, the pressure is hydrostatic and given by $p = p_a + \rho g(h - x_n) \cos \theta$. Because the surface is assumed to be level, $dh/dx_t = 0$; the normal momentum equation simply shows that there is no tangential pressure gradient. Were it not level, variations of the layer depth would produce a pressure gradient (see exercise 8.4).

The tangential momentum equation describes Poiseuille flow driven by gravitational acceleration tangent to the slope. The boundary conditions are no-slip at the wall and no stress at the free surface. That is, $u_t = 0$ at $y = 0$ and $\mu \partial u/\partial y = 0$ at $y = h$. Hence, the tangential equation integrates to

$$u_t = -\frac{\rho g \sin \theta}{2\mu} x_n(x_n - 2h). \tag{8.5}$$

The volume flow rate, per unit width, it just the integral of this with respect to x_n from 0 to h:

$$Q = \frac{g \sin \theta}{3\nu} h^3.$$

Finally, the height of the layer is found to be

$$h = \left(\frac{3Q\nu}{g \sin \theta} \right)^{1/3}. \tag{8.6}$$

Given the ramp angle and flow rate, the free surface will adjust to this height. This idealized result only applies to a very viscous fluid because it assumes that Poiseuille flow establishes on the ramp. Hydraulic engineers are concerned with the opposite extreme: of effectively inviscid flow. Then the free surface will not be level. Some aspects of inviscid free surfaces will be discussed shortly.

8.2 Surface Tension

First, we must note an exception to the condition (8.3) of stress continuity. At the juncture between two immiscible fluids, a force develops in consequence of the dissimilarity of the materials. It is called the surface tension. It is represented mathematically by a discontinuity of stress. Before replacing Eq. (8.3) by a stress jump, let us discuss the origin and nature of surface tension.

Surface tension can be understood as an affinity of molecules for others of their kind. If a small slug of air, or of oil, is introduced at the bottom of a tank of water, it will rise as a bubble. The fluids remain distinct because they are immiscible; they remain distinct because water molecules are more strongly attracted to water than to air or oil. Similarly, the oil molecules draw together, and if the slug is small enough, a spherical bubble of oil rises through the water.

A complementary description is to say that oil is repelled by water. To bring a molecule from inside the oil to the water–oil interface that repulsion must be overcome. Work must be done to add a molecule to the interface. The work required to create the surface of a fluid–fluid interface is stored as free energy. This is a chemical, potential energy that is proportional to the total area of the surface. Denote by Σ_T the coefficient of proportionality between surface energy and surface area. Then the superficial energy is

$$E_{\text{surface}} = \Sigma_T A \qquad (8.7)$$

The coefficient Σ_T is a physicochemical property of the two fluids that can be measured. It is the work required to produce a unit area of interface.

If the interface is deformed from its equilibrium configuration, surface tension provides a restoring force. For instance, a spherical drop that is stretched and released will return to its spherical shape. The restoring force can be related to surface energy by the notion of virtual work. Suppose the area is perturbed to $A + \delta A$. Then the energy is perturbed to

$$E + \delta E = \Sigma_T(A + \delta A).$$

Attribute the perturbation to work done by a force $\boldsymbol{F}$. The work that it does in the course of a displacement $\delta \boldsymbol{x}$ is

$$\boldsymbol{F} \cdot \delta \boldsymbol{x} = \delta E = \Sigma_T \delta A.$$

Consider a few examples. A plane interface is displaced normal to itself. That does not change its area: $\delta A = 0$, so $F_n = 0$. If it is displaced tangentially to itself, δA is again 0. So all components of $\boldsymbol{F}$, whether normal or tangential to the plane surface, are zero. There is no surface tension force at a plane interface.

Now consider a cylindrical surface. Its area is $2\pi r$ times its length, L. If the cylinder is rotated about its axis, or if it is translated parallel to its axis, the surface area is unchanged: $\delta A = 0$. There is no surface tension force in these directions. However, if the cylinder is translated normal to its surface its radius expands to $r + \delta r$ and $\delta A = 2\pi \delta r L$. Now there is a force. The work done by this displacement is

$F_r \delta r$. Equating this to the increase in surface free energy gives

$$F_r = 2\pi L \Sigma_T.$$

Similarly, a spherical surface of area $4\pi r^2$ has a radial force

$$F_r = 8\pi r \Sigma_T$$

associated with its surface energy. Translation of the spherical surface tangentially to itself – that is, rotating it – causes no area change. There is no surface tension force tangential to the sphere. Radial displacement inflates the surface area by $\delta A = 4\pi(r + \delta r)^2 - 4\pi r^2 \approx 8\pi r \delta r$. Equating this times Σ_T to the work $F_r \delta r$ provides the above formula for the radial force component.

It is not hard to infer from these examples that surface tension force is always directed normal to the interface and that it vanishes if the surface is not curved. Displacements normal to a curved surface increase its area; the increased area is associated with the force of surface tension by the principle of virtual work.

In dealing with the continuum fluid, we have invoked stress and pressure in place of force. These are intensive properties – properties that are defined independently of the size of the fluid element used to introduce them. To similarly characterize surface tension, the force on a given area must be replaced by a force per unit area or, equivalently, by a pressure. Recall that pressure produces only a normal force. Surface tension has that same property, so it is appropriate to associate a pressure with surface tension.

Consider, again, the cylinder. The force $2\pi L \Sigma_t$ acts on an area $2\pi r L$. The pressure is the ratio of these:

$$p = \frac{\Sigma_T}{r}.$$

This is the pressure inside the cylinder relative to that outside. For the sphere $F = 8\pi r \Sigma_t$, $A = 4\pi r^2$, and

$$p = \frac{2\Sigma_T}{r}.$$

The factor of 2 disparity between these formulas is not an inconsistency. It arises because the cylinder is curved in only one direction, whereas the sphere is curved in two. Generally, two principle radii of curvature can be identified at each point of a two-dimensional surface. If they are denoted r_1 and r_2 then the surface tension pressure is

$$p = \Sigma_T \left(\frac{1}{r_1} + \frac{1}{r_2} \right). \tag{8.8}$$

For a flat surface, $r_1 = r_2 = \infty$. For a cylinder, $r_1 = r$ and $r_2 = \infty$, whereas for a sphere, $r_1 = r_2 = r$.

Formula (8.8) can be restated as follows: surface tension is proportional to curvature, defined by $\kappa = 1/r_1 + 1/r_2$. That is,

$$p = \Sigma_T \kappa. \tag{8.9}$$

This statement is commonly used in computations because the curvature can be computed without evaluating the two principal radii of curvature. For instance, in the volume of fluid method (§8.4) curvature is evaluated from the divergence of the surface normal:

$$\kappa = \nabla \cdot \hat{n}, \tag{8.10}$$

where $\hat{n}$ is the unit normal to the interface. It can be verified that this gives the correct result for a sphere about the origin. The vector $\hat{n} = (x, y, z)/r$ is normal its surface. The divergence of this is readily found to be $2/r$. Note that $\hat{n}$ is defined to point out from the center of curvature.

To see how surface tension is included in the interface condition, consider a spherical bubble in equilibrium. Surface tension will cause the pressure just inside the bubble to be larger than the pressure just outside the bubble by $2\Sigma_T/r$. Hence, surface tension is modeled as a pressure jump equal to $2\Sigma_T/r$.

In a static fluid the stress tensor (1.10) contains only the pressure. To be general, in the surface tension model a jump is attributed to the normal component of the entire surface stress, rather than just to the pressure per se. Introducing surface tension in this manner alters Eq. (8.3) to

$$[\sigma \cdot \hat{n}] = \Sigma_T \kappa \hat{n}. \tag{8.11}$$

This says that the normal stress experiences a jump, but the tangential stress is continuous. For instance, in two dimensions the stress tensor is the 2×2 matrix

$$\begin{bmatrix} \sigma_{xx} & \sigma_{xy} \\ \sigma_{xy} & \sigma_{yy} \end{bmatrix}.$$

If the surface normal is in the x direction, then $\hat{n} = (1, 0)$. Equation (8.11) gives

$$[\sigma_{xx}, \sigma_{xy}] = (\Sigma_T \kappa, 0).$$

The normal stress σ_{xx} experiences a jump, $[\sigma_{xx}] = \Sigma_T \kappa$; the tangential stress does not $[\sigma_{xy}] = 0$. For a static fluid, the viscous stress is zero; the stress is simply minus the pressure (see page 9). Then Eq. (8.11) reduces to a pressure jump

$$[p] = -\Sigma_T \kappa. \tag{8.12}$$

The minus sign arises because the normal points out from the center of curvature: thus, $[p]$ is the pressure on the convex side of the surface, minus the pressure on the concave side; for instance, it is the pressure outside a sphere minus the pressure inside. The pressure is greater inside because it must counteract the inward force of surface tension.

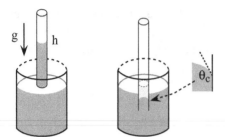

Figure 8.2. The meniscus in a capillary tube for hydrophilic and hydrophobic materials. The liquid rises up a hydrophilic tube and is depressed in a hydrophobic one. The contact angle is defined at the right.

8.2.1 Capillarity and Contact Angle

Surface tension is readily observed via the phenomenon of capillary rise. If a thin,* circular tube is dipped into a glass of water, a column of water will rise up the tube, being capped by a meniscus (see Figure 8.2). The meniscus is a portion of a sphere of radius R. Denote the ambient pressure by p_a. The pressure just inside the meniscus will be $p_a - 2\Sigma_T/R$. It is lower than ambient by the surface tension, if the meniscus curves upward.

A column of water of height h exerts a pressure $\rho g h$ at its bottom. Thus, the pressure at the base of the column of water in the capillary tube is $p_a - 2\Sigma_T/R + \rho g h$. If the glass of water is open to the ambient, the pressure at the surface of the water is p_a. To match this at the base of the column, its height must be such that $\rho g h - 2\Sigma_T/R$ vanishes; that is,

$$h = \frac{2\Sigma_t}{\rho g R}. \tag{8.13}$$

The capillary rise is thereby attributed to surface tension.

One might wonder what mechanism causes the rise. No aspect of surface tension between air and water seems to explain capillarity. Indeed, the meniscus could either rise or fall, as in the cases of Figure 8.2. It would seem that other factors must play a role. Indeed, to some extent, capillarity must be an effect of attraction or repulsion between the liquid and the walls of the tube.

A tube material that has an affinity for water is called hydrophilic; a material that repels water is hydrophobic. The angle at which the meniscus meets the tube wall is called the *contact angle*. As the water is drawn toward a hydrophilic material, its contact angle will be less than 90°; for a hydrophobic material it will be greater than 90°. Figure 8.2 defines the contact angle, θ_{contact}.

The contact angle determines R in Eq. (8.13). R is the radius of the meniscus, not the radius of the tube; they are quite different. Some trigonometry shows that

---

* The formal requirement is that the tube radius satisfy $a^2 \ll \Sigma_T/\rho g$. Then the influence of gravity is small enough for the meniscus to be spherical.

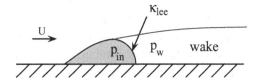

Figure 8.3. The shape of a drop in an airflow. The front side slopes from the wind; the rear is spherical.

the tube radius a is related to the meniscus radius by $a = R \cos \theta_{\text{contact}}$. Hence, the capillary rise formula (8.13) becomes

$$h = \frac{2\Sigma_t}{\rho g a} \cos \theta_{\text{contact}}, \tag{8.14}$$

For a hydrophobic tube, $\theta_{\text{contact}} > 90°$ and h is negative, meaning that the meniscus is depressed below the surface of the water, as in the right side of Figure 8.2.

A similar argument describes a drop of liquid sitting on a flat, hydrophilic surface. The drop will assume the form of of a spherical cap of some radius R. Its base intersects the wall in a circle of radius $r = R \sin \theta_{\text{contact}}$. The volume of a spherical cap is

$$V = 2/3\pi R^3 \left(1 - \cos \theta_{\text{contact}} - 1/2 \sin^2 \theta_{\text{contact}} \cos \theta_{\text{contact}} \right).$$

Given the volume of the fluid in the drop, this determines the radius R in terms of the contact angle. Note that surface tension does not appear in this formula. It does play a role, however. The drop must have a uniform internal pressure if there is no flow of the liquid inside it – and if it is small enough to ignore gravity. That means that the surface curvature must be constant, according to Eq. (8.9). The curvature at any point on the surface of a sphere has the same constant value. Hence, surface tension ensures the spherical-cap shape of the drop.

Place a spherical-cap-shaped drop of water on a sheet of plastic and blow briskly across it. The front of the drop flattens and slopes away from you. The rear retains its spherical form. It looks like the schematic Figure 8.3. This shape can be understood by reference to the computation of flow over a bump in Chapter 4, page 147. The pressure is fairly constant on the lee side of the bump. If the water in the drop is nearly static, it too will be at constant pressure. Hence, the pressure jump across the rear of the drop is nearly constant. Equation (8.12) shows that the curvature must then be constant on the lee side. Let that curvature be κ_{lee}. The rear of the drop has a spherical shape, with radius $2/\kappa_{\text{lee}}$. Let the pressure in the wake be p_w. Then, inside the drop the pressure is

$$p_{\text{in}} = p_w + \Sigma_T \kappa_{\text{lee}}. \tag{8.15}$$

On the front of the hump in Figure 4.16, the pressure is nonuniform and is larger than p_w. Let that external pressure be $p(x)$. At any point, x, on the front of the drop

$$\kappa(x) = \frac{p_{\text{in}} - p(x)}{\Sigma_T} = \frac{p_w - p(x)}{\Sigma_T} + \kappa_{\text{lee}} < \kappa_{\text{lee}}$$

using Eq. (8.15) for the internal pressure. The curvature on the front is less than on the back because of the increased pressure caused by flow stagnation. Indeed, the higher pressure on the windward side is responsible for the form drag on the drop.

At high Reynolds number, the pressure difference between the front and rear scales with the Bernoulli head. Denote $p(x) - p_w$ by $1/2\rho U_\infty^2 C_D$. Then

$$\kappa(x) - \kappa_{\text{lee}} = -1/2 C_D \left[\frac{\rho U_\infty^2}{\Sigma_T}\right]. \tag{8.16}$$

The quantity in brackets has the dimensions of one over length. A nondimensional parameter, called the Weber number, can be defined by introducing the cubed root of volume as a unit of length:

$$We \equiv \frac{\mathcal{V}^{1/3}\rho U_\infty^2}{\Sigma_T}.$$

The magnitude of We characterizes the relative importance of inertial and surface tension forces. The higher the We, the larger the right side of Eq. (8.16) and the more will the drop deform.

The drop will continue to deform until a limit is reached at which it begins to move. For instance, drops on the passenger windows of an airplane begin to stream along the surface as the plane accelerates for takeoff.

The limiting Weber number is determined by another physicochemical property, termed *contact angle hysteresis*. If one draws a slide out of a glass of liquid, the contact angle is observed to be smaller than that seen when the slide is being submerged into the liquid. The difference between the two is the contact angle hysteresis. It characterizes the resoluteness with which the drop is held in place by surface free energy. In broad terms, the limiting Weber number is reached when the drag force exerted by the wind equals the maximum force that can be opposed by contact angle hysteresis.

Properties of contact lines, especially their movement, are far afield of the scope of this book. The reader might consult Probstein (1994). We will leave this topic and return to the discussion of fluid–fluid interfaces.

8.2.2 Disintegration by Capillarity

Turn on a faucet so that the water flows smoothly in a laminar stream. If the flow is sufficiently slow, the jet of water will be seen to break into drops before reaching the sink. This is an instance of capillary disintegration. Disintegration into a very regular stream of drops can be produced by subjecting a thin stream of liquid to periodic forcing. Such controlled breakup is used in ink jet printers to provide a fine, tightly controlled spray for drawing images. Capillary breakup can be studied by computer simulation, as will be seen in §8.2.3.

As a prelude, we review the corresponding instability theory. It dates to the 19th century and the work of Plateau and Rayleigh, among others. The reasoning introduced by Plateau is that a stable equilibrium should minimize the potential energy of the surface. The energy of the interface is proportional to its area. Thus, an interface is unstable if any perturbation exists that lowers its area. If attention is restricted to small perturbations, an assessment can be made of the local stability: if all small disturbances increase the surface area, the interface is locally stable. Whether the area increases or decreases is the question to be addressed.

Consider a straight, circular cylinder of liquid. Let its radius be a_0 and consider a length λ of the cylinder. Its surface area is $2\pi a_0 \lambda$, and its surface energy is therefore $2\pi a_0 \lambda \Sigma_T$.

Now consider a sinusoidal perturbation to the radius,

$$R = a + b\cos(2\pi x/\lambda). \tag{8.17}$$

(The reader may wish to look at the topmost part of Figure 8.4.) This is an axi-symmetric, varicose protrusion on the cylinder. The volume of liquid should be unaltered. Equating the unperturbed and perturbed volumes:

$$\pi a_0^2 \lambda = \pi \int_0^\lambda R(x)^2 dx = \pi(a^2 + 1/2b^2)\lambda.$$

Hence

$$b^2 = 2(a_0^2 - a^2). \tag{8.18}$$

This defines a permissible disturbance.

Now calculate the disturbed surface area to see whether it increases or decreases. The perturbed area is

$$A = \int_0^\lambda 2\pi R(x)\sqrt{(dx)^2 + (dR)^2}.$$

Denote $2\pi/\lambda$ by k. Substituting Eq. (8.17) for the radius into the above integral for the area gives

$$\begin{aligned} A &= 2\pi \int_0^\lambda [a + b\cos(kx)]\sqrt{1 + b^2 k^2 \cos^2(kx)} dx \\ &\approx 2\pi \left(a + 1/4ab^2 k^2\right)\lambda. \end{aligned} \tag{8.19}$$

The approximation is that $(bk)^2 \ll 1$. This is equivalent to requiring the slope dR/dx to be small. That is acceptable if local stability is being examined.

In light of Eq. (8.18), the area is

$$2\pi\lambda \left[a + 1/2a(a_0^2 - a^2)k^2\right].$$

The area of the initial cylinder is $A_0 = 2\pi\lambda a_0$. The perturbation from that is

$$A - A_0 = 2\pi\lambda(a_0 - a)\left[\frac{(a + a_0)a}{2}k^2 - 1\right]. \tag{8.20}$$

As $a < a_0$ by constancy of volume, Eq. (8.18), whether the area increases or decreases depends on the sign of the bracketed term. The disturbance is small if $b^2 \ll a_0$. Then the bracketed factor in Eq. (8.20) is approximately

$$(a_0 k)^2 - 1. \tag{8.21}$$

The area will decrease if $a_0 k < 1$. In terms of the wavelength, the interface is unstable to perturbations with

$$\lambda > 2\pi a_0. \tag{8.22}$$

Oscillations of sufficient length will reduce the surface energy. It may be anticipated that a cylindrical liquid jet will degenerate into a stream of drops if subjected to a long wavelength disturbance. The computer simulation in Figure 8.4 demonstrates how that occurs. The instability criterion is a good guide, although the details of the disintegration are more complex than might be supposed.

The anxious reader may pass over the next few paragraphs to the description of the computation, §8.2.3. For other readers, we discuss the theory further.

The results (8.21) and (8.22) first derived by Plateau, are a step to understanding why a jet of water disintegrates into drops. This long-wavelength, capillary instability is commonly attributed to Lord Rayleigh – although that is somewhat unfair to Plateau.

Rayleigh provided a dynamic theory. The sections of Rayleigh's *Theory of Sound* (1894) on capillary phenomena remain an enjoyable account of early experimental studies, enlightened by succinct, masterful analyses. In addition to instability, types of oscillation driven by capillarity are surveyed. For instance, if the analysis leading to Eq. (8.20) is repeated for nonaxisymmetric disturbances, it is found that the area always increases. Therefore, the interface is stable to perturbations of asymmetric form. These disturbances will produce stable oscillations of the liquid cylinder but are not the mode of its disintegration. Capillarity, in those instances, acts as a restoring force, maintaining the integrity of the cylinder.

A casual analogy to gravity provides a simplified version of Rayleigh's dynamical theory. The hydrostatic pressure produced by a height η of water is

$$p = \rho g \eta. \tag{8.23}$$

The pressure inside a curved interface is

$$p = \Sigma_T \left(\frac{1}{R_1} + \frac{1}{R_2} \right). \tag{8.24}$$

Let the interface be a cylinder subjected to a perturbation $\eta(x)$ of its radius. Then $R_2 = a_0 + \eta$.

The other radius involves derivatives of η. Along the axis the unperturbed radius is infinite and the curvature is initially zero. The perturbed curvature is

$$\frac{1}{R_1} = -\frac{d}{dx} \left[\frac{d\eta/dx}{\sqrt{1 + (d\eta/dx)^2}} \right]. \tag{8.25}$$

This follows from formula (8.10) if it is recognized that the unit surface normal is

$$\hat{n} = \frac{(-d\eta/dx, 1)}{\sqrt{1 + (d\eta/dx)^2}}.$$

If the slope is mild ($|d\eta/dx| \ll 1$), then Eq. (8.25) reduces to

$$\frac{1}{R_1} \approx -\frac{d^2\eta}{dx^2},$$

and if the perturbation is small ($\eta \ll a_0$),

$$\frac{1}{R_2} \approx \frac{1}{a_0} - \frac{\eta}{a_0^2}.$$

Thus, the capillary pressure (8.24) is approximately

$$p - p_0 = -\Sigma_T \left(\frac{d^2\eta}{dx^2} + \frac{\eta}{a_0^2} \right), \tag{8.26}$$

where p_0 is the undisturbed pressure inside the cylinder.

Substituting a disturbance of the form $\eta = b\cos(kx)$, considered previously, gives

$$p - p_0 = \Sigma_T \left(k^2 - \frac{1}{a_0^2} \right) \eta.$$

Comparing to Eq. (8.23), we find the capillary acceleration to be

$$g_T = \frac{\Sigma_T}{\rho a_0^2} \left[(ka_0)^2 - 1 \right]. \tag{8.27}$$

In the next section, the relation

$$\omega^2 = gk$$

between frequency and wavelength of surface gravity waves is discussed [Eq. (8.34)]. Accepting it for now, and applying it to the capillary restoring force, gives

$$\omega^2 = \frac{\Sigma_T}{\rho a_0^3} ka_0 \left[(ka_0)^2 - 1 \right]. \tag{8.28}$$

When $ka_0 < 1$, the right side is negative and the frequency is imaginary. An oscillation with an imaginary frequency grows exponentially because

$$e^{i\omega t} = e^{\sigma t},$$

when $\omega = -i\sigma$. Thus, if $ka_0 < 1$ instead of a frequency, formula (8.28) provides the growth rate

$$\sigma^2 = \frac{\Sigma_T}{\rho a_0^3} \left[ka_0 - (ka_0)^3 \right]. \tag{8.29}$$

This reaches a maximum with respect to k when $k^2 a_0^2 = 1/3$; that is, when the wavelength is

$$\lambda = 2\sqrt{3}\pi a_0 = 10.9 a_0$$

Figure 8.4. Disintegration of a capillary jet. The perturbation wavelength is 25.1, 14.6, 11.8, and 9.2 times a_0 from top to bottom. The last has the largest growth rate according to linear theory. Figure courtesy of F. Albina and M. Peric.

having substituted $k = 2\pi/\lambda$. Rayleigh's more rigorous analysis produced $\lambda = 9a_0$ (Lamb, 1932, p. 473).

If $ka_0 > 1$, the frequency (8.28) is real, and we may call the perturbation a capillary wave. When $ka_0 \gg 1$, the relation between frequency and wavelength is

$$\omega^2 = \frac{\Sigma_T}{\rho} k^3. \tag{8.30}$$

This is called the *dispersion relation* for capillary waves. Wave propagation will be described in the next section. Although capillarity will not factor into that material, Eq. (8.30) enters into exercise 8.6.

8.2.3 Simulation of Capillary Breakup

The Plateau–Rayleigh instability has been computed by subjecting a thin liquid jet to a small, periodic pulsation of the inlet velocity. This simulation was graciously provided by F. Albina and M. Peric. The geometry is axisymmetric. The liquid has $\rho = 1.12\,\mathrm{g/cm^3}$, $\mu = 0.04\,\mathrm{g/cm \cdot s}$, and $\Sigma_T = 56.0\,\mathrm{g/s^2}$. These properties are those of a mixture of glycerine and water. They are selected to emphasize surface tension over viscous forces. The initial diameter of the jet is 2.59 mm, the inlet velocity is 2.13 m/s and it exits into air. The Reynolds number based on nozzle diameter is 1,545. The inlet velocity can be used to convert frequency to wavelength. Thus, if f is the frequency in hertz, the wavelength is $\lambda = U/f$. Theory predicts instability for $\lambda > 2\pi a_0$ and maximum rate of growth when $\lambda = 9a_0$.

The cases of $\lambda = 25.1$, 14.6, 11.8, and $9.2a_0$ are displayed in Figure 8.4. All simulations covered the same time interval. The interface is shown at the end of each simulation. The radial pulsations are seen to grow more rapidly as the wavelength decreases, consistently with Eq. (8.29). The capillary jet at the bottom of the figure has approximately the maximum growth rate, according to linear theory.

Although the initial protrusions are very close to sinusoidal, they grow into nearly spherical beads, connected by a thin string of liquid. That is a consequence of nonlinearity. The perturbed radius is $a + b\cos(2\pi x/\lambda)$, as in Eq. (8.17). At the crest of the cosine, the radius increases; at the trough it decreases. The capillary pressure is higher at the smaller radius section than at the larger radius section. Thus a pressure gradient is formed that drives liquid out of the trough into the crest. The latter fattens to form the main drop. Capillary force drives it toward a spherical shape, stretching the neck between drops. This produces the appearance of beads on a string. The string breaks, setting the bead into an oscillation and forming a much smaller drop in between the beads. These are called the main and satellite drops. This phenomenology corresponds quite closely with experiments.

8.3 Inviscid Free Surface

Capillarity plays a role at small scales. Ripples on a pond are influenced by surface tension. It becomes insignificant for larger waves. A boat motoring across a calm lake will radiate surface gravity waves, much like the Mach waves of a body moving in a compressible gas, discussed in Chapter 7. In the presence of gravity, two fluids are statically stable if the heavier lies below the lighter. If the interface is perturbed by locally raising the heavier fluid and then releasing it, the return of the interface to the stable, static configuration will occur through oscillations. These are just the waves that a boat creates on the lake. They are driven by gravity acting on the density difference.

Gravity waves are not entirely analogous to sound waves in a gas, however. On a thin layer of water the analogy stands: surface gravity waves propagate with speed $\sqrt{gH}$, where H is the depth of the layer and g is the acceleration of gravity. This plays a role similar to the speed of sound.

However, if the water is not very shallow, the analogy fails. If the water is deep, the particular depth becomes irrelevant. To obtain a quantity containing gravity and having the dimension of velocity a characteristic length is needed. The depth is ruled out. The length of the wave serves the purpose. Surface waves propagate at speed $\sqrt{g\lambda/2\pi}$, where λ is their wavelength. This is quite different from sound: sound waves propagate at a fixed speed, irrespective of their wavelength. By contrast, longer gravity waves move faster than shorter ones. With $g = 9.8$ m/s, the wave speed becomes $1.25\,\lambda^{1/2}$ in mks units. Thus, a 1-m-length wave moves with speed 1.25 m/s, whereas one with $\lambda = 4$ m moves at 2.5 m/s.

When speed depends on wavelength, it is called *dispersive propagation*. If two waves, one long, with length λ_1, and one short, with λ_2, are generated at the same position, they subsequently will move apart. After time t, the first is a distance $t\sqrt{g\lambda_1/2\pi}$ from the source, whereas the second is at $t\sqrt{g\lambda_2/2\pi}$. Thus, the waves disperse.

Fourier's theorem states that a smooth function can be decomposed into a sum of sine waves. A pulse in the elevation of a water surface can be regarded as a superposition of sine waves, by this reasoning. But each wavelength moves with a different speed. The pulse will not propagate without distortion; it will spread as the component waves disperse. We will return to this shortly, as it is an important aspect of free-surface flow.

As a shorthand, we again introduce the wave number k, defined as $2\pi/\lambda$. As x increases from 0 to λ, the product kx increases from 0 to 2π. Hence, $\phi \equiv kx$ is a phase function. The wave speed of deepwater waves can be stated

$$c_{ph} = \sqrt{g/k}, \tag{8.31}$$

termed the *phase velocity*. Linear theory provides the formula

$$c_{ph} = \sqrt{\frac{g \tanh kH}{k}} \tag{8.32}$$

for the phase speed on water of arbitrary depth, H. The limit $kH \to 0$ gives the shallow-water formula $c_{ph} = \sqrt{gH}$; the limit $kH \to \infty$ gives the deepwater formula.

A propagating wave is an undulatory pattern in space. A small amplitude, linear wave varies as $\sin kx$. As it propagates, an observer at a fixed position sees the water rise and fall sinusoidally, with frequency ω. It has a corresponding period of $\tau = 2\pi/\omega$. One way to state this combination of spatial and temporal sinuosity is by defining a phase as $\phi = kx - \omega t$. At a fixed position x, the oscillation is periodic in time, varying as ωt. As t increases from 0 to τ the phase varies from 0 to 2π.

If the wave varies as $\sin \phi$, then the crest is at $\phi = \pi/2$. An observer riding with the crest must occupy a position $X(t)$ that moves with the phase velocity c_{ph}; that is, $dX/dt = c_{ph}$. For this observer

$$\phi = kX - \omega t = \pi/2.$$

Differentiating this with respect to time shows that $dX/dt = \omega/k$. Equating this to the phase velocity gives

$$\omega = kc_{ph}. \tag{8.33}$$

For instance, deepwater waves have the frequency

$$\omega = \sqrt{kg}. \tag{8.34}$$

Certain properties of free surfaces can be interpreted in terms of gravity wave propagation. However, another important concept is first needed. If one looks closely at the bow wave in front of a large boat, one might notice that it is not simply an elevated ridge of water. Undulations are seen within the ridge, as in the example of Figure 8.7. The pattern of surface elevation might be described as a modulated undulation. At some distance, the water is calm. Closer to the boat, the surface is disturbed within an approximately wedge-shaped region. The presence of undulations gives the bow gravity wave a different appearance from the bow shock in Figure 7.6. Surface waves on deep water do not produce sharp fronts, as do sound waves. How is the difference explained? In some respect, it must be a consequence of the dispersive property.

Returning to Fourier's theorem – that a pulse can be decomposed into a sum of sinusoids – it is apparent that no unique Mach angle can be defined for the bow wave in front of a moving boat. The sine of the Mach angle is the wave speed divided by the boat speed per Eq. (7.15). But each Fourier wavelength propagates with a different phase velocity, so each has a different Mach angle; no unique angle can be defined. Nevertheless, a V-shaped pattern can be seen to precede a fast-moving boat. What sets its angle? Why do undulations appear within the V-shaped ridge?

8.3.1 Group Velocity and Its Connection to Surface Excrescence

Suppose the boat creates a ridge that rises above the surface of the water and propagates away from the boat. A ridge forms when the sine waves in the Fourier series add constructively to produce an elevation. As time progresses, the ridge will move away from the boat. But shorter waves will be left behind, as they have the lowest velocity. At a distance x, well away from the boat, the form of the ridge will be determined by which waves still superpose constructively at that location. These are waves that have approximately the same phase at the time when they reach x. The phase is

$$\phi = kx - \omega(k)t,$$

where it is noted that for dispersive waves frequency is a function of wave number. The condition that the phase is approximately to same for two waves, k and $k + \delta k$ is that

$$kx - \omega(k)t = [k + \delta k]x - \omega(k + \delta k)t.$$

For small δk, this reduces to

$$\frac{x}{t} = \frac{\partial \omega}{\partial k}.$$

The quantity on the left is termed the *group velocity*:

$$c_g \equiv \frac{\partial \omega}{\partial k}. \tag{8.35}$$

Waves that superpose constructively at position x and time t are those having a wavelength such that $x = c_g(k)t$. The rigorous derivation of this result makes use of the criterion of stationary phase (Lighthill, 1978). That is equivalent to the present, informal, allusion to constructive interference.

The position of constructive interference moves with the group velocity; a single sinusoidal wave moves with the phase velocity. These velocities are unequal. For deepwater waves, the phase velocity is twice the group velocity:

$$c_{ph} = \frac{\omega}{k} = \sqrt{\frac{g}{k}}; \ c_g = \frac{\partial \omega}{\partial k} = \frac{1}{2}\sqrt{\frac{g}{k}}.$$

To the extent that constructive interference produces an excrescence of the water surface, the excrescence moves with the group velocity. To the extent that the energy of a packet of waves moves with the ridge, energy also propagates at the group velocity.

The curves in Figure 8.5 illustrate these ideas. The dashed curves indicate a wave envelope. It moves with the group velocity. The oscillatory solid curve, inside the envelope, indicates the wave. It moves with the phase velocity. In the reasoning

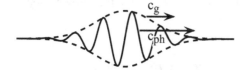

Figure 8.5. A wave packet moves with the group velocity; individual waves move with the phase velocity.

leading to formula (8.35) the envelope is created by constructive and destructive interference of a packet of waves. The oscillation inside the envelope is at the wavelength of the central wave in the packet. Its wave number, k, sets the group velocity, $c_g(k)$, at which the envelope propagates. It may seem odd that the wave moves faster than the packet. In the course of time, the oscillatory curve would be seen to move through the envelope, entering from the left with a small amplitude, growing to the middle, and then decaying as it exits to the right; all the while the dashed curve is also moving from left to right. That may seem odd but is the behavior of a dispersive wave packet.

Now we have some explanation for bow waves on deep water. Consider a boat moving with velocity V in the $-x$ direction. In a frame of reference attached to the boat, the water moves forward with velocity V. To remain fixed relative to the boat, the velocity of a wave must be equal and opposite to the velocity of the water, projected in the direction normal to the wave crest. Thus, a wave propagating at angle θ to the $-x$ axis must have speed

$$c_{\text{ph}} = V \cos \theta. \tag{8.36}$$

The vectors at the left of Figure 8.6 show the speeds of possible waves. The crests are perpendicular to the direction of propagation, as indicated by the dashed lines. The heads of the vectors lie on a circle of diameter V. This is seen by extending the dashed lines until they meet the axis to form a right triangle. Then the requirement that vectors be of length $V \cos \theta$ is met because the diameter of the circle is the hypotenuse of the right triangle, with included angle θ.

Each vector corresponds to the phase speed that can be held stationary at the angle θ. Only one particular wavelength will have just that speed. By Eq. (8.36), each wavelength determines a particular direction, given by

$$\cos \theta = \sqrt{\frac{g\lambda}{2\pi V^2}}. \tag{8.37}$$

If the boat generates energy at a predominant wavelength, that direction will be more noticeable because it has the highest amplitude. A boat of length ℓ may generate disturbances of similar scale. Then $\lambda \sim \ell$ may dominate the wave field. That component propagates in the direction

$$\cos \theta = \sqrt{\frac{g\ell}{2\pi V^2}} \tag{8.38}$$

on deep water. The crests lie perpendicular to this, as denoted by the dashed lines across the arrowheads in Figure 8.6.

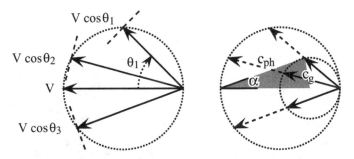

Figure 8.6. Waves with vector wave speed in the direction of any of the arrows will be held stationary in an oncoming stream in the direction $-V$. Their group velocities lie on the smaller circle at right. A line from the ship to the outermost bow wave makes an angle α to the direction of travel.

The quantity

$$\frac{V^2}{g\ell}$$

in Eq. (8.38) is called the *Froude number*. It plays a role in all free-surface flows. Here we see that at high Froude number the dominant disturbance will have crests nearly parallel to the motion ($\theta \approx 90°$). At low Froude number there may be no angle that will satisfy relation (8.38). Then the dominant disturbance cannot be held stationary.

Equation (8.37) sets the direction of the dominant wave crests, but it is not the direction of the corresponding ridge of water elevation in a bow wave. The ridge is created by superposition of a packet of waves. A packet propagates at the group velocity.

Consider any of the vectors in Figure 8.6 such as that at angle θ_1. Its group velocity is in the same direction, with half the magnitude. At the right of Figure 8.6, an arrowhead is marked at the midpoint of each vector. The arrowheads form a circle of $1/2$ the radius of the larger circle. The latter has radius $V/2$; the smaller circle has radius $V/4$.

In a fixed frame of reference, the boat progresses at speed V, continuously generating disturbances on the surface of the water. The components of the disturbance radiate away at various angles. The components with vectors lying on the dotted circles of the Figure 8.6 move with the boat, forming a bow wake that spreads behind the ship. The pattern of surface excrescence is composed from a sum of components with wavelengths corresponding to the various vectors demarcated in Figure 8.6.

In a time t, the ship moves a distance Vt along the $-x$ axis. Waves are continually flowing away at the speed $V\cos\theta$ appropriate to their wave number. The group velocity is one-half this, so the wave packet moves to $1/2Vt\cos\theta$. A line from the boat to the position of that packet will lie at the complementary angle to θ, as in the filled triangle at the right of Figure 8.6. This is analogous to the Mach angle, discussed in Chapter 7. However, unlike that case, each wave has a different Mach angle. It is clear from the construction at the right of Figure 8.6 that all wave packets make an

Figure 8.7. Kelvin wedge and wave crests emanating from the bow of a ship. The photo, taken by Photographer's Mate Second Class Gloria J. Barry, is a U.S. Navy photo and is free for public use.

angle less than α, defined in that figure: for every θ the group velocity and V form a triangle with an included angle less than or equal to α. The angle α is contained in a triangle that is tangent to the smaller circle and has its radius for one side. Hence, it is a right triangle. The side opposing α is the radius, which has length $1/4V$. The hypotenuse has length $3/4V$. Hence $\sin \alpha = 1/3$, giving $\alpha = 19.5°$. This is a famous result obtained by Lord Kelvin (Lamb, 1932).

It can be concluded that all wave energy produced by the moving boat lies in a wedge that makes an angle of 19.5° to the direction of motion. This is called *Kelvin's ship wedge*. The theoretical angle of 19.5° is independent of scale. It is popularly illustrated by photographs of the wave pattern produced by ducks swimming on a still pond. Their bow wave angle is remarkably similar to that seen in photographs of ships plowing through a calm harbor – for example, Figure 8.7. Although the wedge angle is unchanged across such a striking range of scales, the oscillations that are seen inside that bow angle are directly proportional to the size of the body. This is in accord with the theory that group velocity is the speed of energy propagation, distinct from the phase velocity of a pure oscillation. Group velocity determines the angle of the Kelvin wedge. All wave energy lies in that wedge, whether it radiates from a large or from a small body.

The fixed angle of the Kelvin wedge is also independent of speed. That is rather different from Mach angle (page 272). Mach angle decreases with increasing Mach number. That is because sound waves are nondispersive. Deepwater waves have a speed that increases indefinitely with wavelength. The Kelvin angle is independent of speed because the waves that are carried in the bow wake shift to longer wavelength with increasing speed. To be carried with a moving body, waves must have a sufficiently low phase velocity to be held stationary, if they propagate at a suitable angle to the body. In other words, they must meet the condition

$$\sqrt{g\lambda/2\pi} = V\cos\theta.$$

This shows that the lengths of components of the bow wave increase with speed. A large body will produce correspondingly long wavelengths in surface displacement. Only those with $g\lambda/2\pi \leq V^2$ can move with the body. The remainder will radiate away. A slow-moving large body carries little energy in its bow wave. As its speed increases, more wave energy travels with the body. When its speed is high enough for the right side of Eq. (8.38) to be less than unity, most of the wave energy goes into the Kelvin wedge.

The energy deposited in the bow wave must come from the boat. It represents a form of drag. The inability to satisfy condition (8.38) at low Froude number implies that the wave drag is low. It is found experimentally that drag rises quite rapidly as the Froude number increases beyond about 0.2. It is proposed theoretically that this is due to energy lost to maintaining the bow wave.

These considerations apply generally to disturbances at a free surface. Surface disturbances are seen in a stream flowing over a large rock. A rock that nearly spans a stream will produce waves primarily with $\theta = 0$. To be held stationary in front of such a broad object, waves must have a phase speed that satisfies $c_{ph} = V$. That sets their wavelength. Suppose the water is not deep. Then Eq. (8.32) provides the criterion

$$\frac{g}{k}\tanh kH = V^2$$

for a wave at $\theta = 0$ to be held stationary. The largest value of the left side is gH. If the depth is shallow, more particularly if the Froude number V^2/gH is greater than unity, this equation cannot be satisfied. Because $c_{ph} = \sqrt{gH}$ on shallow water, a Froude number greater than unity is analogous to supersonic speed; in the present context, it is called supercritical speed. At supercritical speed, a broad object cannot be preceded by a surface wave. [If it is not broad, it can produce waves at oblique angles, consistent with Eq. (8.37).] When a broad body moves supercritically, the drag is observed to fall. That can be attributed to its inability to produce a bow wave.

The theory of surface gravity waves also applies to the pattern seen in flow over submerged objects (Lighthill, 1978). When water flows over a hump on the bottom of a shallow stream, the surface rides up over the bump. After returning downstream to a flat bottom, the free surface does not become level; rather, it oscillates before flattening out. The oscillations are lee waves, downstream of the obstacle, as in Figure 8.8. They form only on the downstream side because the group velocity is less than the phase velocity. If the hump spans the stream, it will generate plane waves, with crests also spanning the stream. The wave with phase speed equal to the velocity, V, of the stream will be held stationary. That selects the wavelength of the lee wave. The associated wave packet propagates with the group velocity. If the group velocity is less than the phase velocity, the packet will be advected downstream. Hence, the surface distortion caused by the wave packet will be seen on the downstream side.

Figure 8.8. The wavy surface extending downstream, above a submerged body.

It might be noticed that short wavelength ripples are sometimes seen upstream of an obstruction in a stream. Such waves are influenced by surface tension. These capillary waves have a group velocity that is larger than their phase speed, as can be deduced from Eq. (8.30). Hence, the waves that are held stationary by the stream form a surface disturbance on the upstream side. Capillary waves dissipate quickly because their wavelength is short. Hence, they are seen for only a small distance before the obstruction.

8.3.2 Sloshing

When a partially filled container of liquid is accelerated horizontally at a constant rate, the surface is displaced as a whole. It is as if the direction of gravity were rotated from the vertical. If the acceleration changes in time, the liquid may slosh about within the container. Undoubtedly, the reader has experienced the basic phenomenon of sloshing.

Sloshing occurs in railroad tanker cars and in the fuel tanks of rockets. The accompanying shifts of their center of gravity can divert these vehicles from their intended course, possibly causing accidents. Baffles are commonly designed to restrict the amplitude of sloshing that can occur.

An analogy can be drawn between liquid sloshing in a container and the oscillation of a pendulum. The pendulum in a grandfather clock is kept swinging by repeated impulses applied by weights, through a release mechanism. Between the impulses it swings at its natural frequency of $\sqrt{g/\ell}$, where ℓ is length from pivot to the bob. A pendulum can also be forced continually by a motor. If the motor is tuned to the natural frequency, little force is needed to produce a large amplitude swing; if the forcing frequency is out of tune, less amplitude is produced with a given force. Mathematically, the linearized equation of the pendulum is

$$\ell\ddot{\Theta} + g\Theta = 0,$$

where Θ is the angular displacement from the vertical. Θ is assumed to be small. This equation has solution

$$\Theta = A\sin(\omega_n t + \phi),$$

where $\omega_n = \sqrt{g/\ell}$ is the natural frequency, A is an amplitude, and ϕ is a phase. The latter are the two constants of integration.

The equation of a forced pendulum is

$$\ell\ddot{\Theta} + g\Theta = F\sin\omega_f t. \tag{8.39}$$

Figure 8.9. Modes of sloshing in a cylinderical container.

The force, per unit mass, is sinusoidal with frequency ω_f. In addition to the previous homogeneous solution, this has the particular solution

$$\Theta = \frac{F \sin(\omega_f t)}{\ell(\omega_n^2 - \omega_f^2)}.$$

The response goes to infinity when the forcing is at the natural frequency. Of course, the assumption that Θ is small would be violated in that case.

The analogy with sloshing should be apparent. Slide a partially filled glass of water gently along a flat surface, then stop abruptly. The water will be set into oscillations at its natural frequency, much like the grandfather clock. Observe the frequency. Now slide the glass gently back and forth at approximately that same frequency. The water will go readily into regular, large-amplitude oscillations. This is the case of forcing near the natural frequency or of resonant forcing. Increase the frequency a bit. The oscillations are no longer so regular or so large. This is off-resonant forcing. Now oscillate it at a much higher frequency. Try a few; at some higher frequencies, the water will again oscillate violently, perhaps even splashing out of the glass. The splash might form toward the middle of the glass.

The high frequency behavior departs from the simple pendulum analogy. It reveals that the water surface has more than one natural frequency. Various resonances can be hit as the frequency is varied. The pattern of oscillations also varies. At the lowest frequency, the resonant mode is a side to side oscillation. Gently sliding the glass back and forth at a moderate frequency will cause the water to slosh up and down with a pattern similar to that at the left of Figure 8.9: up on one side and down on the other. The maximum amplitude occurs at the walls. This surface displacement is called the lowest natural mode.

Sliding the glass faster back and forth – at slightly less than twice the previous frequency – will excite a mode of oscillation like that at the right of Figure 8.9. The maxima are in the interior. Further modes, at increasingly higher frequencies, are possible theoretically. They are associated with multiple maxima in the interior. In many practical circumstances, only the lowest modes are of concern because they produce large oscillatory forces on the walls of tanks of liquid.

The theory is similar to that of surface waves. Indeed, the slosh can be considered to be a standing surface gravity wave. Its frequency is given by a formula like

Eq. (8.34); however, the wavelength is constrained by the need to fit into the container. In a rectangular container with sides ℓ_x and ℓ_z and depth H, only the modal wave numbers

$$k^2 = [(2n - 1)\pi/\ell_x]^2 + [(2m - 1)\pi/\ell_z]^2 \qquad (8.40)$$

can occur, where n and m are any positive integers. This discrete set of k values, with Eqs. (8.33) and (8.32), gives the natural slosh frequencies. (These are the antisymmetric modes created by side-to-side motion. A symmetric mode, with $2n - 1$ replaced by $2n$ also exists.)

We have largely avoided derivations of results of classical surface wave theory. They can be found in numerous texts (Lamb, 1932; Lighthill, 1978). For the curious reader, classical, linear wave theory will be summarized quite briefly in the following.

Basic wave theory makes the assumption that the water motion is irrotational and described by potential flow (§5.6). The velocity is the gradient of a potential, as in Eq. (5.19): $\boldsymbol{u} = \nabla\phi$. The fluid acceleration is then

$$\frac{d\boldsymbol{u}}{dt} = \nabla\left(\frac{\partial\phi}{\partial t}\right).$$

The linear momentum equation equates this to the pressure gradient and the gravitational acceleration in the $-\hat{y}$ direction:

$$\rho\nabla\left(\frac{\partial\phi}{\partial t}\right) = -\nabla p - \rho g\hat{y}.$$

This is satisfied by

$$p = -\rho g y - \rho\frac{\partial\phi}{\partial t}, \qquad (8.41)$$

which is the linear form of Bernoulli's equation for unsteady, irrotational flow. Although an arbitrary function of time could be added to this formula for pressure, that function can be set to 0 simply by absorbing it into the velocity potential.

Along a free surface the pressure is constant and equal to ambient. Pressure will be measured relative to ambient, so that it is zero on the free-surface. Let the position of the surface be described by the function $y = \eta(x, z, t)$. Then Eq. (8.41) evaluated on the free surface becomes

$$\frac{\partial\phi}{\partial t} = -g\eta. \qquad (8.42)$$

The velocity of the surface equals the normal component of fluid velocity. For small amplitude deformations, y is the normal direction; then $\partial\eta/\partial t = v = \partial\phi/\partial y$. Differentiating Eq. (8.42) with respect to time and using this gives

$$\frac{\partial^2\phi}{\partial t^2} = -g\frac{\partial\eta}{\partial t} = -g\frac{\partial\phi}{\partial y}. \qquad (8.43)$$

Consider a rectangular tank with walls at $x = 0$, $x = \ell_x$, $z = 0$, and $z = \ell_z$. Assume that ϕ is proportional to $\cos(\xi_x x/\ell_x) \cos(\xi_x z/\ell_z)$ in plan form. Let it also be periodic in time, so that

$$\phi = \cos(\xi_x x/\ell_x) \cos(\xi_x z/\ell_z) \sin(\omega t) F(y), \qquad (8.44)$$

where $F(y)$ describes the depth dependence. For a deep tank, substituting this into Laplace's Eq. (5.20) shows that $F = e^{ky}$ with $k^2 = [\xi_x/\ell_x]^2 + [\xi_z/\ell_z]^2$. The disturbance decays exponentially with height below the surface.

The condition that there be no velocity normal to the solid walls is met by equating the normal derivative of ϕ to zero at each of the four lateral boundaries. This gives the antisymmetric modes $\xi_x = (2n - 1)\pi$ and $\xi_z = (2m - 1)\pi$ for the representation (8.44). Equation (8.40) then follows from Eq. (8.43). Substituting the form of ϕ into the latter gives

$$\omega^2 \phi = gk\phi.$$

This is the deepwater dispersion relation $\omega^2 = gk$. The analysis shows that it is the restoring pressure produced by displacement of the interface in a gravitational field that causes oscillatory motion of the liquid.

When sloshing occurs in more complex geometries, analytical solutions for the frequencies are not available. Unsteady computer simulation can provide data on forces on container walls. Viscous action will add damping. That occurs in thin boundary layers, but it can reduce oscillations quite rapidly. Baffles are often added to fuel tanks of rockets to enhance damping.

The reader might have noticed that the oscillations he or she produced in a glass of water damp out quite rapidly. Let us try to estimate an approximate damping time. If it is assumed that the energy of sloshing decays by viscous friction, then an estimate of the rate of energy dissipation per unit volume based on Eq. (1.30), on page 21, might be

$$\mu \overline{\left(\frac{\partial u}{\partial y}\right)^2} \sim \mu \left(\frac{U}{a}\right)^2.$$

The cylinder of radius a is used as the length scale for y variation and U is a velocity that is representative of the oscillation. The total rate of of decay is this dissipation per unit volume times the volume $\mathcal{V}$. The total kinetic energy in the volume is $E = 1/2\rho U^2 \mathcal{V}$. Hence, the rate of decay is estimated to be

$$\frac{dE}{dt} \sim \mu \left(\frac{U}{a}\right)^2 \mathcal{V} \sim \left(\frac{2\nu}{a^2}\right) E. \qquad (8.45)$$

The exponential damping time is $a^2/2\nu$. In a cylinder of radius 4 cm containing water with viscosity of 0.01 cm^2/s, this is 800 s; obviously much longer than is observed.

It was explained in §1.5.2 that dissipation in oscillatory boundary layers can be much larger than this naive estimate. That is because the y variation occurs across a boundary layer that is thin compared to a. Suppose the dissipation occurs near the wall and that Eq. (1.29) describes it. Then the velocity derivative must be estimated as U/δ with $\delta = \sqrt{2\nu/\omega}$. The rate of dissipation per unit of surface area is

$$\frac{\mu U^2}{2\delta}.$$

The total rate of dissipation is this times the area of the wall. For a cylinder filled to height h, the estimate is

$$\frac{\mu U^2}{2\delta} 2\pi a h,$$

based on the area of the cylindrical walls. The bottom wall is ignored. For a sloshing liquid, the velocity U would depend on depth below the surface; however, our purpose is to make an order of magnitude estimate, so that complication is ignored. To conclude the estimate, the last formula is rewritten in terms of the energy $1/2 \rho U^2 = E/V = E/\pi a^2 h$ to obtain the rate of decay

$$\frac{dE}{dt} \sim 2\left(\frac{\nu}{\delta a}\right) E \sim \sqrt{\frac{2\omega\nu}{a^2}} E.$$

The exponential damping time is $\sqrt{a^2/2\omega\nu}$. In a cylinder of radius 4 cm containing water with viscosity $0.01\,\text{cm}^2/\text{s}$ oscillating at a frequency $2\pi/(1/3\,\text{s}) = 18.85$ rads, the damping time is about 6.5 s. This is two orders of magnitude shorter than the previous estimate. It clearly is more consistent with observations. Indeed, experiments on sloshing in tanks show that the damping time is of the proportionate form

$$T_{\text{damp}} \propto \sqrt{\frac{L^2}{2\omega\nu}}, \tag{8.46}$$

where L is a characteristic dimension of the tank. The coefficient of proportionality depends on the depth of liquid but is generally of order 1. High frequencies decay quickly. That is another reason why only the lowest modes are usually of concern.

8.3.3 Numerical Simulation of Sloshing

A railroad tanker car is partially filled with oil. The car suddenly sways. A wave of oil is sent crashing into one wall. What are the forces on the wall? Will the car topple over?

The linear theory just described is a starting point, but it will not be accurate when the slosh is large. A full computer simulation provides a far better accounting of the extreme forces that could cause the tanker to topple over.

In a stationary frame of reference, the container tank moves and initiates the slosh. For computational purposes, it may be preferable to work in a frame attached to the tank; then the liquid is subjected to an imposed force that represents the sway. The two frames of reference produce equivalent solutions. In the container frame, the geometry and grid are constant, independent of time.

Computations are presented of a rectangular container, rocked periodically from side to side. If the forcing frequency is near resonance, a large sloshing flow is expected. The container is 1.2 m wide and 0.6 m high. The liquid is water, filled to a

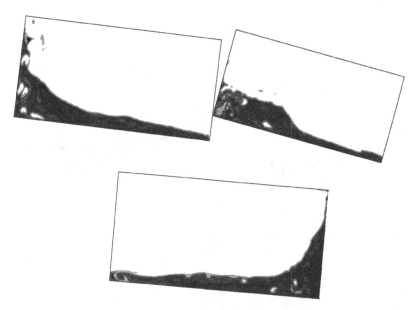

Figure 8.10. Sloshing induced in a swaying container. Figure courtesy of M. Peric.

height of 0.12 m. The amplitude of side-to-side rolling is 10° with a period of 2.25 s. This is slightly above the resonant frequency of linear theory.

Three instants in the sloshing motion are displayed in Figure 8.10. They are vaguely reminiscent of the side-to-side mode of Figure 8.9. However, the liquid is thrown up on impact with the side wall. As it slams into the wall, the water produces an impulsive pressure, as seen in Figure 8.11. This figure records the pressure at the lower-left corner of the tank. As the water settles back down the wall, a secondary maximum occurs before the pressure relaxes back to hydrostatic. Validations against experiment have shown the impulsive forces to be quite accurately predicted by this computation. Figure 8.11 includes an experimental trace, demonstrating the degree of fidelity.

It takes only the merest experimenting with a container of water to recognize that the liquid riding up the side wall might overturn and splash as it comes back

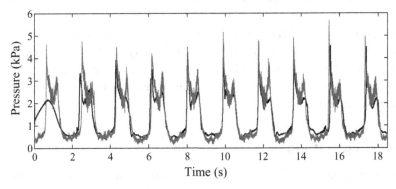

Figure 8.11. Pressure beneath sloshing liquid. Dark line is computation; light line is experiment. Figure courtesy of M. Peric.

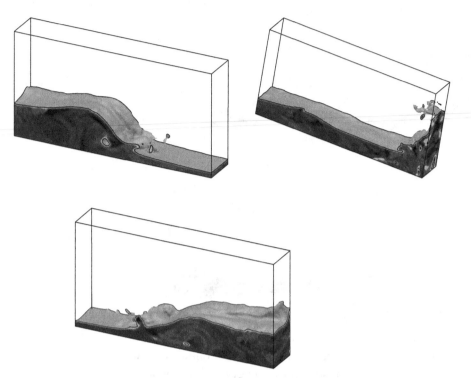

Figure 8.12. Splash of liquid in a swaying container. Figures courtesy of M. Peric and T. Zorn.

down. The surface will then be irregular and three dimensional. It is a challenge to free-surface computation to reproduce this. The ability to do so is demonstrated in Figure 8.12.

The three-dimensional, 204,800 cell, hexahedral mesh utilized in this simulation was judged adequate to capture the main features of the splash. This computation was provided by M. Peric and T. Zorn. They judged that the main forces on the container walls could be obtained from these simulations. A turbulence model was invoked to represent effects of small-scale eddies, although irregularities of the interface are resolved in the simulations. An analogy to DES, §6.4.4, might be drawn. In place of shear flow eddies, splashing against the end walls is the source of the resolved turbulence.

The liquid–gas interface was represented by the volume of fluid (VOF) method, described in the next section. That method does not explicitly follow the interface; it interpolates between computational nodes to define the interface. Some degree of numerical diffusion occurs. Despite that, the VOF technique can simulate the interfacial dynamics with fair accuracy.

8.4 Volume of Fluid Method

Computation of a free boundary presents several new issues. The interface position is unknown. It is found as part of the solution. Without knowing the position of the boundary in advance, it is not possible to fit a mesh to the geometry. The most accurate method would be to generate the mesh dynamically so that it conforms to the interface. In a time-dependent solution the mesh would evolve as a function of time. Such methods exist, but a simpler approach has been developed, which is used in most practical CFD applications.

Introduce an indicator function $\mathcal{I}(x)$ which has a value $\mathcal{I}_a$ in fluid a and $\mathcal{I}_b$ in fluid b. The interface is midway between these values. Although we could define $\mathcal{I}_a = 0$, $\mathcal{I}_b = 1$ and the interface as $\mathcal{I} = 1/2$, the more general definition allows density to be used in an incompressible flow as the indicator. That avoids the need for the indicator to be a new variable.

The interface is tracked by solving for the field $\mathcal{I}(x)$ and locating the contour that identifies it. If $\mathcal{I}$ satisfies the pure convection equation

$$\frac{\partial \mathcal{I}}{\partial t} + \boldsymbol{u} \cdot \nabla \mathcal{I} = 0$$

level contours of $\mathcal{I}$ are carried with fluid elements. Thereby, condition (8.2) is enforced. However, in any real computation numerical diffusion will spread the contours of $\mathcal{I}$. Effectively, it satisfies a convection–diffusion equation. This produces an inaccuracy. However, the midpoint value $(\mathcal{I}_a + \mathcal{I}_b)/2$ that identifies the interface will usually retain a second-order accuracy.

The midpoint contour will cut across cells, dividing them into portions of fluid a and portions of fluid b. Cells that are not divided lie entirely in one fluid or the other. Those that are divided must be treated specially. In the VOF method, the material properties are set to volume-weighted average values. For instance

$$\mu = \frac{(I - I_a)\mu_b + (I_b - I)\mu_a}{I_b - I_a},$$

where I is the value of the indicator averaged over the cell. This is a rather simple approach. The Navier–Stokes equations are solved for a variable property fluid. In each cell, the properties are of one, the other, or a weighted average of the two components. Interface conditions (8.1) and (8.3) are met automatically. These are conditions of continuity and the fluid is treated as a single continuum; hence, all fields, including velocity and stress, are continuous.

But, in the presence of surface tension the stress should be discontinuous, in accordance with Eq. (8.11). An effective method to incorporate surface tension forces, within the VOF framework, is by converting it to a body force. This is called the continuum surface force model. The interface is a level surface of the indicator function. Hence, the gradient of the indicator function is directed normal to the interface. The unit normal is simply

$$\hat{n} = \frac{\nabla \mathcal{I}}{|\nabla \mathcal{I}|}. \tag{8.47}$$

The curvature is then evaluated by Eq. (8.10).

The Navier–Stokes momentum equation contains the gradient of the stress tensor. We represent the surface tension force by something similar. This can be accomplished by the formal device of a δ function: the surface force becomes a body force of infinite magnitude, located at the position of the interface. Let us consider a less formal rationale.

Consider two points, one in each of the two fluids, and separated in the normal direction. Let the stress at these positions be σ_a and σ_b. If the points are a distance h apart, the stress gradient is approximated by

$$\frac{\sigma_b - \sigma_a}{h}.$$

If h is very small, the points are very near to the interface. They differ in consequence of immiscibility. Invoking Eq. (8.11) for the stress jump, the above becomes

$$\frac{\Sigma_T \kappa \hat{n}}{h},$$

where it has been recognized that the body force is in the normal direction. In the expression (8.47) for the normal direction, the denominator is approximately $(\mathcal{I}_b - \mathcal{I}_a)/h$. Thus, we obtain the formula

$$F_\Sigma = -\frac{\nabla \mathcal{I}}{\mathcal{I}_b - \mathcal{I}_a} \Sigma_T \kappa \tag{8.48}$$

for the body-force model of surface tension. This is added to the right side of the Navier–Stokes equations in cells containing the interface. The minus sign occurs because formula (8.48) is the force exerted by surface tension, on the fluid.

8.4.1 Pouring a Viscous Liquid

A very viscous liquid pours in a smooth stream, as it leaves a spout and falls under the acceleration of gravity. An everyday example is thick syrup pouring onto a plate. As the syrup accelerates, the stream of liquid narrows. This is simply conservation of mass: given the volumetric flow rate, the cross-sectional area decreases inversely with the accelerating velocity. If the stream becomes sufficiently thin, it will waver back and forth before reaching the plate. It might fold on itself as it falls into the pool of syrup already on the surface.

In one of his delightful papers on fluid phenomena, G. I. Taylor shows how this can be understood by analogy to the buckling of an elastic rod. When a thin, flexible rod is squeezed along its axis, it will bow outward at its middle, if the compressive load is sufficient. For instance, if the rod is held vertically against a surface and a weight is placed on its free end, so as to exert a load in the axial direction, the rod will buckle if the weight is too large. Taylor argued that the liquid column is subjected to a compressive force as it slows on approaching the surface. The compressive load initiates an instability in the viscous, liquid stream that is analogous to the buckling of a beam. The thread of liquid must be thin for this to occur. That is what the analogy suggests and that is what is observed.

Having argued that the wafting that is seen as the stream settles onto the plate is an instability caused by compressive loading, Taylor designed an experiment to

Figure 8.13. A stream of glycerine pouring onto a surface, computed with the volume of fluid method. Views at 1.45 and 1.5 s.

magnify the behavior. Instead of falling through air onto a solid surface, a glycerine stream was allowed to fall through a tank of water. The density difference between glycerine and water times gravity provided the acceleration. At some depth, the water was replaced by a fluid of density higher than water but less than glycerine. The downward gravitational acceleration decreased abruptly. The stream was subjected to compressive strain at the density contrast. Shortly thereafter the stream was seen to waft erratically back and forth. The origin of instability in the compressive strain was convincingly demonstrated.

The wavering seen in the fluid stream poses a challenge to CFD. A time-accurate, unsteady computation is required, with good grid resolution around the pouring liquid. The entire zone where it oscillates must be adequately gridded. Simulations were done in two dimensions. The fluids were glycerine ($\mu = 8.0\,\mathrm{g/cm} \cdot \mathrm{s}$, $\rho = 1.26\,\mathrm{g/cm^3}$) and water ($\mu = 0.01\,\mathrm{g/cm} \cdot \mathrm{s}$, $\rho = 1\,\mathrm{g/cm^3}$). Surface tension was set to zero. The glycerine pours under gravity after it is injected at $v = 2.5\,\mathrm{cm/s}$. It takes approximately 1.5 s for the jet to reach the wall. This gives us some idea about the necessary time-step: Δt was set to $10^{-3}\,\mathrm{s}$.

A 334×291, structured grid was used for the simulations. The grid was constructed so that the majority of the grid lines was concentrated in the jet region and in the vicinity of the wall. Because the jet becomes unstable and begins to oscillate as it approaches the wall, extra refinement was required in a region surrounding the initial inflow location.

As the stream accelerates, its width necks down to a minimum, as seen in the part of the full view at the left of Figure 8.13. Then, as it decelerates near the lower wall, it is seen to waver as a function of time. A pool of glycerine is accumulating on the lower wall. The tails of the pool are remnants of the start-up of the computation. As the glycerine is first injected into the water, vortices form on either side of the stream. They advect down with the initial mass of glycerine, as it impinges on the lower wall.

A closeup is shown at the right of Figure 8.13. The viscous fluid is piling into the pool accumulated on the surface. The stream above the pool is somewhat irregular. Instability is triggered as the stream begins to widen on approaching the wall. This evidences the compressive strain in Taylor's reasoning. Presumably, the seed for the irregular motion is numerical noise created by the iterative solution method. Taylor's study shows that the wavering is irregular in the physical experiment as well.

This example provides evidence that the volume of fluid method is up to the demands of simulating pouring of a viscous liquid.

EXERCISES

8.1 *Fluid–fluid interface.* The boundary condition at the interface between two viscous fluids is that the velocity and stress are continuous. In the case of pipe flow, governed by Eq. (E1.1), the relevant stress is $\mu\, du/dr$.

An engineer at a chocolate factory has suggested that liquid chocolate could be pumped more efficiently if it were surrounded by a layer of less viscous liquid. Analyze flow through a straight, circular pipe of two liquids, the less viscous forming an annular ring around the more viscous central liquid (chocolate).

Let the pipe radius be R and let the liquid chocolate lie in $0 \leq r \leq R_1$. Assume that the outer liquid has one-tenth the viscosity of chocolate. Determine the ratio

R_1/R that will give the greatest volume flow of chocolate for a given pressure drop.

8.2 *Capillarity.*

    (a) A thin, waxy tube is immersed partway into a vial of water. The contact angle is 110°. What is the displacement of the meniscus from the ambient water level? The density and surface tension of water are $1\,\mathrm{g/cm^3}$ and $74\,\mathrm{g/s^2}$ in cgs units. The radius of the tube is $1\,\mathrm{mm}$ and the acceleration of gravity is $980\,\mathrm{cm/s^2}$. What is the pressure just inside the meniscus, relative to ambient?

    (b) A short slug of liquid is inserted into a thin tube made of hydrophilic material. The tube is lying horizontally and the slug sits in between its two ends. One side of the tube is exposed to atmospheric pressure, the other is initially at atmospheric pressure too. Sketch the shape of the slug of liquid. What is the formula for the pressure inside the liquid?

Now the pressure is increased a bit on one side of the tube but remains at atmospheric on the other. The pressure difference is too small to dislodge the slug. Sketch the shape of the menisci on either end of the slug. If the liquid is static, its internal pressure is uniform. How can that be consistent with the interface condition at either end of the slug?

8.3 *Compute a meniscus.* Compute the capillary rise of water in a narrow tube by applying the conditions of surface tension and wall adhesion. The surface tension coefficient for the water–air interface is $\Sigma_T = 0.0756 \, \text{N/m}$. Specify the contact angle for water at solid walls as $\alpha = 30°$.

Define the problem as axisymmetric, unsteady, laminar, two-phase flow, using the VOF method to track the interface. The flow domain is shown in the accompanying figure. The diameter of the capillary tube is $d = 5 \, \text{mm}$, whereas the length of the tube is $L = 50 \, \text{mm}$. The dimensions of the container are $D \times H = 200 \times 100 \, \text{mm}$. Use a fine grid within the tube (at least 10 cells along the radius). The time-step should be small, $\Delta t \leq 1.e-4 \, \text{s}$ say, but integrate to steady state to obtain the capillary rise.

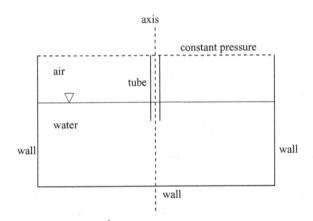

Compare the simulation results with the analytical solution for the steady state height h.

Repeat for contact angles of $\alpha = 90°$ and $\alpha = 150°$.

8.4 *Spreading, viscous drop.* An axisymmetric drop is placed on a smooth surface and spreads under the action of gravity alone; ignore surface tension. Show that the radius of the wetted surface increases as

$$R(t) = \left[\frac{2^{10} g V^3}{3^5 \pi^3 \nu} \right]^{1/8} t^{1/8}$$

(i.e., as t to the 1/8 power), where V is the volume of the drop. Note that the volume of the drop is constant in time. Use a lubrication approximation. Here are steps to the answer:

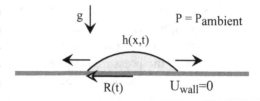

(a) $v = u(h)\partial_r h + \partial_t h$ on the surface $y = h(r, t)$ of the drop. This is simply the condition that the fluid flow relative to the drop surface is in the tangential direction. That is equivalent to the condition (8.2), which states that the fluid velocity relative to the surface is zero in the normal direction.

(b) For the y-momentum equation, invoke the hydrostatic approximation

$$\frac{\partial p}{\partial y} = -\rho g.$$

What does this give for $\partial_r p$, in terms of $h(r, t)$?

(c) What is the $u_r(y)$ profile assumed by the lubrication approximation? The continuity equation, in cylindrical coordinates, is

$$r\frac{\partial v}{\partial y} = -\frac{\partial r u_r}{\partial r}.$$

Substitute the $u_r(y)$ profile on the right, and integrate with respect to y, using $v(0) = 0$ and $v(h)$ from item a above.

(d) Rearrange this into the evolution equation for h. Show that if $h(r, t) = At^{-1/4}H(\eta)$, where $\eta = Brt^{-1/8}$, with $A = (3vV/16g)^{1/4}$ and $B = (3v/16gV^3)^{1/8}$ this evolution equation becomes

$$2\partial_\eta \left(\eta^2 H\right) + \partial_\eta \left[\eta H^3 \partial_\eta H\right] = 0. \tag{E8.1}$$

This change of dependent and independent variables simply neatens the equation.

(e) At this point, the 1/8 power has been derived: the edge of the drop is where $H(\eta) = 0$. Call this η_0. Then $R = \eta_0 t^{1/8}/B$. Solve (E8.1) for the case of a constant volume drop to find η_0.

8.5 *Group velocity.* Plot

$$\sin(0.95x - 0.9t) + \sin(1.05x - 1.1t)$$

for $0 < x < 60\pi$, at times $t = 0, 10, 20, 30$. Rewrite this as a product of two trigonometric functions. What is the group velocity? What is the phase velocity?

8.6 *Stone in a pond.* The dispersion relation for capillary–gravity waves is the sum of Eq. (8.30) and (8.34):

$$\omega^2 = gk + \frac{\Sigma_T}{\rho}k^3.$$

Rewrite this in terms of the nondimensional frequency and wave number

$$\tilde{\omega} = \omega \left(\frac{\Sigma_T}{\rho g^3}\right)^{1/4} \quad \text{and} \quad \tilde{k} = k \left(\frac{\Sigma_T}{\rho g}\right)^{1/2}.$$

Evaluate the dimensional factors for water in s and cm ($\rho = 1\,\text{g/cm}^3$ and $\Sigma_T = 74\,\text{g/s}^2$, $g = 980\,\text{cm/s}^2$).

Plot the phase velocity versus wavelength, in nondimensional form. Do the same for the group velocity.

A small stone is dropped into a pond. A ring of capillary–gravity waves radiates out from the point of entry. Why is the surface of the pond calm inside the ring? What is the outward speed of the ring? Write your answer in dimensional form and then evaluate the speed for water.

Bibliography

ANDERSON, J. D. 1982. *Modern Compressible Flow with Historical Perspective*, McGraw-Hill.

BATCHELOR, G. K. 1967. *An Introduction to Fluid Dynamics*, Cambridge University Press.

DÉLERY, J. M., LEGENDER R., AND WERLÉ H. 2001. "Toward the elucidation of three-dimensional separation," *Annu. Rev. Fluid Mech.* **33**, 129–54.

DRAZIN, P., AND REID, W. H. 1995. *Hydrodynamic Stability*, Cambridge University Press.

DURBIN, P. A., AND PETTERSSON REIF, B. A. 2001. *Statistical Theory and Modeling for Turbulent Flow*, John Wiley & Sons.

FERZIGER, J., AND PERIC, M. 2002. *Computational Methods for Fluid Dynamics,* 3rd ed., Springer-Verlag.

FLETCHER, C. A. J. 1991. *Computational Techniques for Fluid Dynamics,* 2nd ed., Springer-Verlag.

HAMROCK, B. J., SCHMID, S. R., AND JACOBSON, B. O. 2004. *Fundamentals of Fluid Film Lubrication*, 2nd ed., Marcel Dekker.

LAMB, H. 1932. *Hydrodynamics*, 6th ed., Cambridge University Press.

LIGHTHILL, M. J. 1975. *Mathematical Biofluiddynamics*, SIAM.

LIGHTHILL, M. J. 1978. *Waves in Fluids*, Cambridge University Press.

MILNE-THOMSON, L. M. 1968. *Theoretical Hydrodynamics*, Dover.

PANTON, R. L. 1997. *Incompressible Flow*, Wiley Interscience.

POPE, S. B. 2000. *Turbulent Flow*, Cambridge University Press.

PROBSTEIN, R. F. 1994. *Physico-chemical Hydrodynamics*, Wiley Interscience.

RAYLEIGH, J. W. S. 1877. *The Theory of Sound*, Dover, 1945 edition.

ROTT, N. 1990. "Note on the history of the Reynolds number," *Annu. Rev. Fluid Mech.* **22**, 1–11.

SAAD M. A. 1997. *Thermodynamics: Principles and Practice*, Prentice Hall.

TANNEHILL, J. C., ANDERSON, D. A., AND PLETCHER, R. H. 1997. *Computational Fluid Mechanics and Heat Transfer*, Taylor & Francis.

TAYLOR, G. I. 1971. *Scientific Papers of G. I. Taylor*, G. K. Batchelor, ed., Cambridge University Press.

THOMPSON, J. F., SONI, B., AND WEATHERILL, N., eds. 1999. *Handbook of Grid Generation*, CRC Press.

VANDYKE, M. 1982. *An Album of Fluid Motion*, Parabolic Press.

VANDYKE, M. 1975. *Perturbation Methods in Fluid Mechanics*, Parabolic Press.

WHITE, F. M. 1991. *Viscous Fluid Flow*, 2nd ed., McGraw-Hill.

Index

Printed in the United States
By Bookmasters